The Biology of Blood-Sucking in Insects
Second Edition

Blood-sucking insects transmit many of the most debilitating diseases in humans, including malaria, sleeping sickness, filariasis, leishmaniasis, dengue, typhus and plague. In addition, these insects cause major economic losses in agriculture both by direct damage to livestock and as a result of the veterinary diseases, such as the various trypanosomiases, that they transmit. The second edition of *The Biology of Blood-Sucking in Insects* is a unique, topic-led commentary on the biological themes that are common in the lives of blood-sucking insects. To do this effectively it concentrates on those aspects of the biology of these fascinating insects that have been clearly modified in some way to suit the blood-sucking habit. The book opens with a brief outline of the medical, social and economic impact of blood-sucking insects. Further chapters cover the evolution of the blood-sucking habit, feeding preferences, host location, the ingestion of blood and the various physiological adaptations for dealing with the blood meal. Discussions on host–insect interactions and the transmission of parasites by blood-sucking insects are followed by the final chapter, which is designed as a useful quick-reference section covering the different groups of insects referred to in the text.

For this second edition, *The Biology of Blood-Sucking in Insects* has been fully updated since the first edition was published in 1991. It is written in a clear, concise fashion and is well illustrated throughout with a variety of specially prepared line illustrations and photographs. The text provides a summary of knowledge about this important group of insects and will be of interest to advanced undergraduate and to postgraduate students in medical and veterinary parasitology and entomology.

M̄ike L̄ehane is Professor of Molecular Entomology and Parasitology in the Liverpool School of Tropical Medicine.

The Biology of Blood-Sucking in Insects

SECOND EDITION

M. J. Lehane

Liverpool School of Tropical Medicine

CAMBRIDGE
UNIVERSITY PRESS

CAMBRIDGE
UNIVERSITY PRESS

Shaftesbury Road, Cambridge CB2 8EA, United Kingdom

One Liberty Plaza, 20th Floor, New York, NY 10006, USA

477 Williamstown Road, Port Melbourne, VIC 3207, Australia

314–321, 3rd Floor, Plot 3, Splendor Forum, Jasola District Centre, New Delhi – 110025, India

103 Penang Road, #05–06/07, Visioncrest Commercial, Singapore 238467

Cambridge University Press is part of Cambridge University Press & Assessment, a department of the University of Cambridge.

We share the University's mission to contribute to society through the pursuit of education, learning and research at the highest international levels of excellence.

www.cambridge.org
Information on this title: www.cambridge.org/9780521543958

First published 2005

A catalogue record for this publication is available from the British Library

ISBN   978-0-521-83608-1   Hardback
ISBN   978-0-521-54395-8   Paperback

Contents

Tables

Boxes

Preface

Blood-sucking insects are the vectors of many of the most debilitating parasites of humans and their domesticated animals. In addition they are of considerable direct cost to the agricultural industry through losses in milk and meat yields, and through damage to hides, wool and other products. So, not surprisingly, many books of medical and veterinary entomology have been written. Most of these texts are organized taxonomically, giving details of the life cycles, bionomics, relationships to disease and economic importance of each of the insect groups in turn. I have taken a different approach. This book is topic-led and aims to discuss the biological themes common to the lives of blood-sucking insects. To do this I have concentrated on those aspects of the biology of these fascinating insects that have been clearly modified in some way to suit the blood-sucking habit. For example, I have discussed feeding and digestion in some detail because feeding on blood presents insects with special problems, but I have not discussed respiration because it is not affected in any particular way by haematophagy. To reflect this better I have made a slight adjustment to the title of the book in this second edition. Naturally there is a subjective element in the choice of topics for discussion and the weight given to each. I hope that I have not let my enthusiasm for the particular subjects get the better of me on too many occasions and that the subject material achieves an overall balance. The major changes in this second edition most often reflect the revolutionary influence that molecular biology has had on the subject in the past 12 years.

Although the book is not designed as a conventional text of medical and veterinary entomology, in Chapter 9 I have given a brief outline of each of the blood-sucking insect groups. This chapter is intended as a quick introduction for those entirely new to the subject, or as a refresher on particular groups for those already familiar with the divisions of blood-sucking insects. There are several introductory textbooks of medical and veterinary entomology available to those requiring more information.

The book is primarily intended for advanced undergraduate and for postgraduate students, but because it looks at topics that cut across the normal research boundaries of physiology and ecology, behaviour and cell biology, I hope it may also be useful for more established scientists who want to look outside their own specialism. I have tried to distil this broad spectrum of information, much of which is not readily available to the non-specialist, into a brief synthesis. For those who want to look further into a particular area I have included some of the references I found most useful in writing the text, and these will provide an entry into the literature. Clearly the subjects covered by the book encompass a vast number of publications and I am sure to have missed many important and interesting references for which I apologize in advance both to the reader and my fellow scientists. Many of the topics discussed in the different chapters are interrelated. To avoid repetition, and still give the broadest picture possible, I have given cross-references in the text which I hope the reader will find useful.

From a comparative point of view it is an unfortunate fact that most of the work on blood-sucking insects has been carried out on a few species. Consequently, tsetse flies and mosquitoes pop up on every other page. In many instances it remains to be seen how widely the lessons we have learned from these well-studied models can be applied. Where possible I have tried to point to general patterns that fit whole groups of blood-sucking insects. To help me in this I have divided the blood-sucking insects into three convenient but artificial categories: temporary ectoparasites, permanent ectoparasites and periodic ectoparasites. These categories are based solely on the behaviour biology of the blood-feeding stadia in the lives of these insects. Temporary ectoparasites are considered to be those largely free-living insects, such as the tabanids, mosquitoes, blood-feeding bugs and blackflies, that visit the host only long enough to take a blood meal. I also include insects such as the tsetse here, even though the male may be found in swarms closely associated with the host for large parts of its life. Permanent ectoparasites are considered to be those insects that live almost constantly on the host, such as lice, the sheep ked and tungid fleas. Finally, periodic ectoparasites are considered to be those insects that spend considerably longer on the host than is required merely to obtain a blood meal, but that nevertheless spend a significant amount of time away from the host. Insects that fall into this category include many of the fleas and Pupipara. These categories are no more than a useful generalization in the text; I make no claims for their rigour and I realize that it could be argued in several instances that an insect will sit as easily in one category as another.

Acknowledgements

I gladly and gratefully acknowledge the help I have received from many people during the writing of this book or the previous edition, particularly P. Billingsley, A. Blackwell, J. Brady, H. Briegel, I. Burgess, E. Bursell, R. Dillon, J. D. Edman, R. Galun, A. G. Gatehouse, M. Gaunt, M. Gillies, R. H. Gooding, M. Greaves, C. Green, M. Hafner, J. Hogsette, H. Hurd, A. M. Jordan, K. C. Kim, J. Kingsolver, M. Klowden, A. M. Lackie, B. R. Laurence, E. Levashina, A. G. Marshall, P. Mellor, D. Molyneux, P. Morrison, W. A. Nelson, G. O'Meara, G. S. Paulson, G. Port, N. A. Ratcliffe, J. M. Ribeiro, P. Rossignol, M. Rothschild, W. Rudin, C. J. Schofield, M. W. Service, J. J. B. Smith, W. Takken, S. Torr, G. A. Vale and J. Waage. I also thank Paula Hynes, Maria Turton, Paula Dwyer and Dafydd Roberts for help with the illustrations. Finally, most thanks go to my family, particularly Stella, without whose encouragement, support and practical help this book would never have been finished.

Acknowledgements

1

The importance of blood-sucking insects

Insects are the pre-eminent form of metazoan life on land. The class Insecta contains over three-quarters of a million described species. Estimates for the total number of extant species vary between 1 and 10 million, and it has been calculated that as many as 10^{19} individual insects are alive at any given instant (McGavin, 2001). That gives about 200 million for each man, woman and child on Earth! It is estimated that there are 14 000 species of insects from five orders that feed on blood (Adams, 1999) but, thankfully, only 300 to 400 species regularly attract our attention. These blood-sucking insects are of immense importance to humanity.

Humans evolved in a world already stocked with blood-sucking insects. From their earliest days insects would have annoyed them with their bites and sickened them with the parasites they transmitted. As humans evolved from hunters to herders, blood-sucking insects had a further impact on their wellbeing by lowering the productivity of their animals. It is reasonable to assume that, because of their annoyance value, humanity has been in battle with blood-sucking insects from the very beginning. In recent years this battle has intensified because of an increasing intolerance of the discomfort they cause, our fuller understanding of their role in disease transmission and the demand for greater agricultural productivity. But despite considerable advances in our knowledge of the insects and improvements in the weapons we have to use against them, there is still no sign of an eventual winner in this age-old battle.

Many keen observers of nature suspected that insects were in some way involved with many of the febrile illnesses of humans and their animals well before confirmatory scientific evidence was available. The explorer Alexander von Humboldt recorded such a belief amongst the tribes of the Orinoco region of South America. The great German bacteriologist Robert Koch reported the belief of the tribes of the Usambara Mountains of East Africa that the mosquitoes they encountered when they descended to the plains were the cause of malaria (Nuttal, 1899). Sir Richard Burton, in his travels in East Africa, recorded the similar belief of Somaliland tribes that mosquitoes were responsible for febrile illnesses (Burton, 1860). Many of the peoples living near the tsetse fly belts of East and West Africa associated tsetse flies with sleeping sickness of humans and nagana of animals. In our

Table 1.1 *An outline of the early investigations that laid the foundations of medical and veterinary entomology.*

Date	Source	Subject
1878	Manson	Development of *Wuchereria bancrofti* in a mosquito
1893	Smith and Kilbourne	*Babesia bigemina,* the causative agent of Texas cattle fever, transmitted by the tick, *Boophilus annulatus*
1895	Bruce	Transmission of nagana by tsetse fly
1897	Ross	Malaria parasites seen to develop in mosquitoes
1898	Ross	Transmission of avian malaria by mosquitoes
1898	Simond	Transmission of plague from rat to rat by fleas
1899	Grassi, Bignami and Bastianelli	*Anopheles* spp. are the vectors of human malaria
1900	Reed *et al.*	Transmission of yellow fever by the mosquito *Aedes aegypti*
1902	Graham	Transmission of dengue by mosquitoes
1903	Bruce and Nabarro	Sleeping sickness in humans transmitted by tsetse fly
1903	Marchoux and Salimbeni	Transmission of fowl spirochaetes, *Borrelia conserina,* by the tick *Argus persicus*
1907	Mackie	Spirochaete causing relapsing fever transmitted by lice
1909	Chagas	*Trypansoma cruzi,* causative agent of Chagas' disease, transmitted by reduviid bugs

own western tradition North American stock ranchers held the belief that Texas cattle fever was transmitted by ticks (in the class Arachnida, not Insecta) well before this was confirmed experimentally.

The fact that insects are vectors of disease was only confirmed scientifically at the end of the nineteenth century. The key discovery was made in 1877 (reported in 1878) by a Scottish doctor, Patrick Manson, working for the customs and excise service in China. He found that larval stages of the filarial worm, *Wuchereria bancrofti*, developed in the body of a mosquito, *Culex pipiens quinquefasciatus* (Manson, 1878). This was the start of an avalanche of investigations that laid the foundations of medical and veterinary entomology. Some of the key discoveries of this era are outlined in Table 1.1. The main insects involved in the transmission of all the most important vector-transmitted diseases (Table 1.2) are now well known. The list of diseases transmitted is an impressive one and includes the medical scourges malaria, sleeping sickness, leishmaniasis, river blindness, elephantiasis, yellow fever and dengue, and the veterinary diseases nagana, surra, souma, bluetongue, African horse sickness and Rift Valley fever

Table 1.2 *Rounded estimates for the prevalence of disease, the number at risk and the disability adjusted life years (DALYs) for major vector-borne diseases. Figures in millions (M). (DALYs were introduced in the World Bank Development report of 1990 as an estimate of the burden a disease causes to the health of the population. They are often used for comparative purposes and for use in prioritization.)*

Disease	Prevalence	At risk	DALYs	Major distribution	Major vectors
Malaria	273M	2100M	42M	Tropics and subtropics	Anopheline mosquitoes
Onchocerciasis (river blindness)	18M	120M	1M	Tropical Africa, Yemen, Latin America	Blackflies (*Simulium* spp.)
Lymphatic filariasis (elephantiasis)	120M	1100M	5.6M	Africa, Asia and South America	Various mosquitoes
African trypanosomiasis	0.5M	50M	2M	Sub-Saharan Africa	Tsetse flies
Chagas' disease	16–18M	120M	0.7M	Central and South America	Triatomine bugs
Leishmaniasis	12M	350M	2M	Africa, Asia and Latin America	Sandflies
Dengue	50M	3000M	0.5M	Asia, Africa and Americas	Various mosquitoes

Data largely from World Health Organization web pages as of 11 December 2002: http://www.who.int/tdr/media/image.html.

(the piroplasms being tick-borne). Gauging the extent of these diseases is much more problematical, even for human disease. One reason is that health statistics are a moving target, particularly for those diseases such as yellow fever that occur as epidemics. But the greatest problem is that the heartland of these vector-borne diseases is in the under-developed world where, for a variety of reasons, accurate statistical data are often difficult or impossible to gather. For this reason, figures given for the extent of a disease are often not based entirely on hard data, but are an estimate founded largely upon the experience of an expert. Table 1.2 gives an estimate of some of the major vector-transmitted diseases of humans, but obviously, as just indicated, care needs to be taken in the interpretation of the figures.

Blood-sucking insects cause very serious losses to agriculture (Table 1.3). One way this happens is through the transmission of parasites. The

Table 1.3 *Estimated losses in agricultural production caused by blood-sucking insects.*

Insect	Year	Animal mainly affected	Estimated losses (millions US$)	Geographical region
Haematobia irritans (horn fly)	1991	Cattle	800	USA
Stomoxys calcitrans (stable fly)	1965	Cattle	142	USA
Tabanids	1965	Cattle	40	USA
Mosquitoes	1965	Cattle	25	USA
Melophagus ovinus (sheep ked)	1965	Sheep	9.4	USA
Lice	1965	Cattle	47	USA
		Sheep	47	
		Swine	3	
		Goats	0.8	
Tsetse fly	1999	Cattle	4500	Sub-Saharan Africa
Insects, ticks, mites	1994		3000	USA

Information from: Budd, 1999; Geden and Hogsette, 1994; Kunz *et al.*, 1991; Steelman, 1976.

most celebrated case is trypanosomiasis, transmitted by tsetse flies across 9 million km^2 of Africa (Hursey, 2001), and estimated to cause agricultural losses of about US$4.5 billion a year (Budd, 1999). The counter argument has also been proposed that the tsetse has prevented desertification of large areas of land by overgrazing, and has been the saviour of Africa's game animals. The debate has been clearly outlined by Jordan (1986). Other examples of spectacular losses caused by insect-transmitted disease are the death in 1960 of 200 000 to 300 000 horses in Turkey, Cyprus and India caused by African horse sickness, transmitted by *Culicoides* spp. (Huq, 1961; Shahan and Giltner, 1945); and the estimated deaths in the USA, between 1930 and 1945, of up to 300 000 equines from Western and Eastern equine encephalitis transmitted by mosquitoes (Shahan and Giltner, 1945).

In the developed countries it is usually direct losses caused by insects themselves that are of greatest concern. In exceptional circumstances the insects may be present in such numbers that stock are killed; for example, 16 000 animals died in Romania in 1923 and 13 900 in Yugoslavia in 1934 because of outbreaks of the blackfly *Simulium colombaschense* (Baranov,

1935; Ciurea and Dinulescu, 1924). More usually losses are caused not by death but by distress to the animal. Good examples are the reductions in milk yields, weight gains or feed efficiencies that are commonly caused by the painful bites of the tabanids and biting flies. Estimated losses in the USA have been calculated (Steelman, 1976). More recent estimates suggest insects, ticks and mites cost the US livestock producer in excess of $3 billion annually (Geden and Hogsette, 1994). The horn fly is perhaps the major pest in the USA, with an estimated loss in excess of $800 million annually (Kunz *et al.*, 1991). Losses are caused by reduced feed conversion efficiency, reduced weight gains and decreased milk production and are the result of blood loss, annoyance, irritation and behavioural defensive responses on the part of the host.

The sheer annoyance that blood-sucking insects cause to us can easily be overshadowed by their importance in medical and veterinary medicine. In some parts of the world, at certain times of the year, there may be so many blood-sucking insects that any activity outside is difficult or impossible without protective clothing. For example, the biting activity of the midge *Culicoides impunctatus* is thought to cause a 20 per cent loss in working hours in the forestry industry in Scotland during the summer months (Hendry and Godwin, 1988). Such disruption is common during the summer blooms of insects at many of the higher latitudes, and also in many of the wetter areas of the tropics. These levels of annoyance are still rare for most people, and for this reason the concept of nuisance insects is much more difficult to grasp than that of a vector or an agricultural pest causing economic damage. Perhaps the best way to view annoyance caused by insects is as a tolerance threshold. It can then be viewed as a variable with widely separated upper and lower limits; a handful of mosquitoes may be a minor inconvenience to the beggar in the street but intolerable to the prince in the palace.

I suggest that, in the developed world at least, we are increasingly intolerant of nuisance insects. There are several underlying reasons: the increased awareness in the general population of the importance of insects in the spread of disease (sometimes over-exaggerated); the growing stress placed on hygiene and cleanliness; and increasing urbanization, so that for many people blood-sucking insects are not the familiar, everyday things that they were once to our grandparents working in a rural economy. This reduction in our tolerance of nuisance insects causes problems. The extended leisure time and mobility of many people in the developed world means that they spend more time in increasingly distant places. The countries involved are often anxious to promote and develop their tourist industries, and this has led to pressure to control nuisance insects. This can be seen in places such as the Camargue in southern France, the Scottish

Highlands (Blackwell, 2000), the Bahamas, New Zealand (Blackwell and Page, 2003), Florida and many parts of the Caribbean (Linley and Davies, 1971). In addition, population growth has put increased pressure on marginal land which in the past may have been left alone because of nuisance insect problems. Development of this land for leisure, commerce or housing with no insect control input can be disastrous for the developer, user or purchaser.

2

The evolution of the blood-sucking habit

It is believed that haematophagy arose independently at least six times among the arthropods of the Jurassic and Cretaceous periods (145–65 million years ago) (Balashov, 1984; Ribeiro, 1995). The very patchy nature of the insect fossil record means that discussion of the evolution of the blood-sucking habit has until now relied heavily on detective work, with the major clues lying in the diversity of forms and lifestyles seen in modern-day insects, and in some cases in the details of their relationships with vertebrates. From careful interpretation of this evidence quite credible accounts of the likely evolution of the blood-sucking habit can be made. From this starting point it has been convincingly argued that the evolution of the blood-sucking habit in insects has occurred on several occasions, in each case along one of two main routes (Waage, 1979), and these are discussed below. Insect molecular systematics is beginning to emerge from its 'Tower of Babel' stage (Caterino et al., 2000) and it will make a major contribution in defining the detail of the evolutionary routes taken by haematophagous insects (Esseghir et al., 1997; Hafner et al., 1994; Lanzaro et al., 1998; Mans et al., 2002; Sallum et al., 2002). The proposed population bottleneck suffered by phlebotomines in the late Pleistocene and the subsequent radiation of the species out from the eastern Mediterranean sub-region is a good example of what we can expect (Esseghir et al., 1997).

2.1 Prolonged close association with vertebrates

In the first route it is suggested that haematophagous forms may have developed subsequent to a prolonged association between vertebrates and insects that had no specializations immediately suiting them to the blood-sucking way of life. The most common association of this type is likely to have centred around the attraction of insects to the nest or burrow of the vertebrate host. Insects may have been attracted to the nest for several reasons. The humid, warm environment would have been very favourable to a great many insects. In some circumstances, such as the location of the nest in a semi-arid or arid area, the protected habitat offered by the nest may have been essential to the insects' survival. For many insects the nest would also have proved attractive for the abundant supply of food to be found there. Certainly many current day insects such as the psocids are attracted to the high concentrations of organic matter to be found in nests. Indeed, psocids may become so intimately associated with this habitat that they develop a phoretic association with birds and mammals, climbing into fur and feathers, to be translocated from one nest site to another (Mockford, 1967; Mockford, 1971; Pearman, 1960).

Initially feeding on dung, fungus or other organic debris, the insects attracted to the nest would also have encountered considerable quantities of sloughed skin, hair or feathers. The regular, accidental ingestion of this sloughed body covering probably led to the selection of individuals possessing physiological systems capable of the efficient use of this material. Behavioural adaptations may then have permitted occasional feeding direct from the host itself. It is easy to see how this may have gone hand in hand with the adoption of a phoretic habit. Morphological and further behavioural adaptations would have allowed the insect to remain with the host for longer periods with increasingly efficient feeding on skin and feathers.

The mouthparts developed for this lifestyle, in which the insect feeds primarily on skin and feathers, were almost certainly of the chewing type, such as those seen in the present-day Mallophaga. While these mouthparts are not primarily designed to pierce skin some mallophagans do feed on blood. *Menacanthus stramineus*, a present-day mallophagan, feeds at the base of feathers or on the skin of the chicken. The insect often breaks through to the dermis, giving it access to blood on which it will feed (Emmerson *et al.*, 1973). Blood has a higher nutritional value than skin and is far easier to digest. This is reflected in the increased fecundity of blood-feeding Anoplura compared to skin-feeding Mallophaga (Marshall, 1981). Once blood was regularly encountered by insects, it is likely that its high nutritional value favoured the development of a group of insects that regularly exploited blood as a resource. This would have developed progressively, through physiological, behavioural and morphological adaptations, first to facultative haematophagy and eventually, in some insects, to obligate haematophagy. One way in which the progression from skin feeding to blood feeding may have occurred is seen in members of the mallophagan suborder the Rhynchophthirina, such as the elephant louse, *Haematomyzus elephantis*. This insect possesses typical mallophagan biting-type mouthparts (Ferris, 1931; Mukerji and Sen-Sarma, 1955) which are not primarily adapted for obtaining blood. By holding the mouthparts at the end of an extended rostrum (Fig. 2.1) the insect manages to use them to penetrate the thick epidermal skin layers of the host to get to the blood in the dermis.

It is thought that haematophagous lice developed from an original nest-dwelling, free-living ancestor (Kim, 1985) along the pathway described above. We do not know when the change occurred from free-living nest

Figure 2.1 Despite having the chewing mouthparts typical of mallophagans, *Haematomyzus hopkinsi* is unusual in feeding on blood. The chewing mouthparts are held on the end of an unusual, elongated rostrum, which may well be an adaptation helping the insect reach the blood-containing dermis through the thick skin of its wart-hog host. (Courtesy of Vince Smith)

dweller to parasite, but it may well go right back to the appearance of nesting or communal living in land-dwelling vertebrates, which is thought to have happened during the Mesozoic (225–65 million years ago). So lice may have predated the emergence of mammals and birds and been parasitic on their reptilian ancestors (Hopkins, 1949; Rothschild and Clay, 1952). It is highly likely that ancestral forms were parasitic on primordial mammals and that from there they radiated along the lines of mammalian evolution. Helping drive this rapid speciation of the permanent ectoparasites was the reproductive isolation they suffered from being confined on specific vertebrate hosts, which may well have enhanced the effects of classical geographic reproductive isolation. Co-evolution of the host and permanent (and to a lesser extent temporary) ectoparasites probably led to rapid speciation in lice and other ectoparasitic forms. The evidence for co-speciation in lice is strong. Sequence analysis of mitochondrial cytochrome oxidase I genes suggests co-speciation in the pocket gophers *Orthogeomys*, *Geomys* and *Thomomys* and their chewing lice (Fig. 2.2) (Hafner *et al.*, 1994). Co-speciation also predicts temporal congruence between chewing lice and gopher speciation. This is borne out by analysis of the molecular data, in which the synonymous substitution rate is approximately an order of magnitude greater in the lice compared to the gophers. This roughly parallels the differences in generation times of the two groups, suggesting equal rates of mutation per generation. While the case for pocket gophers and their chewing lice is strong, the extent to which co-speciation is generally the case is unclear. Classical taxonomy, which has tended to group species with origins on the same host, may be misleading. Molecular studies are showing this is a dangerous practice and that not all species have stuck to the co-evolutionary model mentioned above (Johnson *et al.*, 2002a; Johnson *et al.*, 2002b).

Some beetles also appear to be developing along the evolutionary highway described above. Several hundred species have been reported from nests and burrows (Barrera and Machado-Allison, 1965; Medvedev and Skylar, 1974). Most of these are probably free-living, feeding on the high levels of organic debris to be found at these sites. Some of these beetles have developed a phoretic association with the mammal which allows them to transfer efficiently between nest sites. Many of these phoretic forms also feed on the host by scraping skin and hair, and some have progressed to the stage when they will occasionally take blood (Barrera, 1966; Wood, 1964).

The prolonged association of the insect with the vertebrate, which is the cornerstone of this first route for the evolution of the blood-feeding habit, may not always have relied on encounters in the nest habitat. Free-living ancestral forms with few, if any, clear adaptations for the blood-sucking way of life may have also developed prolonged associations with

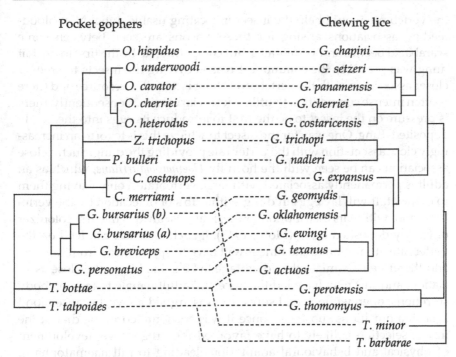

Figure 2.2 Phylogenies of pocket gophers and their chewing lice based on nucleotide sequence data (Hafner *et al.*, 1994). The figure shows composite trees based on multiple methods of phylogenetic analysis. Branch lengths are proportional to inferred amounts of genetic change. Pocket gopher genera are *Orthogeomys*, *Zygogeomys*, *Pappogeomys*, *Cratogeomys*, *Geomys* and *Thomomys*. *Geomys bursarius* is represented by two subspecies (a = *G. b. halli*; b = *G. b. majusculus*). Chewing louse genera are *Geomydoecus* and *Thomomydoecus*. The program COMPONENT was used to document significant similarity in branching structure between these trees. Because the host and parasite trees were based on DNA sequences from the same gene (cytochrome c oxidase subunit I), rates of DNA evolution could be compared in the two groups. Based on these data, Hafner *et al.* (1994) estimated that chewing lice were evolving approximately ten times faster than pocket gophers in this gene region, which is in line with the predictions if co-evolution is occurring (see p. 10). But see Page *et al.* (1996).

the vertebrate at some point distant from the nest. This type of association may have had several different underlying reasons, such as attraction to feed on vertebrate secretions, or the use of the vertebrate as a basking or swarming site. But probably the most important factor was the use of the host's dung as a larval habitat.

The vertebrate may live in a harsh environment where dung rapidly dries up or it may bury its dung. In either case, closely associating with the vertebrate would help the insect in locating usable dung. Non-blood-feeding associations arising for these reasons are seen between some scarabaeid beetles and vertebrates, but competition to be the first to exploit dung was probably the commonest reason driving the insects to an ever closer association with the vertebrate. Dung is a limited resource and there is often intense competition to utilize it as a larval site. Consequently, there is pressure on the insect to be the first to introduce its eggs into the newly deposited dung. One way for the insect to achieve this is to form an increasingly close association with the vertebrate providing the dung. Such a close association can be seen with the horn fly, *Haematobia irritans*, which as an adult is permanently associated with large vertebrates, only leaving them to oviposit. It will lay eggs in dung within 15 s of defecation by the vertebrate and, within its distribution range, is almost always the first colonizer of freshly deposited dung (Mohr, 1943). If such an insect can feed on the vertebrate, it minimizes the time it will have to spend away from the host and therefore maximizes the advantage to be gained from the close association, and this may have led the ancestral adult female to feed on body secretions, open wounds and sores. As was argued above, because blood is such a nutritious substance, once it was encountered in the diet of the insect, selection is likely to have favoured the progressive development of physical and behavioural adaptations leading to full haematophagy. Selection will have led to the progressive development of mouthparts, allowing the insect to dislodge scabs, open up old sores and eventually to penetrate unbroken skin. The evolution of organisms capable of breaking the skin and obtaining a blood meal may very well have been an accelerated process. The wounds produced by the first blood-suckers will have provided regular blood-feeding opportunities for other organisms that could not break the skin by their own efforts. Once given access to blood these too may well have followed the evolutionary pathway outlined above.

The spasmodic appearance of blood feeding among male insects is a difficult issue to explain, but the close or permanent association of females with the vertebrate as discussed above is one factor that may explain it in some insects. Under these circumstances it may become advantageous for the male to become associated with the vertebrate because of the greater likelihood of success in finding a mate. This may then lead to blood feeding in the male, as it would minimize the time spent away from its host and therefore maximize its chances of successful mating. Similarly, if blood-sucking females are irregularly and/or widely dispersed in a habitat, then the male may gain a mating advantage by staying with the vertebrate and waiting for the female to arrive to feed. This is seen in the 'following swarm' of the tsetse flies but also in the mosquito *Aedes aegypti*, the males of which do not take blood (Teesdale, 1955).

2.2   Morphological pre-adaptation for piercing

The second route for the evolution of the blood-sucking habit suggests that blood feeding developed in some insect lineages from ancestral insects that were morphologically pre-adapted for piercing surfaces (Beklemishev, 1957; Downes, 1970; Waage, 1979). Entomophagous insects are strong candidates for such a conversion. The Rhagionidae are a good example: most of the group are predacious on other insects, but a few species have turned to blood feeding. How could this changeover have come about? Entomophagous insects would have been attracted to nests and burrows by the accumulation of insects to be found there and so would have encountered vertebrates. Away from the nests they would have been attracted to vertebrates by the accumulation of insects around them, or the vertebrates may have regularly congregated in the wet areas that are the breeding sites for many of the 'lower' Diptera. The vertebrates involved may have been permanently resident amphibians or reptiles, or larger vertebrates that regularly visited such sites for drinking or bathing purposes. In each of these cases it is easy to see how entomophagous insects could have made repeated and possibly prolonged contact with vertebrates. These predatory insects would have physiological and morphological adaptations (such as efficient protein-digesting enzymes and piercing mouthparts) facilitating the switch to haematophagy. Haematophagy in these individuals was at first probably an occasional, chance event which led to full haematophagy through continued close association with the vertebrate host. It is thought that haematophagy developed along these lines in the ancestors of the blood-feeding bugs and in blood-feeding rhagionids and possibly in some blood-feeding Diptera. There is some doubt about which came first, larval feeding on nest debris or adult feeding on other insects in the nest habitat. Fleas may also have evolved along this pathway from free-living mecopteran stock (Hinton, 1958; Tillyard, 1935). The Mecoptera, or scorpion flies, contain a modern-day group, the Boreidae, which are apterous and are capable of jumping. They live in moss and feed on insects. Similar insects may well have been the ancestors of the fleas, a view receiving support from molecular systematics (Whiting, 2002).

The lifestyles of several present-day insects support the idea that entomophagous insects gave rise to some blood-sucking insect groups (Waage, 1979). Many personal experiences in Britain (as many as three in one day) with the flower bug *Anthocoris nemorum* show that this insect is willing and able to pierce human skin. This insect is entomophagous, living on and around flowers where it pounces on small insects visiting the flowers to feed. While its probings of my skin cause a sharp pain, I have yet to find one that has obviously ingested any blood; however, it still establishes the fact that entomophagous insects will often show an interest in vertebrates as potential sources of a meal. The hemipteran bug *Lyctocoris campestris* is also an entomophagous species, but it takes matters further. It can live in birds' nests, where it feeds on other insects, but it will also take blood meals from vertebrates (Stys and Daniel, 1957). Evidence from blood-sucking insects themselves also points to the close links between entomophagy and the blood-feeding habit. The mosquitoes *Aedes aegypti* and *Culex tarsalis* will take body fluids from insect larvae presented to them under laboratory conditions (Harris *et al.*, 1969). Indeed, this form of feeding is so successful that these mosquitoes go on to produce viable eggs as a result, which opens up the intriguing possibility that this may occur naturally in the field.

It has also been argued that haematophagy may have arisen in some insect groups (including the mosquitoes; Mattingley (1965)) from plant-feeding ancestors. This is certainly a possibility as many plant-feeding insects possess piercing and sucking mouthparts that would pre-adapt them for haematophagy. This is seen in the moth *Calpe eustrigata*, which is one of a group of noctuiids that possess an unusually modified, sharp proboscis used in most species for the penetration of fruit rinds. But *C. eustrigata* uses it to penetrate vertebrate skin for the purposes of blood-feeding. It is probable that plant-feeding ancestors of modern-day blood feeders would have developed haematophagy only if they were in a position of continual association with the vertebrate host. This may have occurred through mechanisms similar to those already outlined. Attraction to free-living vertebrates may have occurred in order to feed on bodily secretions or to use dung as a larval medium. Or insects may have been attracted to nests to feed on fruits or seeds stored there by the vertebrate. In this context it is interesting to note that the hemipteran bugs are exceptional in using cathepsin-like digestive proteinases. That is consistent with a proposed evolutionary path for bugs from sap-sucking (Billingsley and Downe, 1988; Houseman *et al.*, 1985; Terra, 1988) or seed-feeding ancestors. Sap feeders, not needing proteases, may have lost their trypsins. If they then moved to blood feeding they would need to reacquire proteolytic activity. Seeds are often sources of powerful anti-serine protease molecules produced to protect the seed from insects. Seed feeders may have moved to cathepsins to avoid these inhibitors. It is argued that having lost their trypsins they had to make use of the cathepsins contained in the lysosomes of all cells, re-routing them for extracellular digestion. Molecular evidence suggests Triatominae from the Americas and Asia are monophyletic with an origin in northern areas of South America, in Central America, or in the southern region of North America about 95 million years ago (Gaunt and Miles, 2002; Lyman *et al.*, 1999).

3

Feeding preferences of blood-sucking insects

3.1 Host choice

Blood-sucking insects feed from a range of different host animals; because of the bites we receive we are acutely aware of the fact that many of them feed from humans, but many other animals are also exploited, including other mammals, birds, reptiles, amphibians and fish, and even insects, arachnids and annelids (Hocking, 1971). Any one insect does not feed equally well from all of these potential resources; it displays host choice. For some insects, particularly some permanent ectoparasites, the host choice may be very specific. Occasionally, for example for human lice, just a single species. For other blood-sucking insects host choice is clearly not as restricted as introduced exotic hosts (e.g. those in zoos) quickly become incorporated into the diet of local blood-sucking insects.

Let us consider what is meant by host choice. In its main sense it denotes the species of host animal or animals from which blood-sucking insects obtain their blood meals. But host choice can go beyond particular species of host chosen. Insects often choose to feed on particular individuals from among preferred species, which may well have implications for disease transmission (Burkot, 1988; Kelly, 2001; McCall and Kelly, 2002).

Although most blood-sucking insects in their undisturbed, natural surroundings show a preference for feeding from a particular group or species, or even a primary cohort of their chosen species, the degree of host specificity shown varies greatly from one type of insect to the next. Some are entirely dependent on a single species of host while others are willing to feed from a wide range of hosts. As a rough rule of thumb, it has often been said that there is a direct relationship between the locomotory capabilities of an insect and the number of different hosts that it utilizes. Thus, the permanent ectoparasites, with their limited capacity for movement away from the host, contain most cases of precise dependence. For example, the louse *Haematomyzus elephantis* is confined to elephants, *H. hopkinsi* to the wart-hog (Clay, 1963) and *Pediculus capitis* to humans. Considering the more mobile periodic ectoparasites, we can find many examples such as the 'human' flea, *Pulex irritans*, that have a narrow preferred range (badgers and foxes) on which most insects are found, but that can occasionally be discovered on a wide range of other hosts (in this case humans, pigs and other large mammals) (Marshall, 1981). When considering the largely free-living, temporary ectoparasites we find that many have catholic tastes. The mosquito *Culex salinarius*, for example, has a wide range of potential hosts that includes birds (45 per cent of the blood meals tested), equines (17 per cent) and canines (15 per cent) (Cupp and Stokes, 1976). This mosquito will demonstrate its cosmopolitan tastes further if it is disturbed during the meal by moving willingly from one host species to another to complete this single, full blood meal. This is a regular and natural occurrence for this insect. In one investigation 13 per cent of the blood meals tested were found to be from a mixture of hosts (Cupp and Stokes, 1976).

The above rule of thumb works well in a broad sense but often breaks down when particular insects are considered, as can be seen with the help of two examples. The amblyceran louse, *Menacanthus eurysternus*, has poor locomotory abilities off the host and according to our rule of thumb it should be limited to a small number of hosts. In fact it has been recorded from 123 different species of bird (Price, 1975)! In contrast, each of four closely related species of winged streblid were found to be specific to a different genus of bat and this relationship held true even in caves containing three of the bat genera in crowded conditions (Maa and Marshall, 1981).

Before looking at some of the factors that underpin host choice, let us make some general comments of particular relevance to temporary ectoparasites. The commonest hosts are probably large, social herbivores, which present an abundant and easily visible food source for many ectoparasites. They are also reliable food sources because normally they move only slowly from one pasture to another. Carnivores are less abundant than their prey, often solitary, less visible and less predictable, occupying a large home range. For these reasons carnivores are less likely to be a primary element in the host choice of a blood-sucking insect. Also large vertebrates are likely to be preferred over smaller ones. This is because small vertebrates, which suffer greater losses from the attention of a blood-sucking insect, use their agility to develop efficient defences.

Moving from these generalizations to consider the specific choices made by particular insects is a more complex issue. Sometimes a reasoned case for the evolution of a particular pattern of host choice can be made, but on most occasions we are in the dark. The choice may be determined by a large number of factors (probably acting in combination), including behavioural, physiological, morphological, ecological, geographical, temporal and genetic considerations. To give an idea of the complexity of the issue, let us look at a small number of examples, showing how each of these factors may affect host choice.

Although it is not always the case (Canyon *et al.*, 1998; Charlwood *et al.*, 1995; Prior and Torr, 2002), host defensive behaviour can have a considerable effect on the number of blood-sucking insects successfully feeding on that host (Torr *et al.*, 2001). The anti-mosquito behaviour displayed by a range of ciconiiform birds is a good demonstration of the protective effects host defensive behaviour can have. At roost some of these birds perform up to 3000 defensive movements per hour! But the green heron and the crowned night heron are far less active, performing as few as 650 defensive movements per hour: as a consequence they receive far more bites than their more active relatives (Webber and Edman, 1972). If no other factors were in operation it is easy to see how natural selection would bear on such a case and would lead a blood-sucking insect to select and eventually specialize in feeding from the easiest available target. The fact that under natural conditions mosquitoes still feed from a wide range of ciconiiform birds, including those displaying the highest levels of defensive behaviours, is a testament to the complexity of the factors that determine host choice. None the less, grooming and host defensive behaviour in general are likely to be highly efficient agents of natural selection. For humans the concept of defensive behaviour needs to be extended to include such practices as sleeping under bednets, screening houses and the use of repellents (modern preparations containing active ingredients such as di-ethyl toluamide (DEET) or traditional concoctions such as mixtures of cow dung, urine and ash) (MacCormack, 1984).

The intensity of defensive behaviour often correlates with the number of blood-sucking insects attacking the host. Thus is it possible that the seasonal changes in the numbers of blood-sucking insects may play a part in the seasonal changes in host choice because of changes in the amount of host defensive behaviour (Edman and Spielman, 1988). Such seasonal changes in host choice are important in the transmission of some zoonoses to humans. For example, in the USA the arbovirus eastern equine encephalitis is enzootic in birds and the mosquito *Culiseta melanura* is its vector. During the spring and early summer, when mosquito numbers are comparatively low, this insect feeds almost exclusively on passerine birds. Later in the season mosquito numbers increase and so does defensive behaviour by the birds. At this time mosquito host choice becomes more catholic and includes other bird groups and mammals. It is also in this later part of the season that epizootics of the virus occur in birds and horses and epidemics occur in humans. All this is enabled by the greater range of hosts chosen by the mosquitoes, driven by changing levels of host defensive behaviour. Mathematical models of disease transmission have been produced that incorporate the assumption that different individual levels of host defensive behaviour lead to a non-homogeneous pattern of host choice within a host species (Kelly and Thompson, 2000).

I suggest that physiological factors are mainly important in determining host choice in the sense that once the insect has become associated with a

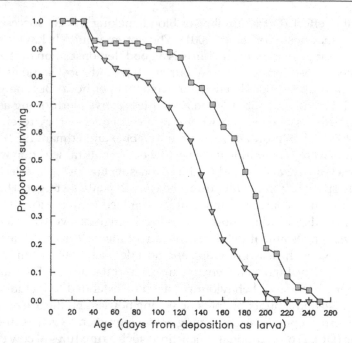

Figure 3.1 The tsetse fly, *Glossina morsitans morsitans*, shows greater longevity when it feeds on rabbits (■) rather than goats (▼) (Jordan and Curtis, 1972).

narrow range of hosts then specialization in its physiology will occur that may limit the range of other hosts it can exploit (Krasnov *et al.*, 2003). This is because natural selection will ensure that all of the insect's systems will become tuned to the exploitation of the resources of its major host, which may restrict the insect's ability to deal with unusual situations. In some cases choice of the wrong host can, for physiological reasons, lead to the death of the insect. For example, a blood meal from the guinea-pig may form oxyhaemoglobin crystals which will rupture the intestine of several blood-sucking insects (Krynski *et al.*, 1952).

The physiological consequences of moving on to an unusual host can have a less obvious, but a nevertheless damaging, effect on the success of the insect. Several experimental studies have shown that the fecundity of a blood-sucking insect depends on the host on which the insect feeds. Reduced fecundity can be generated by a reduced rate of development of the insect, reduced longevity (Fig. 3.1), a skewed sex ratio, or reduced food intake or rate of digestion (Chang and Judson, 1977; Nelson *et al.*, 1975; Rothschild, 1975). We can look at the tsetse fly as an example. *Glossina austeni* fed on rabbit blood showed a consistently higher fecundity than flies of the same species fed on goat blood (Jordan and Curtis, 1968). In a further series of experiments it was shown that both *G. austeni* and

G. *morsitans morsitans* fed on pig blood produced the heaviest puparia, closely followed by flies fed on goat blood, while the puparia produced from flies fed on cow blood were considerably lighter (Mews *et al.*, 1976). Clearly under natural conditions it would be a selective advantage to the insect to develop mechanisms that avoided such a restriction on its reproductive success by careful choice of hosts. It is also of interest to note that the source of the blood meal or its quality may also influence the course of an infection in the insect. If tsetse flies feed on goats or cattle they will develop a much higher infection rate with *Trypanosoma vivax* than if they feed on mice (Maudlin *et al.*, 1984). Similar results are also found in more natural vector combinations (Nguu *et al.*, 1996).

As with the physiological restrictions on host choice just discussed, morphological factors are also important as a limitation on the range of available hosts once a certain degree of specialization for a particular host has taken place. With permanent and periodic ectoparasites especially, restriction in host range usually involves morphological specialization of the insect's mouthparts, ovipositor or locomotory attachment apparatus. Specializations of any of these can easily restrict the host range of the insect concerned; indeed, in many cases specialization may even limit the area of the body of the preferred host that the insect can utilize. This is discussed in more detail in Section 7.1.

It has also been suggested that morphological characteristics of the host are important in determining host choice. One example is the frequently reported preference of *Anopheles gambiae* for human adults rather than children. It is suggested that the choice is a direct consequence of the difference in size of these two potential hosts, the number of bites received by an individual being in direct proportion to the surface area that individual contributes to the total immediately available to the mosquito (Port *et al.*, 1980). Interestingly this is not the case for adult cows and calves; here it is differences in levels of defensive behaviour that determine that adults are bitten more than calves by tsetse flies (Torr *et al.*, 2001).

A clear example of ecological influence on host choice is furnished by the fleas. The immature stages of most fleas live on the detritus in the homes of the host animal. In other words, host choice for fleas is largely determined by the ecology of the host, most fleas being restricted to hosts with long-term homes, or at least seasonal lairs that are used each year. Because they so commonly use long-term homes, mammals are the main hosts for fleas. Rodents are unrivalled as home builders and as a consequence are the most afflicted mammals, harbouring 74 per cent of all known flea species (Marshall, 1981). Humans are also habitual home dwellers and because of this we are the only primates that are regularly flea-ridden. Interestingly the 5000-year-old late-Stone-Age Tyrolean ice man was infested with *Pulex irritans* (Spindler, 2001).

A second example of ecological influence on host choice, of considerable importance in the health of Europeans, can be seen in the changing feeding patterns of the anopheline mosquitoes of Europe. Until the middle of the nineteenth century benign tertian malaria (caused by *Plasmodium vivax*) was endemic in Europe as far north as Scandinavia, with *P. malariae* as far north as Holland and the UK, and *P. falciparum* extensive in southern Europe. At this time malaria began to retreat from Europe and transmission of the disease now rarely occurs. Why did this happen? Ronald Ross described the full cycle of malaria in 1898 and soon after this, in 1901, it was realized that there were areas of Europe that abounded with anopheline mosquitoes, and into which malaria carriers routinely passed, but in which the disease was not transmitted (Falleroni, 1927). Several theories were put forward to account for this apparent anomaly, but it was not until after the introduction of the precipitin test, for the identification of mosquito blood meals (see Pant *et al.*, 1987), that relevant evidence

---

Box 3.1 The importance of human-biting rates of mosquitoes for the transmission of malaria

The vectorial capacity C of a population of mosquitoes transmitting malaria can be described by the equation

$$C = m \times a^2 \times \frac{P^n}{-\ln(P)}$$

where m is the number of mosquitoes per person and a is the proportion of these mosquitoes that bite humans (biting once to acquire the infection and transmitting it on the occasion of a subsequent meal on a human, hence a^2) – the quantity a is usually estimated using techniques such as those described in Box 3.2.

The value $P^n/-\ln(P)$ describes the expectation of the infective life of the vector population (Garrett-Jones and Shidrawi, 1969).

If we take the example described in the text we can see that in those parts of Europe where malaria was being transmitted $a = 1/8$, whereas in those areas where mosquitoes were present but no malaria was being transmitted $a = 1/400$.

If we assume that m and $P^n/-\ln(P)$ are constants, then the equation tells us that humans in the malarious areas are 2500 times more likely to contract malaria than those in the non-malarious areas. The difference in this example is entirely due to the different proportions of mosquitoes biting humans in the two areas. A good account of these mathematical models is given by Dye (1992).

Source: From Burkot (1988)

Box 3.2 Identification of the source of a blood meal

The ability to identify the source of an insect's blood meal accurately is a powerful tool for the investigation of a wide range of questions relating to the life of blood-sucking insects. For example, it allows us to investigate the nature of host choice and it can help us to understand whether parasites manipulate the host-seeking behaviour of their vectors (Koella et al., 1998b). It may help in monitoring the efficacy of various control or surveillance tools such as repellents, bednets and traps and to estimate the degree of coverage needed for emerging malaria vaccines (Mukabana et al., 2002a).

Until recently blood meal identification has been achieved through immunological means (Pant et al., 1987). Increasingly molecular techniques are taking over (Mukabana et al., 2002a). These are usually 'fingerprinting' techniques that look at the number of tandem repeats (VNTRs) and/or short tandem repeats (STRs), both of which occur abundantly in nuclear DNA in most eukaryote genomes. The number of tandem repeats at these loci is highly variable with alleles differing in length by an integral number of repeat units. As well as being highly polymorphic these loci are stably inherited and so are informative genetic markers. The markers can be easily accessed because they can be amplified in vitro by polymerase chain reaction and the size of the products easily determined. They have been used to identify successfully the blood meal source of mosquitoes (Koella et al., 1998a; Mukabana et al., 2002a; Mukabana et al., 2002b), crab lice (Lord et al., 1998) and tsetse flies (Torr et al., 2001). Not only can they identify the species that was the source of the blood meal, they can be so specific that they can identify the particular individual within a species from which the blood meal was taken.

---

began to accumulate. The key factor was that anopheline mosquitoes in the malaria-free areas were 400 times as likely to feed on animals other than humans, while in malarious areas 1 in 8 feeds were from humans (Hackett and Missiroli, 1931). The importance of these rates of biting for the transmission of malaria is explained in Box 3.1. Immunological tests are now being replaced by more sensitive and specific molecular techniques for the identification of vector blood meals (Mukabana et al., 2002a) (Box 3.2).

Why should the mosquitoes of one area bite humans so regularly and mosquitoes of the same species in another area bite them so infrequently? Although it was suggested that the mosquitoes were actually changing in their relative attractions for humans and domesticated animals (particularly pigs, horses and cattle), no such change could be demonstrated experimentally. It is now widely accepted that the switch from humans to animals was a response to the changing proportions of human and non-human hosts available. At this time animal husbandry practices in Europe were changing so that far more animals were being kept and were available to the mosquitoes as food sources. A contributory factor was the improving standard of living for the human population, which led to larger and brighter housing that was much less attractive as a resting site for the mosquito. The mosquitoes were diverted to the darker animal sheds around the house, considerably reducing the likelihood of their biting humans (Harrison, 1978; Takken *et al.*, 1999). Also of importance was the decreasing birth rate among the population, reducing the numbers of human hosts available, a factor accentuated by the increased mechanization of agriculture which also led to a reduction in the populations in rural communities. So we can see that host choice is not only a behavioural, decision-making process on the part of the insect, but can also be strongly influenced by the relative availability of different host species.

Geographical considerations are also important in host choice because if there is no overlap in the range of insect and potential host then the host is not available as a food source. Global climate change may be affecting these overlaps (Sutherst *et al.*, 1998). Geographic overlap of host and insect can work on a variety of scales. The effect is obvious from a global viewpoint, but not so obvious when more localized issues are considered. The tsetse fly can be used as an example, one that has considerable economic importance. On the largest geographical scale we can see that tsetse flies, apart from a small fly belt in Saudi Arabia (Elsen *et al.*, 1990), are only found in Africa, where they are widely distributed from the southern borders of the Sahara to Mozambique. Within this region they transmit trypanosomes causing sleeping sickness in humans and nagana in animals. The game animals of Africa have some degree of immunity to nagana, but normally it quickly kills introduced domesticated animals. This disease and its transmission by the tsetse fly have had a tremendous influence on the history of tropical Africa. The combination largely prevented the invasion from the north of armies dependent on horses. It also impeded Africa's development by limiting the use of draft animals and prevented (and indeed still prevents) the use of large tracts of land for ranching. In other words large herbivores entering the tsetse's geographical range, including horses and cattle, become available as hosts and are often chosen as food sources. But geographical overlap is also important in tsetse fly biology on a much smaller scale. Long before Bruce (1895) demonstrated that the tsetse fly transmits the trypanosomes that cause illness and death, the herdsmen of southern Africa had learned by bitter experience that the fly was deadly. This fact became obvious to them because the tsetse flies of this region are not evenly spread throughout the land, but are restricted to certain areas known as fly belts. The herdsmen learned that if they avoided these belts, and the flies that lived there, then their stock did not acquire nagana (McKelvey, 1973).

Geographical effects on host choice can be seen on an even smaller scale. The tree hole mosquitoes *Aedes triseriatus* and *Ae. hendersoni* are sibling species living sympatrically in woodlands in the eastern and midwestern USA. It might be expected that two such closely related species living in the same woodland would be feeding on similar hosts, but geographical separation again affects host choice. *Aedes triseriatus* feeds mainly at ground level and consequently feeds on ground-dwelling animals such as deer and chipmunk. In contrast, *Ae. hendersoni* feeds mainly in the canopy of these woodland trees on animals such as tree squirrels (Nasci, 1982). An important example, because it has consequences for malaria transmission, occurs with the malaria vectors within the *Anopheles gambiae* complex (see Section 3.2). It has often been reported that *An. arabiensis* is much more zoophilic that *An. gambiensis s.s.* but in a carefully designed experiment that controlled for equal accessibility to hosts, which were presented outside houses, it was found that both species showed statistically similar levels of human- and cattle-biting activity (Diatta *et al.*, 1998). So an unwillingness to bite indoors may be the basis of previous reports of zoophily in *An. arabiensis* rather than a real difference in host preference. Clearly geographical overlap on even the smallest scale is very important in determining host choice; indeed, host abundance and proximity may be the ultimate arbiter of host choice in many instances.

As we have just seen, availability of hosts is often a prime factor in determining host choice. For the host to be available there must not only be geographic overlap between the insect and host but also a temporal overlap because most temporary ectoparasites feed only during a well-defined period of the day. Let us look at two contrasting ways in which temporal overlap could occur. Some hosts show such efficient defensive behaviour that temporary ectoparasites can feed only when the host is at rest. The degree of overlap between the activity period of the insect and the resting period of such a host will then be important in determining host choice. In contrast, animals living in deep burrows are unlikely to be hosts for exophilic insects when at rest, but may be hosts if their activity period coincides with that of the insect.

Seasonal variation in the choice of host has been recognized in some blood-sucking insects, and it can have serious consequences for disease transmission to humans. In North America the arbovirus St Louis encephalitis is transmitted during the summer months by the mosquitoes *Culex nigripalpus* and *C. tarsalis*. Both these mosquitoes show a marked seasonal change in their feeding patterns, switching from bird feeding in the winter and spring to mammal feeding in the summer, when arbovirus transmission occurs (Edman, 1974; Edman and Taylor, 1968; Tempelis and Washino, 1967) (see discussion above concerning eastern equine encephalitis).

Evidence is beginning to appear to suggest that memory can play a role in host choice. Thus many blood-fed *Anopheles arabiensis* that had been artificially transported returned to the houses where they had previously obtained a blood meal. Some probably flew more than 400 m to do so, which implies a considerable spatial memory (McCall and Kelly, 2002; McCall *et al.*, 2001). In addition, as well as returning to successful feeding sites, vectors may choose their next host on the basis of previous success- ful feeding experiences. Such imprinting was seen in *Culex* spp. in field cages where the tendency was for individual mosquitoes to choose the same host on which they had previously fed successfully (Mwandawiro *et al.*, 2000). Significantly this preference was not retained by the offspring, which is what would be expected if this is indeed learned behaviour. Such learned behaviour in host selection, if it proves to be widespread, can have a considerable impact on disease transmission (Kelly and Thompson, 2000; McCall and Kelly, 2002). For example, a major component of host selection within a group of hosts may be the level of individual host defensive behaviour (see above and Chapter 7). Sick hosts tend to display fewer defensive behaviours and are fed upon most often (Day and Edman, 1983). If this were absolute then no uninfected hosts would ever become infected. Clearly that is not the case, but nevertheless this aspect of host selection may be a factor determining the rate of disease transmission in a population.

It has often been suggested that there are genetically determined behavioural traits that produce different feeding patterns in different pop- ulations of a single insect species (rather than the well-recognized phe- nomenon of different feeding patterns seen among members of species complexes – see Section 3.2). Most of the support for this is anecdotal but convincing evidence has been reported for *Aedes simpsoni* and *Ae. aegypti* (Mukwaya, 1977). These species showed direct evidence of both zoophilic and anthropophilic populations in the laboratory and in the field. It was established that single species were involved when no crossing or hybrid sterility appeared during crossing and backcrossing experiments of the populations studied.

3.2   Host choice and species complexes

For most blood-sucking insects the spectrum of host choice differs with changing place and season and given different proportions of available hosts. Selective adaptation to such changes in local circumstance is the fac- tor driving speciation. This process is continual and as different species are in the process of emerging they are not always easily distinguishable by their morphological characteristics. Indeed, several blood-sucking insect 'species', including *Anopheles gambiae sensu lato* and *Simulium damnosum s.l.*, are known to be species complexes made up of reproductively isolated sibling species, having differing biological characteristics. These differences commonly include host choice. Let us consider in more detail the *An. gambiae* complex, in which the differences and the distribution of the different members of the complex can be very important in disease control terms (Besansky, 1999; Coetzee *et al.*, 2000; Coluzzi *et al.*, 2002).

Anopheles gambiae s.l. is an enormously important malaria vector throughout Africa south of the Sahara. A major filariasis vector, it also transmits the arboviral disease O'nyong-nyong. As such an important vector it has received considerable attention from entomologists. Before the realization that *An. gambiae s.l.* was a complex of species, the conflicting nature of the biological data gathered was a source of puzzlement and controversy. It is now known that the *An. gambiae* complex is made up of seven very similar sibling species and even within those divisions variations are known to occur (Gentile *et al.*, 2002; Lanzaro *et al.*, 1998; Powell *et al.*, 1999). Pre-copulatory isolation barriers are believed to minimize sibling hybridization (Favia *et al.*, 1997) which would result in sterile male progeny. These barriers are probably the major factors driving speciation processes within the complex, but the precise nature of the barriers is not known. Four of the seven sibling species breed in fresh water: *An. gambiae sensu stricto*, which is distributed throughout sub-Sahelian Africa, particularly the more humid regions; *An. arabiensis*, which is also found widely distributed in Africa but shows some preference for drier areas; and *An. quadriannulatus* A and B, which show a more restricted range, being found only in Ethiopia, Zanzibar and parts of southern Africa. A fifth species, *An. bwambae*, is only known from mineral springs in the Semliki forest of Uganda; it is a malaria vector, but because of its narrow geographical range is of minor importance. Finally there are two species that breed in brackish water: *An. melas*, which is found along the west African coast, and its east African equivalent, *An. merus*.

All four freshwater species and *An. bwambae* are morphologically indistinguishable and require cytotaxonomic, biochemical or molecular methods for identification (Coluzzi *et al.*, 2002; Munstermann and Conn, 1997). The saltwater species can be identified from the careful use of morphological criteria. Because of their differing biological characteristics, the seven sibling species have very different vectorial capacities. The principal vector of human malaria and one of the two major vectors of filariasis in the complex is *An. gambiae s.s.*, which is primarily an endophilic species biting humans. In many parts of this mosquito's range humans are the major available host and, in these circumstances, blood meal identifications often reveal that virtually 100 per cent of the meals are of human origin (Davidson and Draper, 1953). Interestingly, when cattle or other animals are housed next to, or in, the dwelling place, the proportion of feeds from humans can fall to 50 per cent or less (Diatta *et al.*, 1998; Killeen *et al.*, 2001; White, 1974; White and Rosen, 1973), which brings to mind the European malaria story told in Section 3.1. In complete contrast, *An. quadriannulatus* is a strongly zoophilic species and is not a vector of human disease. Over most of its range it is exophilic, but in the highlands of Ethiopia, probably as an adaptation to the cold nights, it is endophilic. The endophilic form may rarely feed on humans but does not transmit human disease. The fundamental difference in host choice between *An. gambiae s.s.* and *An. quadriannulatus* has an innate olfactory basis (Dekker *et al.*, 2001). *Anopheles arabiensis* falls between the two: ecologically and behaviourally it is an extremely plastic species; across its geographic range, exophilic and endophilic, anthropophilic and zoophilic forms are to be found. Generally, *An. arabiensis* is exophilic and zoophilic, but in the absence of other hosts it can live quite happily on humans and may rest inside their dwellings. It is a malaria vector but is less efficient than *An. gambiae s.s.* This can be seen in the sporozoite rates (the proportion of the mosquito population carrying sporozoites) of *An. arabiensis* which are commonly about $1/15$ of those of *An. gambiae s.s.* (White *et al.*, 1972). The higher sporozoite rates in *An. gambiae s.s.* are not a reflection of different susceptibilities to malaria parasites – the two species are equally susceptible – but arise because *An. gambiae s.s.* lives longer and feeds more often on humans (White, 1974). Despite being an important vector of malaria, *An. arabiensis* is not a major filariasis vector as it tends to occur seasonally which precludes it from maintaining the high transmission rates required for the establishment of endemic foci of this disease (White, 1974). The two saltwater species feed primarily on non-human hosts but in the absence of these they can survive quite happily on humans. Of the two species, *An. melas* feeds more readily and regularly on humans and is therefore the more important vector of both malaria and filariasis, particularly in coastal areas where alternative hosts to humans are scarce. So, even with seven very closely related insects, innate host preference can vary greatly.

4

Location of the host

The difficulty that hungry blood-sucking insects have in locating their next blood meal depends upon the closeness of their association with the host. At one extreme we have the permanent ectoparasites which are in the happy position of having food continually 'on tap'. Only by accident will they find themselves more than a few millimetres from the skin of the host and the blood that it holds. At the other extreme are those temporary ectoparasites, such as blackflies and tabanids, that do not remain permanently in the vicinity of the host. When these insects are hungry their first problem is to locate the host, often a difficult and complex behavioural task. These differences in lifestyle are reflected in the number of antennal receptors different types of blood-sucking insect possess (Chapman, 1982). Not surprisingly the more independent, host-seeking insects possess the most receptors. Thus, lice have only 10 to 20 antennal receptors and fleas about 50, but the stablefly, which spends most of its time at some distance from the host, has nearly 5000 antennal receptors. Considering two bugs, we see that *Cimex lectularius* has only 56 antennal receptors compared to 2900 on the more adventurous *Triatoma infestans*.

The level of reliance on blood is also an important factor in host location. So for obligate haematophages such as the tsetse fly and triatomine bugs regular host location is absolutely essential. In contrast, facultative haematophages can often overcome periods when they cannot find a host by feeding on other foods such as nectar – mosquitoes and stableflies are examples.

Most of the detailed information on host finding is restricted to a small number of temporary ectoparasites and the discussion that follows will concentrate largely on these.

4.1   A behavioural framework for host location

The location of the host is an integrated, but flexible, behavioural package that gathers momentum as the host is tracked down. The behaviour patterns involved are not arranged in a strict hierarchy, that is they do not occur in a strict sequence with behaviour one always being followed by behaviour two, followed by three, and so on. This versatility allows a flexible response on the part of the insect to the differing circumstances in which it will encounter hosts. However, it is probable that insects mostly encounter host-derived stimuli in a particular sequence. The insect often makes use of this predictability by permitting the current behavioural pattern in the host location sequence to lower the response threshold for subsequent host-related stimuli. For example, an insect that would not normally respond to a certain visual stimulus may respond strongly if it has just been exposed to an increase in carbon dioxide levels. In this way a behavioural momentum is built up during host finding. This behavioural momentum is further enhanced by the wide range of increasingly strong host stimuli that the insect encounters as host location proceeds (Sutcliffe, 1987).

From observations of blood-sucking insects both in the laboratory and the field, and from the clear evidence on the discrimination and selectivity they can show, we can predict that a variety of host signals are used in host finding. Information on what signals are used and the processes involved is still far from complete. In general, visual and olfactory stimuli, aided by anemotactic and optomotor responses, are the most important signals when the insect is still at some distance from the host. Nearer to the host different stimuli become important, particularly humidity and heat.

For the purposes of explanation, the various behaviour patterns involved in host location can be conveniently divided into three phases (in reality the whole process is a continuum with one behaviour pattern dovetailing into the next) (Sutcliffe, 1987):

(a)   *Appetitive searching* – driven by hunger the insect indulges in non-oriented behaviour likely to bring it into contact with stimuli derived from a potential host. This usually occurs at specific times of the day regulated by the insect's internally programmed activity cycle.

(b)   *Activation and orientation* – upon receipt of host stimuli (activation) the insect switches from behaviour patterns driven from within (appetitive searching) to oriented host location behaviour driven by host stimuli. The insect uses these host-derived stimuli to track down the host. These stimuli are of increasing variety and strength as the insect and host come closer together.

(c)   *Attraction* – the final phase, in which host stimuli are used to bring the insect into the host's immediate vicinity, and in which the decision of whether or not to contact the potential host is made.

Categorization of host location in this way has another benefit. It clearly indicates that host location is a series of behavioural events, rather than just one. This may help explain the conflicting reports on insect behavioural responses to certain host stimuli, and at the same time give us a clear warning of the need for great care in the design of behavioural experiments and in the interpretation of results from them. The effect of reflected ultraviolet (UV) light on the tsetse fly can be used to illustrate this point. Reflected UV light is a deterrent to the fly during the orientation phase of host finding, but will increase the number of flies landing during the attraction phase. It would be easy to confuse the two effects in a poorly designed experiment.

Throughout the rest of the chapter, to avoid confusion over which phase of host location is being discussed, I will use the words appetitive searching, activation, orientation and attraction only to refer to the different phases of host finding outlined above.

4.2   Appetitive searching

Blood-sucking insects usually have a delay period between their emergence from the egg, or previous developmental stage, and their first blood meal. The reasons for this are not clear, especially as other activities, such as mating and dispersal, commonly occur during this period. One reason for the delay may be that, after adult emergence, the reproductive system of many female blood-sucking insects undergoes a maturation period lasting several days. Blood meals taken before maturation has passed a certain point do not add to the reproductive success of the insect, but visiting the host to get them will greatly increase the insect's chances of being damaged or killed. For example, in the cat flea, *Ctenocephalides felis felis*, the reproductive system takes about four days after emergence to mature and a further two days after the blood meal to produce eggs. It will be selectively advantageous for the insect to remain off the host (and unfed) for the first four days after emergence, as this will minimize the danger from the host's grooming activity without affecting the insect's reproductive success (Osbrink and Rust, 1985). Another reason for the delay in blood feeding may lie in the progressive thickening and hardening of the cuticle that takes place during the teneral period. This thickening may mean the mouthparts are insufficiently hard to permit efficient skin penetration. This occurs in mosquitoes, which for approximately the first 24 hours after emergence cannot pierce skin. As the post-emergence delay period progresses, or as the time since the last blood meal lengthens, the insect becomes increasingly hungry and more likely to begin host seeking. Bouts of activity will be mainly restricted to particular times of the day (Gibson and Torr, 1999) because activity in blood-sucking insects, as in other animals, occurs in set patterns during each 24-hour (circadian) cycle. The timing of these activity bouts is internally programmed and the time of day at which they occur is characteristic for each species (Fig. 4.1). The patterns are not inviolable; species commonly show variations in periodicity when collected from different habitats or at different times of the year. As hunger increases these periods of activity intensify (Fig. 4.2) and also occupy longer periods of time

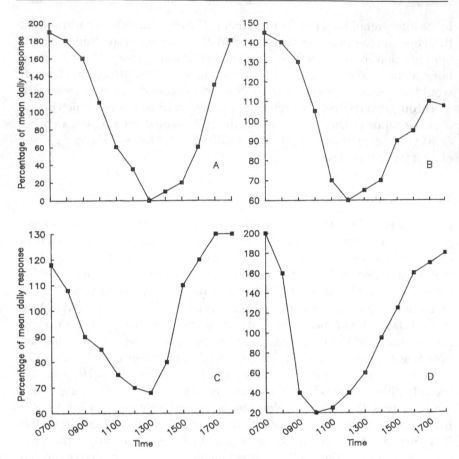

Figure 4.1 Insects will commonly show just a single peak of activity in a day (Lewis and Taylor, 1965), but there are plenty of exceptions. The tsetse fly, for instance, shows two endogenously controlled peaks in activity, one at dawn and one at dusk (Brady, 1975). The key behaviours in the life of the fly most frequently take place at these times. Some of these are shown in the graph, expressed as percentages of mean daily response: A. field biting activity; B. spontaneous flight in actographs; C. optokinetic responsiveness; D. olfactory responsiveness (Brady, 1975).

(Brady, 1972). Host location during the day or night each has its particular advantages and disadvantages (Table 4.1).

Much of the activity seen in the hungry insect is appetitive behaviour, that is behaviour that maximizes the chances of the insect contacting a signal derived from a host animal. The simplest appetitive behaviour pattern is to sit still and wait for a host stimulus to arrive. This may sound a very chancy business, but providing the insect chooses the resting site carefully, it could be a sound strategy, combining maximum energy

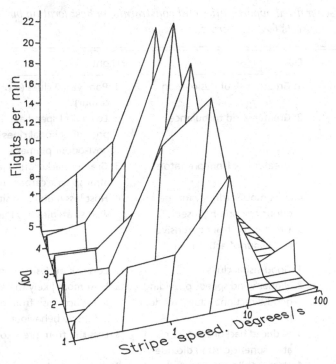

Figure 4.2 Hunger will lead to an increase in the overall activity of the insect. This can be demonstrated by monitoring various behavioural responses. Here the changing, kinetic responsiveness of male tsetse flies to a moving 'target' (stripe speed) is recorded over a five-day period. As the fly becomes hungrier it shows increased flight activity in response to a given stimulus. It is also interesting to note that the pattern of response stays the same throughout the period, showing that flies are 'tuned' to particular patterns of movement (Brady, 1972).

conservation with a strong chance of encountering a host. This strategy is likely to be employed to a greater or lesser degree by virtually all blood-sucking insects, but it is likely to be used more by forest-dwelling species than by those living in more open terrains (Sutcliffe, 1986). Insects almost certainly also engage in active appetitive behaviour. The evidence for this is that non-activating, non-orientating, non-attracting sampling devices, such as suction traps sunk below ground or electric nets, will often catch large numbers of hungry female insects that are impregnated but not ready for egg laying. As these insects are often trapped at peak biting times, it is quite possible that they are engaged in non-oriented appetitive activity.

We know virtually nothing of the search patterns used by insects undertaking active appetitive behaviour. Based on the reasonable assumption that the insect will attempt to optimize its chances of encountering a host

Table 4.1 *Generalized opportunities and constraints on host location by blood-sucking insects feeding during the day or night.*

	Day	Night
Disadvantages	1. Greater risk of desiccation	1. Poor visual clues (especially colour)
	2. Greater wind turbulence	2. Low wind speed and hence poor directional clues in host-odour plumes
	3. Greater risk from predators	3. Greater background levels of atmospheric carbon dioxide
	4. Host mobile (disadvantage for odour-responding insects?)	4. Host less mobile, so sit-and-wait strategies less feasible
	5. Greater risk from defensive behaviour of active host	
Advantages	1. Good visual clues	1. Less risk of desiccation
	2. Higher wind speeds providing good directional clues in odour plumes	2. Host more likely to be at rest so reduced risk from host defensive behaviour
	3. Reduced background levels of atmospheric carbon dioxide	3. Less risk from predators
	4. Host mobile, making a sit-and-wait strategy feasible	4. Less atmospheric turbulence and hence more continuous odour plumes

Gibson and Torr, 1999.

while minimizing its energy expenditure, theoretical work has been carried out to determine such patterns for flying insects. In a wind blowing consistently from one direction the optimum strategy is for the insect to fly across the wind, allowing the maximum number of air streams to be monitored for a particular energy expenditure (Linsenmair, 1973). In the field, winds often veer rapidly from one direction to another. If they veer by more than 30° from the mean, then downwind flight becomes the most energy-efficient method of sampling the maximum number of airstreams (Sabelis and Schippers, 1984). Whether these theoretical conclusions relate to the appetitive behaviours displayed by blood-sucking insects in the field remains to be seen.

4.3   Activation and orientation

Activation occurs when the insect comes into contact with a suitable signal from a potential host animal. Such a signal may simply change the behavioural awareness of the insect without causing any observable activity on the insect's part. In such a case, orienting behaviour would be released by a subsequent stimulus, for which the insect has now been primed. Alternatively, the stimulus may directly cause the insect to switch over from endogenously driven appetitive searching to oriented host location behaviour. The insect then uses the information contained in host-derived signals to orientate towards the host. A range of stimuli are used by insects in activation and orientation.

Olfaction

There seems to be an olfactory component to host finding in virtually all blood-sucking insects (Takken, 1996). It might be assumed that it is of most importance to forest-dwelling insects, because direct visual contact with the host is most restricted for them. But the evidence from tsetse flies suggests that the species showing the clearest response to odours are those (*Glossina morsitans*, *G. pallidipes* and *G. longipennis*) living in fairly open situations. The forest-dwelling *palpalis* group show little response at a distance. This is probably explained by the fate of the odour plume in the two situations, with a plume at the flight height of insects acting as a much more continuous, relatively linear guide for insects in an open situation compared to a forest (David *et al.*, 1982; Elkinton *et al.*, 1987). Field experiments on plume structure show that an insect following an odour plume could be flying $>90°$ away from the host for up to 25 per cent of the time when following an odour plume passing through vegetation, even at 5 m from the source (Brady *et al.*, 1989). So odour is probably most important in orientation for night-feeding forms living in relatively open situations.

Olfactory stimuli implicated in host location to date include carbon dioxide, lactic acid, ammonia, acetone, butanone, fatty acids, indole, 6-methyl-5-hepten-2-one and phenolic components of urine (Geier *et al.*, 1999; Klowden *et al.*, 1990; Knols *et al.*, 1997; Meijerink *et al.*, 2000). For tsetse flies the most potent chemicals affecting behaviour are carbon dioxide, acetone, octenol, butanone and various phenols (Gibson and Torr, 1999), but when these are dispensed in the field at natural dose rates they attract only about 50 per cent of the tsetse a natural host would attract (Hargrove *et al.*, 1995; Torr *et al.*, 1995). This suggests that other kairomones remain to be identified.

It is generally accepted that carbon dioxide can be involved in both the activation and orientation of virtually all blood-sucking insects. It is normally present in the atmosphere at between about 0.03 per cent and 0.05 per cent, occasionally rising to 0.1 per cent in dense vegetation at night. It is secreted by the skin of hosts, but the major emissions occur in exhaled breath which, in humans, contains about 4.5 per cent carbon dioxide. So

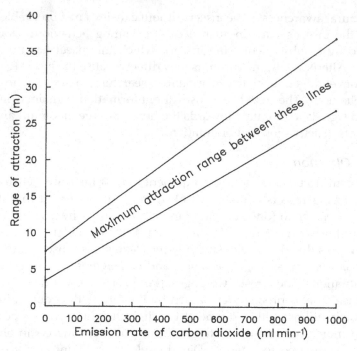

Figure 4.3 The effect of varying the emission rate of carbon dioxide on its drawing power for mosquitoes has been measured by various authors under field conditions. The lower edge of the 'attraction range' shown in the figure is the furthest trapping point at which an effect of the carbon dioxide was noted. The upper edge of the zone is the nearest trap at which no effect of the carbon dioxide was seen (Gillies, 1980).

the carbon dioxide in an odour plume produced from a solitary human would remain above background concentration until the exhaled breath had been diluted by a factor of about 100 (Gillies, 1980). Field recordings of tsetse receptor responses to odours support the idea that odour plumes break up into filaments and packets under field conditions, with the odour concentration and frequency of packets decreasing with distance from the host (Voskamp et al., 1998). Predicting how dilution of the plume will occur is no easy task. Packets of relatively undiluted odour are likely to travel for quite some distance downwind, and the distance from the host at which activation and orientation are still likely to occur will vary according to local meteorological conditions, with windspeed being particularly important (Brady et al., 1995; Griffiths and Brady, 1995). Perhaps the easiest way of dealing with the problem is to look directly at the effects of odour in the field (Fig. 4.3). The range over which a host animal can activate and orientate an insect, on the basis of odour alone, has been calculated for some blood-sucking insects. An ox draws the tabanid *Philoliche zonata* at 80 m, but this is reduced to only 30 m for species of *Tabanus* (Phelps and Vale, 1976). Calves draw a number of mosquito species at distances between 15 and 80 m (Gillies and Wilkes, 1969; Gillies and Wilkes, 1970; Gillies and Wilkes, 1972). For the tsetse flies *Glossina morsitans* and *G. pallidipes* the maximum odour-based attraction distance for an ox is estimated to be about 90 m (Vale, 1977; Vale, 1980). The number of insects attracted is dependent on the amount of odour released. With tsetse again, an increase in the amount of odour released equivalent to an increase of 10-fold in the body mass of hosts resulted in a 2.5-fold increase in number of insects caught (Hargrove *et al.*, 1995). It seems likely that at least a component of this increased catch rate is that the larger odour dose is drawing flies from further distances than the 90 m for a single ox.

Carbon dioxide on its own has been shown to cause activation in several blood-sucking insects (Bursell, 1984; Bursell, 1987; Omer and Gillies, 1971; Warnes, 1985) and its involvement has been suggested from circumstantial evidence in others (Compton-Knox and Hayes, 1972; Nelson, 1965; Roberts, 1972). In mosquitoes it is the change in concentration of carbon dioxide rather than the level of carbon dioxide encountered that is the important factor eliciting behavioural responses (Kellogg and Wright, 1962; Wright and Kellogg, 1962). For example, it has been shown in mosquitoes that carbon dioxide causes continued upwind flight only when received at continually varying concentrations (Omer and Gillies, 1971). This is probably true for all blood-sucking insects. To utilize odour pulses mosquitoes are sensitive to very small changes in carbon dioxide levels. Changes as small as 0.05 per cent will elicit behavioural responses in a wind tunnel (Mayer and James, 1969) and electrophysiological recordings from the carbon dioxide receptors on mosquito palps show responses to changes as small as $+0.01$ per cent (Grant *et al.*, 1995; Kellogg, 1970). The antennal receptors of *Stomoxys calcitrans* are slightly less sensitive, responding to an increase in carbon dioxide levels of 0.023 per cent, but starvation lowers the response threshold of these receptors (Warnes and Finlayson, 1986). It has been shown that this sensitivity to host odours is strongly down-regulated following the blood meal and, in *Aedes aegypti* at least, that this is hormonally controlled. The down-regulation in sensitivity is at the level of the receptor of the peripheral nervous system rather than in the central nervous system (Davis, 1984; Denotter *et al.*, 1991; Fox *et al.*, 2001; Klowden and Lea, 1979; Klowden *et al.*, 1987; Takken *et al.*, 2001). It has also been reported that the activation response to carbon dioxide in *S. calcitrans* rapidly habituates. The stimulus had to be increased from 1.04 to 2.04 per cent carbon dioxide to induce the same activation response in habituated flies (Warnes, 1985). As mentioned above, odours break into pulses and filaments the further they move from the source (Murlis *et al.*, 2000), so the flying insect will be exposed to pulses of odour, of varying strength and spacing rather than a continuous plume of odour. Such pulses will militate against habituation. In support of this view that habituation is less likely in the field because of the pulsed nature of the odour, it is worth noting that in moths the responses to pulsed pheromone are considerably different from the responses obtained after exposure to an homogeneous pheromone cloud (Baker, 1986).

As well as acting as an activating agent, carbon dioxide also acts as an orientation stimulus guiding blood-sucking insects to hosts (Omer and Gillies, 1971; Warnes and Finlayson, 1985). This can be neatly shown by filtering carbon dioxide from the breath of a host (Laarman, 1958) or by using two sets of inanimate traps, only one of which is baited with carbon dioxide (Fallis and Raybould, 1975). In both cases fewer insects arrive at the source without carbon dioxide.

In the natural situation, carbon dioxide is only one of several host stimuli that the insect receives. Carbon dioxide can act in concert with other stimuli giving a response that is different from that of either stimulus given alone (Gillies, 1980). There is a spectrum of responses that can be seen to these dual stimuli, ranging from synergism (in which the two stimuli give an overall reaction that is greater than the sum of the two stimuli given separately) to an interaction in which one stimulus primes the insect to respond to the second which, if given alone, has no effect (Bar-Zeev et al., 1977; Bos and Laarman, 1975; Laarman, 1958). For example, lactic acid is an activating and orientating stimulus for some mosquitoes, but only if carbon dioxide is also present in the airstream (Price et al., 1979; Smith et al., 1970). Other components of host breath can also act in concert with carbon dioxide. Octenol, which is a component of ox breath, enhances the catches of tsetse flies in traps, especially when carbon dioxide is also released (Hall et al., 1984; Vale and Hall, 1985a). This enhancement is possibly explained by the fact that octenol, which clearly can be used by tsetse for orientation, does not cause activation of tsetse flies. Acetone, another constituent of ox breath, further enhances the effectiveness of these traps.

An interesting aspect of the interaction of host stimuli is that even closely related species may respond differently to combinations of stimuli. Looking at the levels of efficiency of a synthetic odour consisting of 1.2 l of carbon dioxide per hour, 5 mg of acetone per hour and 0.05 mg of octenol per hour for two tsetse species, we find that it is almost as effective in drawing *Glossina morsitans morsitans* to a field trap as natural ox odour, but is only half as effective as the ox for *G. pallidipes*. Similarly, if we look at the drawing power of these bait components for muscoid biting flies, we see that they are lured strongly by carbon dioxide alone, or carbon dioxide–acetone mixtures, but that octenol has little, if any, effect on catches when released at the levels given above (Vale and Hall, 1985b). Perhaps the clearest example of the species-specific nature of the response to odour mixtures is seen in the drawing power of the bovine urine components, 4-methylphenol and 3-n-propylphenol, for different tsetse species. Used singly, 3-n-propylphenol drew roughly equal numbers of *G. pallidipes* and *G. m. morsitans*. When used in combination, trap catches increased by up to 400 per cent for *G. pallidipes* (Vale *et al.*, 1988), in contrast to trap catches for *G. m. morsitans*, which decreased.

Why have insects developed distinctive responses to mixtures of different host-derived stimuli? Multiple stimuli are likely to be a far surer guide to the presence of a host than one stimulus received alone. This is because while one stimulus alone may be of non-host origin, this is very unlikely for combinations of stimuli, especially when they are received in particular proportions. In other words, responding to stimuli received in combination is likely to maximize the chances of host encounter while minimizing energy consumption. Also, insects show preferences for particular hosts (see Section 3.1). Responding to particular combinations of host signals may permit a degree of selection for a particular host that is still some distance away – different animals have different odours. Carbon dioxide is released by all potential hosts and, although it may be good for alerting the insect to the possibility of a meal, it does not allow the insect to discriminate between hosts. Body odours, however, may be characteristic for particular groups of host animal or even for particular species. An extreme example of selection at a distance is seen in the blackfly, *Simulium euryadminiculum*, which is drawn to a unique, non-polar product of the uropygial gland of a particular water fowl, the loon (Fallis and Smith, 1964; Lowther and Wood, 1964). This is probably an exceptional case, based as it is on a very specific chemical signal. In most instances, it is more likely that the insect discriminates between hosts on the basis of the varying proportions of a number of less specific stimuli. The *Anopheles gambiae* complex of mosquitoes (see Chapter 3) provides an example of this sort. *An. quadriannulatus*, which feeds on a broad range of bovids, responds strongly to carbon dioxide (Dekker and Takken, 1998; Knols *et al.*, 1998). In contrast the highly anthropophilic *An. gambiae s.s.* responds only weakly to carbon dioxide but is strongly attracted by human foot odours (Dekker *et al.*, 1998). These foot odours are probably generated by the action of *Coryneform* skin bacteria on human secretions (Braks *et al.*, 2000) and fatty acids may be an important component (Bosch *et al.*, 2000; Knols *et al.*, 1997; Takken and Knols, 1999). Responding to such particular, species-specific odours rather than more commonly produced odours such as carbon dioxide is a sensible strategy for such a specialist feeder (Takken and Knols, 1999). It is interesting to note that people differ in their attractiveness to mosquitoes and blackflies (Ansell *et al.*, 2002; Brady *et al.*, 1997; Lindsay *et al.*, 1993; Schofield and Sutcliffe, 1997; Schofield and Sutcliffe, 1996). Work in Tanzania showed that with *An. gambiae* this differential attraction was entirely odour based

(Knols *et al.*, 1995). This raises the interesting possibility that differential attractiveness of individuals for mosquitoes may be due to the status of the bacteria they harbour (Takken and Knols, 1999).

It seems that the response to combinations of stimuli can also be used to detect the presence of hosts the insect wishes specifically to avoid. This is seen in the reduced catches of tsetse flies using stationary bait animals if humans remain in the vicinity of the bait (Vale, 1974a). In this case humans can be recognized by tsetse flies not only by their characteristic odour, but also visually.

How are synergistic responses to host odours generated? Two possibilities present themselves. First, each odour may be treated separately by the peripheral nervous system with integration and amplification of the signals occurring in the central nervous system. Or second, the response of peripheral receptors to a substance may be significantly higher in the presence of the second chemical. Electrophysiological recordings from mosquito receptors are consistent with the former theory (Davis and Sokolove, 1975; Mayer and James, 1970).

Although it is clear that blood-sucking insects use olfactory clues to trace the host animal, it is still uncertain how they achieve this feat. The odour plume is not a smooth, homogeneous cone of molecules fanning out from the source. In field conditions the plume is a series of loosely gathered lamellae and filaments of odour that are all continually mixed and dispersed by the vagaries of the wind (Wright, 1958). As a consequence, the plume is a heterogeneous patchwork of odours of differing concentrations. In addition, the wind often veers from its mean direction, with the effect that the odour plume meanders as it moves away from the source, possibly breaking into discrete packets. Wind direction within each meander or packet of odour remains remarkably consistent over considerable distances (David *et al.*, 1982). Tracing the source of such a plume is a major task. Several hypotheses have been proposed to account for the ability of insects accurately to find the source of such an airborne odour trail. Early theories tended to stress the importance of features of the odour plume, such as the increasing concentration of odour as the source is approached (Sutton, 1947), the pulse frequency of the odour in the plume (Wright, 1958), or the boundaries of the plume where odoriferous air meets odour-free air (Farkas and Shorey, 1972). Currently, the most widely held theories suggest that the concentration of the odour, or other features of the plume, are largely irrelevant and that, above a threshold level, odour presence is merely used as a switch to release particular behavioural responses (Kennedy, 1983). The best-worked examples of flying insects following an odour trail are male moths homing in on a pheromone source (Baker, 1990). Within an odour plume these insects continually fly in an internally programmed, upwind, zig-zagging pattern. This mainly serves to hold them within the plume and to move them closer to the odour source. To be able to fly

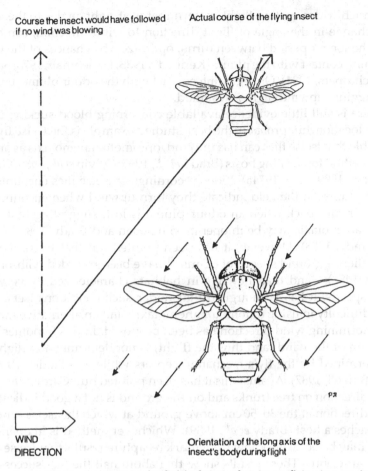

Course the insect would have followed if no wind was blowing

Actual course of the flying insect

WIND DIRECTION

Orientation of the long axis of the insect's body during flight

Figure 4.4 Fixed objects in the field of view of an insect flying in a straight line in still air will apparently move backwards and parallel to the insect's flight plane. If wind is blowing at an angle to the insect's flight plane, then the insect will be blown off course. The fixed objects in the insect's field of view will now apparently move at an angle to a plane through the long axis of the body (short arrows). Flying insects can use this information to determine wind direction (Kennedy, 1940).

upwind (positive anemotaxis) the insect must know in which direction the wind is blowing. The flying insect is of low inertia and is blown along by the wind such that, to all intents and purposes, it seems to be in still air except for the movements it generates in flight. For this reason the flying insect is unable to sense wind direction by using air movements as a guide. Instead it determines wind direction by observing the apparent movement of fixed objects in its environment (this is called optomotor anemotaxis and is explained further in Fig. 4.4). When contact with the plume is lost, narrow, zig-zagging, upwind flight at a small angle to wind direction gives way to a much wider, casting flight pattern that is at right angles to the wind. This change in the angle of flight direction to wind direction, combined with the longer period between turns, optimizes the chances of the insect regaining contact with the plume (Kennedy, 1983; Linsenmair, 1973; Sabelis and Schippers, 1984). On regaining contact with the odour plume, narrow, zig-zagging, upwind flight is resumed.

There is still little evidence available concerning blood-sucking insect odour location, but probably the best-studied example is the tsetse fly. It is probable that tsetse flies can use upwind, optomotor anemotaxis as an efficient method for locating hosts (Brady *et al.*, 1995; Colvin *et al.*, 1989; Gibson and Torr, 1999; Vale, 1974a). Video recordings of tsetse flies encountering odour plumes in the field indicate they turn upwind when encountering odour or turn back when an odour plume is lost, suggesting that optomotor anemotaxis may be in operation (Gibson and Brady, 1985; Gibson and Brady, 1988). However, it has been pointed out that tsetse flies are rapid fliers – ground speeds of $6.5\,\mathrm{ms}^{-1}$ have been recorded (Gibson and Brady, 1988) – and that they live in habitats characterized by very low wind speeds. It has been argued that under such conditions tsetse may have difficulty in using optomotor anemotaxis, and an alternative method for determining wind direction has been proposed. In this hypothesis the direction of the wind (and thus the flight) is not determined in flight but is determined by the fly's mechanoreceptors while it is landed (Bursell, 1984; Bursell, 1987). Against this it has been pointed out (Carde, 1996) that wind direction on tree trunks and on the ground is not a good indicator of wind direction at the 30–50 cm above ground at which the tsetse typically approaches a host (Brady *et al.*, 1989). Whichever method is in operation, and it may be both, we know from mark recapture results that tsetse track odour efficiently. These results show that about half the flies successfully tracking the odour source flew directly upwind to it, probably in a single flight (Griffiths *et al.*, 1995) suggesting an 'aim and shoot' strategy like that suggested by Bursell. The other half of the flies dribbled in to the source over a 20-minute period, suggesting they lost the odour at some stage and either landed and waited for another odour clue before continuing upwind or that they were engaging in flights employing optomotor anemotaxis. Modelling various strategies for odour tracking by tsetse flies suggests that upwind anemotaxis, even if it achieves only a modest upwind bias of 20 per cent to 40 per cent, will result in virtually all insects tracking an odour source from 100 m away in about 300 seconds (Williams, 1994). This biased random walk approach to host location, whether wind direction is determined when the fly is in flight or landed, may be an optimal strategy for fast-flying insects like tsetse (Gibson and Torr, 1999).

Wind speed influences the efficiency of host location. Thus tsetse flies are more efficient at tracking odour sources as wind speeds increase to

$0.5\,\text{ms}^{-1}$, but take longer to track odour sources as wind speed increases above $1.0\,\text{ms}^{-1}$. Here the probable reason is the initial relative straightening of the odour plume followed by its breaking into complex lamellae and plumes as wind speed increases (Brady *et al.*, 1995). For weaker flying mosquitoes very low wind speeds may be required to enable them to fly effectively and therefore to track down hosts. This can be seen from the negative correlation between suction trap catches of mosquitoes in Florida and wind speed. Wind speeds of $0.5\,\text{ms}^{-1}$ reduced trap catches by 50 per cent and wind speeds of $1.0\,\text{ms}^{-1}$ trap reduced catches by 75 per cent. There appears to be no wind speed threshold below which this phenomenon ceases to occur (Bidlingmayer *et al.*, 1995). So for mosquitoes, relatively low-velocity winds may be impairing the flight ability of the insect that can only fly at approximately $1.0\,\text{ms}^{-1}$ and so effective host location must occur at wind speeds below this.

It is interesting to note that many flying temporary ectoparasites, including tsetse, tabanids, mosquitoes and simuliids, all approach the host while flying close to the ground, or slightly above the top of the predominant vegetation. The advantages of this are experimentally unproven, but the nearer you fly to the ground the greater the perceived angular rate of change of fixed objects on the ground. This will increase the efficiency of optomotor anemotaxis, which would be of particular use when wind speeds are low. A low-flying insect is also able to resolve finer detail in the visual field and this may also be of importance. It has been pointed out that shear forces generated by the frictional drag of the moving air with the ground increase rapidly near to the ground. It has been suggested that the low approach to the host used by many blood-sucking insects may allow the insect to use these shear forces to determine wind direction, but no experimental evidence has been produced to test this hypothesis. While the above is probably one factor determining this ground-hugging approach to the host, there are probably also other factors involved. The insect may adopt a compromise between, on the one hand, the height giving the most favourable wind speed for host odour detection, optomotor anemotaxis and upwind flight and, on the other hand, a height permitting a stealthy approach to the host and one giving maximum protection from any predators.

There have been reports that some blood-feeding insects release odours while they are feeding that attract other insects to the feeding-permissive host (Ahmadi and McClelland, 1985; Alekseev *et al.*, 1977; Charlwood *et al.*, 1995; McCall and Lemoh, 1997; Schlein *et al.*, 1984). It is hard to see any selective advantage to the female in expending energy to help other females find a host. In addition, drawing more insects to the host is likely to reduce the feeding success of the original feeder through density-dependent effects (see Section 7.5). Even if she escapes with a full meal, she may well have impaired her offsprings' chances of success because of the increased

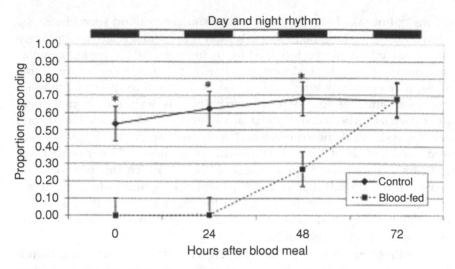

Figure 4.5 Blood-fed mosquitoes are not attracted to hosts, as can be seen here in the response of unfed and blood-fed *Anopheles gambiae* to a human hand held in a wind tunnel-olfactometer at different times after the blood meal (Takken *et al.*, 2001).

competition for larval resources that will take place. These arguments do not exclude the possibility that insects are drawn by substances being released by successfully feeding insects, they merely suggest that the release of such substances does not have the specific purpose of bringing others to the host. Because it may have implications for the distribution of bites among a host population, this phenomenon may be germane to patterns of disease transmission and deserves more study.

Work on odour reception in blood-sucking insects is entering the molecular phase, with the description of a family of odorant receptors in *Anopheles gambiae* (Biessmann *et al.*, 2002; Fox *et al.*, 2001; Hill *et al.*, 2002; Xu *et al.*, 2003). Interestingly, one of these, a female-specific antennal receptor, is down-regulated 12 hours after blood feeding (Fox *et al.*, 2001) when the mosquito shows substantial reduction in response to human odours (Takken *et al.*, 2001) (Fig. 4.5). This evidence adds weight to the possibility that down-regulation of responsiveness in the peripheral rather than the central nervous system may underlie the well-studied phenomenon of decreased host-seeking behaviour in fed mosquitoes (Davis, 1984).

Vision

Vision is also important in the activation and orientation of many blood-sucking insects. Not surprisingly, it is most widely used by diurnal insects that live in open habitats. This can be seen in blackflies: savannah and open-country species, unlike their forest-dwelling relations, are drawn strongly to silhouette traps even in the absence of carbon dioxide or other odour sources (Fredeen, 1961; Peschken and Thorsteinson, 1965; Thompson, 1976). It is likely that in most instances vision is not used alone in the search for the host, but is integrated with information from the other senses. A common pattern of events is that the insect is initially activated by host odour, and then uses smell to track the host from a distance (to give an impression of scale, let us say from about 100 m – see above). As it closes on the host, visual contact is made and visual information is then used in the final stages of orientation (Sutcliffe *et al.*, 1995). This explains results such as the five- to seven-fold increase in efficiency of odour-baited traps for tsetse flies over the same traps relying on visual appeal alone (Politzar and Merot, 1984).

At what distance does visual orientation become important? Clearly this will vary with the visual acuity of the insect involved and the size and visibility of the object concerned. The more contrast a bait has with the background, and the larger and more mobile it is, the more visible it will be over greater distances. Estimates for the range of visible effectiveness are available for some insect–object combinations. For a range of mosquito species visual orientation to stationary, unpainted, thin plywood traps, 2.4 m long by 1.5 m high, begins between about 5 and 20 m (Bidlingmayer and Hem, 1980). For blackflies a cow silhouette trap 1.1 m long becomes visible at about 8 m (Sutcliffe *et al.*, 1995). Tsetse flies respond to a moving screen, 1.3 m long by 1.0 m high, at distances up to about 50 m (Chapman, 1961) and to a herd of moving cattle at up to 183 m (Napier Bax, 1937).

The main visual organs in insects are the compound eyes. They provide the sensory input used in the discrimination of pattern and form, movement, light intensity, contrast and colour. In addition to their compound eyes, insects possess other light receptors called ocelli. There is still speculation about their function and it seems probable that this varies among different insects. For example, ocelli are rapidly responding horizon detectors that play an important role in maintaining a level attitude in flight in some insects (Stange, 1981; Wilson, 1978), but in the blood-sucking bug *Triatoma infestans* they play a significant part in the negative response to light (Lazzari *et al.*, 1998).

The initial detection of an object depends on differences in colour contrast, relative brightness (intensity contrast) or relative movement between the object and the background. Once detected, it is possible that the shape of the object may be of importance in deciding whether it is worth pursuing. However, shape discrimination at a distance seems to be poor – for example, tsetse quite happily pursue vehicles or a variety of other inanimate objects dragged through the bush. The relative importance of each visual capacity is likely to vary with the habits of the insect. For example, utilization of colour information is unlikely to be well developed in insects active at night (Wen *et al.*, 1997), but these insects may be particularly sensitive to intensity contrast. For example, the eyes of nocturnal mosquitoes are very much more sensitive to light than those of diurnal mosquitoes, whose eyes have much more resolving power (Land *et al.*, 1997; Land *et al.*, 1999), and this may be a general phenomenon in blood-feeding insects. Also, nocturnal mosquitoes have eyes that can function over a wide range of light intensities, including very low light intensities such as star-lit conditions, whereas the eyes of diurnal mosquitoes are probably restricted to effective use in daylight only (Land *et al.*, 1999). The visual sensitivity in nocturnal mosquitoes is sufficient to permit optomotor anemotaxis (see above) (Bidlingmayer, 1994). In open-country, day-biting flies, utilization of colour information and the ability to detect movement may be particularly highly developed (Gibson and Young, 1991; Green, 1986; Vale, 1974b).

Although insects have eyes and are sensitive to much of the light we can see, it is important to appreciate that they do not see the world as we do. Behavioural work shows us that blood-sucking insects utilize colour information (Table 4.2) (Allan and Stoffolano, 1986b; Bradbury and Bennett, 1974; Green, 1989). It also shows us that insects are sensitive to ultraviolet light, which is invisible to us, and, conversely, that most insects are probably insensitive to parts of the spectrum, particularly the red end above 650 nm, to which we are sensitive. In other words, the coloured patterning of objects seen by insects is not necessarily that seen by humans. Electroretinogram responses in the compound eyes of those blood-sucking insects studied show peaks in two or more areas of the spectrum (Table 4.2). In common with most other insects, blood feeders show peaks of sensitivity in the near ultraviolet at about 355 nm and in the blue–green part of the spectrum between 450 and 550 nm. In addition, in the higher Diptera, another shoulder of responsiveness appears in the red–orange part of the spectrum around 620 nm. But crude sensitivity measured in electroretinograms is not always a good guide to behavioural responses. Let us look at one example of the complexity of behavioural responses to a visual signal. Tsetse flies are drawn to ultraviolet light sources (Green and Cosens, 1983), but are repelled by ultraviolet-reflecting surfaces as shown by the reduced efficiency of traps covered with them (Green, 1989). So whether the light is presented as a point source or a surface can be important. Things are even more complicated because the response to the same stimulus can vary under different behavioural conditions. So while ultraviolet-reflecting surfaces are repellent during orientation of the tsetse fly to a trap, the same surfaces promote landing if the fly arrives in the trap's immediate vicinity (Green, 1989). Underpinning all of these differing behavioural responses, but giving no clue to the complexity of their nature, would be a strong electroretinogram reaction to ultraviolet light.

Table 4.2 *Different blood-sucking insects respond in different ways to spectral information, but, as we can see in this table, a range of flies are generally attracted to blue/black targets while being repelled by yellow ones.*

Spectral sensitivity	Ultraviolet	Blue	Green	Orange-red	Most attraction	Least attraction or repulsion
Mosquitoes	+	+	+		Blue, black, red	White, yellow
Blackflies	+	+	+		Blue, black, red	White, yellow, ultraviolet
Tabanids	+ (350 nm)	+ (477 nm)	+ (520 nm)		Dark colours, blue, black, black, red. Some species only to white	Green, silver, yellow
Stableflies	+ (360 nm)	+ (450–550 nm)		+ (625 nm)	Low-intensity colours, blue, ultraviolet	High-intensity colours
Hornflies	+ (360 nm)	+ (460 nm)			ultraviolet	
Tsetse	+ (350–365 nm)	+ (450–550 nm)		+ (600–625 nm)	Ultraviolet, blue, black, white	Yellow, green

Theoretically, there are two different types of visual information available to blood-sucking insects. The first set of information comes from the intensity contrast between the target and the background. If the insect is not using colour information, the maximum contrast will be between black and white or between grey and either black or white, depending on the shade of grey used. The ease with which the target background combination is seen by the insect also depends on the intensity ('brightness') of the objects. The second set of information comes from colour. Sensitivity of the insect to colour can be shown in experiments demonstrating either an increased or decreased response to a colour compared to an achromatic series (the spectrum of greys running from black to white) of equal intensity and contrast with the background. In a similar way to the intensity contrast information discussed above, the information to be gained from colour depends on both its intensity and contrast with the background. The colour combination giving maximum contrast will depend on the visual system of the insect, which may be 'tuned' to detect red/green divisions or blue/green divisions, or possibly both, as in humans. In the real world the two sets of information (from intensity contrast and colour) are effectively intermingled. The experimental separation of the responses to the two different sets of information is, at best, extremely difficult and requires careful measurement of the visual characteristics of the targets involved.

The importance of colour and intensity contrast has been illustrated in work on the tabanid *Tabanus nigrovittatus*. Large numbers of these flies are drawn to low-intensity blue panels presented against grey backgrounds, irrespective of the intensity of those backgrounds. This strongly suggests that the flies can respond to colour alone, but the most efficient targets are those showing not only colour contrast but also maximum intensity contrast with the background. In the case of the high-intensity blue panels used in these experiments, this was achieved by presenting them against low-intensity, grey background panels (Allan and Stoffolano, 1986b). High-intensity contrast is also an efficient draw for stableflies (La Breque *et al.*, 1972; Pospisil and Zdarek, 1965), simuliids (Bradbury and Bennett, 1974; Browne and Bennett, 1980) and mosquitoes (Browne and Bennett, 1981). The responsiveness of blood-sucking insects to targets showing strong intensity contrast with the background is not surprising as large homeotherms are low-intensity objects that will appear dark in contrast to high-intensity vegetation (Allan *et al.*, 1987). Work such as this has important implications for the design of traps that are economically as well as biologically efficient (Gibson and Torr, 1999).

For humans, the colour giving the greatest contrast against a green background is blue (Hailman, 1979). We know that blue targets are highly efficient for tsetse flies, tabanids and muscids (Allan and Stoffolano, 1986a; Allan and Stoffolano, 1986b; Challier *et al.*, 1977; Holloway and Phelps, 1991) but we do not know why. Only when further work has determined the spectral discrimination functions and colour matching functions of the eyes of these higher Diptera will we know if the high efficiency of these traps is because for the insect, as for humans, blue gives maximum colour contrast with the green background. The spectral sensitivity of the adult female mosquito *Aedes aegypti* has been determined and ranges from ultra-violet (323 nm) to orange-red (621 nm), with sensitivity peaks in the ultra-violet (λ (max) $= 323$–345 nm) and green (λ (max) $= 523$ nm) wavelengths, and this probably accounts for preferences of this mosquito for traps of a particular colour (Muir *et al.*, 1992). Other studies on mosquitoes show colour preference is species-specific (Burkett *et al.*, 1998).

Patterning of the target normally decreases its power as an activating and orienting agent, as can be seen from the work on Diptera. If the pattern is a simple contrasting one then considerable numbers of tabanids still arrive, but as the patterning becomes more complex, progressively fewer insects are caught (Bracken and Thorsteinson, 1965; Browne and Bennett, 1980; Hansens *et al.*, 1971). Similarly, the most effective tsetse targets are of a uniform colour; increasing the complexity with chequered patterns, stripes or complex edges lowers their appeal (Gibson, 1992; Turner and Invest, 1973). It is likely that patterned targets are less effective because of reduced visibility of the target from a distance. While considering patterning, it is interesting that not all zebras are striped; only those sympatric with tsetse flies bear the characteristic stripes, suggesting that striping may be a means of reducing the attention zebras receive from tsetse flies and the trypanosomes they transmit (Gibson, 1992; Waage, 1981).

The insect compound eye can be very sensitive to movement. A measure of movement detection is the flicker fusion frequency of the eye (the rate of flicker that the eye is just unable to distinguish). Flicker fusion frequency is a function of the recovery time of the photoreceptors. In humans the flicker fusion frequency lies somewhere between 20 and 30 flashes per second (20 to 30 Hz). Fast-flying, diurnal insects typically have a flicker fusion frequency of 200 to 300 Hz, while in slow-flying or nocturnal insects this drops to between 10 and 40 flashes per second. The ommatidia of the tsetse fly *Glossina morsitans morsitans* are sensitive to frequencies of over 200 Hz (Miall, 1978), enabling the insect to detect rapid movements in its visual field. Tsetse flies are also sensitive down to 1 flash per 25 s (0.04 Hz) (Turner and Invest, 1973), suggesting that the insect may also be responsive to very slow movements across its visual field. Another gauge of movement detection is angular velocity sensitivity, which is a measure of responsiveness to the rate of movement of an object across the visual field. For tsetse flies, activation is optimal at angular velocities of about 3 to 7° per second. This is the equivalent of an antelope trotting at 10 km per hour at 20 to 60 m from the resting fly, or a vehicle travelling at 30 km per hour at 60 to 180 m from the fly (Brady, 1972).

How widespread the use of movement detection is in the finding of hosts is unclear. A complication, when designing experiments to look at the effect of movement, is that a moving animal breathes faster and sweats more, increasing the output of odour components which themselves draw blood-sucking insects. In consequence targets are commonly substituted for hosts in these experiments. Tsetse flies, males in particular, are drawn by large, dark, moving objects (Brady, 1972; Gatehouse, 1972; Vale, 1974b), but the situation here is also complicated. Under field conditions, it seems that most tsetse drawn to moving objects mate rather than feed, while those drawn to stationary objects are almost exclusively interested in feeding, not mating (Owaga and Challier, 1985; Vale, 1974b). Superficially, this seems to be a useful strategy acting to minimize wasted effort in tsetse flies. Inseminated female flies moving to stationary targets do not have to fend off males, while the males (grouped in the following swarm around moving animals) spend less time chasing previously inseminated females. Movement is also important for host finding in some mosquito species (Sippel and Brown, 1953), but the evidence for its use in host finding by tabanids (Bracken et al., 1962; Bracken and Thorsteinson, 1965; Browne and Bennett, 1980) and simuliids (Thompson, 1976; Underhill, 1940) is equivocal.

For many species of blood-sucking insect, the evidence on shape preference is also conflicting. One reason for the confusion is that orientation and attraction are often confused in these studies. Flies may well show a preference for a particular shape to orientate towards but, on close approach to the target, they are not attracted to it and sheer off. Unless specific measures are taken to collect or count these insects, no useful information on orientation will come from the experiment (Vale, 1974b). Another reason for confusion is that much of the work on shape preference has been carried out using two-dimensional rather than three-dimensional targets. There are reasons for this. Often the work has had the aim of producing cheap targets that could be used for the control of the insects and the work has often shown that insects have preferences. For example, tsetse are drawn to two-dimensional targets in the ranked order: circle > square > horizontal oblong = vertical oblong (Torr, 1989). For work aimed at investigating the natural host-finding mechanisms of insects, insights are more likely to be gained from the use of three-dimensional targets. Tabanids, for example, show no preference for two-dimensional targets of different shape (Browne and Bennett, 1980; Roberts, 1977), but when three-dimensional targets are employed, they show considerable preference for spheres over cubes or vertical cylinders (Bracken et al., 1962). Clearly, the three-dimensional target comes closest to resembling the host and it is reasonable to conclude that shape is important in drawing tabanids to hosts. Mosquito species can also distinguish between three-dimensional targets (a pyramid and a cube), with different species showing a preference for each (Browne and Bennett, 1981), but some blood-sucking insects, most simuliids studied for example, appear to show no preference for visual targets of a particular shape (Bradbury and Bennett, 1974; Browne and Bennett, 1980; Fredeen, 1961).

The size of traps is important: large traps draw or catch more tabanids (Bracken and Thorsteinson, 1965; Thorsteinson and Bracken, 1965; Thorsteinson et al., 1966), simuliids (Anderson and Hoy, 1972) and tsetse flies (Hargrove, 1980). In addition three-dimensional traps catch more tabanids than two-dimensional screens (Bracken et al., 1962; Thorsteinson et al., 1966). The greater drawing power of large traps is probably explained in part by the wider area over which large and/or three-dimensional traps are visible. However, alighting responses of tsetse flies may also increase with increasing trap size (Hargrove, 1980).

Odours and visual clues are probably the most important factors in the orientation of most insects to the host, but there may also be other mechanisms. It has often been speculated that invisible parts of the electromagnetic spectrum (infrared radiation in particular) may be involved in host location. Sound is a factor guiding the mosquitoes of *Corethrella* spp. to their tree frog hosts (McKeever, 1977; McKeever and French, 1991).

4.4 Attraction

Vision, as well as being important in activating and orientating insects to the host, is also important in their decision as to whether and where to land. In general, the colour preferences displayed by alighting insects are the same as those involved in their activation and orientation, but not always (see the discussion of tsetse and ultraviolet light above). To generalize, host-seeking insects prefer to land on dark, low-intensity colours similar to those of many host animals.

Host animals are contoured and shaped, and blood-sucking insects can recognize different parts of the animal. For example, different tabanid species select markedly different landing sites on cattle (Fig. 7.3) (Mullens and Gerhardt, 1979). This is probably visual recognition of the preferred site because flying insects show definite preferences for particular parts of complex inanimate targets and also for targets of particular shapes. For example, many simuliids and mosquitoes show a preference for the extremities of a target (Browne and Bennett, 1981; Fallis et al., 1967). Simuliids, attacking sticky animal silhouettes, are mainly caught on the projections (for example 'ears') (Wenk and Schlorer, 1963), with the inner ('body') portions of the target catching few insects (Browne and Bennett, 1980). In other studies with cow silhouette traps the head proved the important landing site for simuliids (Sutcliffe *et al.*, 1995). To give another example, the blackfly, *Simulium euryadminiculum*, is activated by the odour of the uropygial gland of the loon which it also uses to orientate towards the host, but it only lands if the target has a prominent extension mimicking the neck of the bird, which is the normal feeding site of the fly (Bennett *et al.*, 1972; Fallis and Smith, 1964). A neat example of the effect of shape on the alighting response is given in the five-fold increase in tsetse flies landing on horizontally elongated targets over the same target presented vertically (Vale, 1974b). Given a target of the right shape, blood-sucking insects will often choose to alight at a colour or intensity border, or at the edge caused by the confluence of two angled planes (Allan and Stoffolano, 1986b; Bradbury and Bennett, 1974; Brady and Shereni, 1988; Browne and Bennett, 1980; Fallis *et al.*, 1967; Turner and Invest, 1973). The size of the target may also be important in determining landing rates for tsetse flies (Hargrove, 1980).

In many cases visual stimuli alone are not sufficient to enable the insect to alight. Mosquitoes are activated and orientate towards large visible objects, but on closing with an inanimate target (at about 30 cm), and in the absence of other host-associated signals, they sheer away from it and do not land (Bidlingmayer and Hem, 1979). Odours are important, and in general those odours that were involved in activation and orientation are also of significance in the attraction phase of host location (de Jong and Knols, 1995). Similar findings can be seen in many tsetse and stableflies reacting to unbaited traps (Hargrove, 1980; Vale, 1982). Tsetse flies tend first to fly around a target (Vale, 1983). Whether or not they land is strongly influenced by the target's shape, size, colour and pattern and whether carbon dioxide or other odours are associated with it (Brady and Shereni, 1988; Gibson, 1992; Green, 1986; Hargrove, 1980; Torr, 1989; Vale, 1974b; Vale and Hall, 1985b; Warnes, 1995). Clearly a range of host-associated stimuli is important to the insect in the attraction phase of host location.

In addition to vision and smell, new stimuli are also becoming available to the insect now it is close to a host. One of the most important of these is heat. Thus, for some mosquitoes and sandflies, at least, addition of heat significantly increases trap catches (Kline and Lemire, 1995; Nigam and Ward, 1991). Blood-sucking insects can be very sensitive to heat. The mosquito *Aedes aegypti* shows maximal spike frequency changes in the cold and hot receptors of their antenna 1 sensilla coeloconica, in response to a temperature change of $+0.2\,^{\circ}\text{C}$ (Davis and Sokolove, 1975). Little information is available on the distances over which heat is effective. It has been demonstrated that there are marked convection currents, with local thermal differences of $1\,^{\circ}\text{C}$ or more, at up to and beyond 40 cm from a human arm (Wright, 1968). Clearly *Ae. aegypti* could detect such a target thermally from a considerable distance. The more sedentary bedbug appears to be less sensitive to heat, requiring a temperature differential of 1–2 °C (Marx, 1955; Overal and Wingate, 1976). The bedbug is only drawn to warm (37 °C) objects when they are presented at a range of about 5 cm. Heat may act in various ways as a clue to the presence of the host. The insect may respond to the radiant heat emitted by the host, to temperature gradients between the host and insect (convective heat), or directly to the body heat of the host once it has been contacted (conducted heat). Unfortunately it is often difficult to establish which heat stimulus is being utilized by the insect. Both the hemipteran *Rhodnius prolixus* (Wigglesworth and Gillett, 1934) and the yellow fever mosquito *Aedes aegypti* (Peterson and Brown, 1951) use convective heat in attraction. The cat flea, *Ctenocephalides felis felis*, is drawn to warm targets with a maximum response to objects at 40 °C, but only if the target is moving or otherwise creates air currents (Osbrink and Rust, 1985). Whether this is a response to radiant or convective heat is unknown. Heat is not a short-range orientation factor for all blood-sucking insects. Thus the drawing power of sticky targets for the blackfly *Simulium venustum* (Fallis *et al.*, 1967) and horseflies (Bracken and Thorsteinson, 1965) is not enhanced by heating the target.

Heat, as well as acting as an attracting agent, can also be used to gain information about the host animal. The louse *Pediculus humanus* rarely leaves the host, but it will do so if the host dies or suffers from a serious febrile illness. The louse is responding to temperature changes, leaving the host if its temperature falls or rises substantially from normal (Buxton, 1947). The louse can then move as far as 2 m in quest of a new host, using heat as one of the locating signals (Wigglesworth, 1941). It has occasionally been suggested that hosts with fever, due perhaps to a parasite transmitted by a vector insect, may show increased drawing power for a blood-sucking insect, compared to an uninfected host (Gillett and Connor, 1976; Mahon and Gibbs, 1982; Turell *et al.*, 1984). This suggestion, if correct, would have considerable epidemiological consequences. It was not found to be the case for mosquitoes feeding on malaria-infected small mammals, where hyperthermia had no significant impact on the numbers of feeding insects (Day and Edman, 1984). This was confirmed in a study showing mosquitoes displayed no preference for infected over uninfected humans (Burkot *et al.*, 1989). In addition, direct experiments using different skin temperatures within the physiological range showed no effect on mosquito attraction (Grossman and Pappas, 1991).

As with other host location stimuli (Fig. 4.2), the response of the insect to heat can be influenced by its physiological state. This is shown well by the streblid *Trichobius major*, which lives in association with the bat *Myotis velifer*. In summer, the bat's temperature rises to about 33 °C before it takes flight. Fed flies are repelled by these temperatures and migrate off the bat onto the wall of the roost. In contrast, hungry flies are drawn by the heat and migrate onto the bat (Overal, 1980). In winter, when the bats are not active, body temperature remains below 10 °C and the flies are found permanently on the bat.

Most of the studies on water vapour perception by blood-sucking insects have been physiological or structural (Altner and Loftus, 1985) rather than behavioural, and therefore the degree of involvement of water vapour in short-range orientation-attraction is unclear. When it has been reported as important it is usually as a synergistic agent with another stimulus. Thus the upwind movement of the mosquito *Ae. aegypti* in a stream of warmed air is enhanced if the relative humidity of the air is between 40 and 60 per cent (Bar-Zeev *et al.*, 1977; Eiras and Jepson, 1994), and similar results are found for sandflies (Nigam and Ward, 1991).

4.5   Movement between hosts

There is considerable variation in the extent of the association between the blood-sucking insect and the host. At one extreme the permanent ectoparasites are continuously associated with the host, and at the other extreme some temporary ectoparasites, including many of the blood-sucking Diptera, visit the host only for long enough to obtain a blood meal. Clearly insects in the latter category, which actively seek the host, must be capable of a considerable degree of movement, but even permanent ectoparasites must disperse from one host to another and so must also have an efficient means of locomotion.

For those insects that live within the pelage or feathers of the host, having wings is an encumbrance. For this reason permanent ectoparasites, and most periodic ectoparasites, have secondarily lost their wings, relying on their legs to enable them to move around on the host and to transfer between hosts. When hosts are in close bodily contact, such as during copulation or suckling, these ectoparasites walk from the current host onto a new one. Permanent and periodic ectoparasites commonly have other adaptations that make movement within the covering of the body easier. These include the modification of the tarsal claws to enable them to grip the hairs or feathers of the host efficiently. A common modification is in the effective diameter of the claw and a loose correlation is found between the size of the claw and the diameter of the host's hairs, etc., when a range of species is considered (Hocking, 1957). As well as tarsal claws, the legs of ectoparasitic insects are often well endowed with backward-pointing setae. These may help the insect to move through and/or maintain its position in the host's covering; for example, polyctenid bugs, such as *Eoctenes* spp., swim rather than walk in the fur of their hosts, and extensions of the limbs act as flippers to help this type of movement (Marshall, 1981). A modification seen in the limbs of several highly active bat ectoparasites is the presence of pseudojoints (unsclerotized rings of cuticle), which increase the flexibility of the limb (Marshall, 1981) and presumably aid the movement of the insects on or between host animals.

To return to the question of wings, some periodic ectoparasites, such as keds of *Lipoptena* spp., have the best of both worlds. They emerge with wings, which enable them to move rapidly and effectively onto a host, but once the insect successfully reaches the host the wings are shed. Presumably the loss of flight capability is more than offset by the advantage gained by being able to move freely in the host's pelage. Increased mobility in the pelage, enabled by the loss of the wings, may be a crucial factor in permitting the insect to escape from the host's grooming activities. Interestingly, keds can still transfer between hosts even after shedding their wings, using the classic permanent ectoparasite's method of walking from one to the other when the hosts are in close bodily contact (Samuel and Trainer, 1972). Not all periodic ectoparasites find wings a significant encumbrance in their movements around the host animal. For example, many adult hippoboscids and streblids retain their wings throughout their life, but in these cases the wings are considerably modified, being either toughened or constructed so that the insect can fold them in a way that avoids abrasion by the host's covering (Bequaert, 1953; Marshall, 1981).

Temporary ectoparasites tend to take very large blood meals, thus limiting the danger from the host by minimizing the number of visits paid to it, and acting as an insurance policy in case hosts are difficult to find in future. Natural selection has ensured that these insects have optimized their locomotory systems to maximize the size of the meal they can take, while at the same time minimizing the risks attendant in carrying it out. In tsetse flies the meal is often two to three times the fly's unfed body weight (Langley, 1970). Such a meal seriously impairs the insect's mobility, and, after feeding, flight speed falls from 15 to only 3–4 miles per hour (Glasgow, 1961). Clearly this is a dangerous time for the fly. Maximum generation of lift by the tsetse fly is attained at 32 °C. As tsetse flies normally feed around dawn and dusk, ambient temperature will usually be below this optimum. To compensate for this the fed fly generates heat endogenously. Immediately following the blood meal the fly raises its thoracic temperature towards the optimum of 32 °C by 'buzzing' (that is, producing the characteristic sound after which the tsetse is named). Buzzing is caused by the rapid contraction of the flight musculature while the wings are uncoupled from the flight motor mechanism. By raising the thoracic temperature in this way, the tsetse maximizes its lift and flight speed, allowing it to take an enormous blood meal while retaining the maximum chance of avoiding host defensive responses and escaping predators (Howe and Lehane, 1986).

The fleas have lost their ancestral wings, but many have found a way of putting the thoracic musculature to good use – it helps them to jump. Jumping fleas represent a halfway house between insects that rely on walking to transfer from host to host and those that fly. For their size fleas can jump amazing distances. For example, the cat flea, *Ctenocephalides felis*, which is only about 3 mm long, can jump a height of 330 mm, that is 110 times its own length (Bossard, 2002; Rothschild *et al.*, 1972). These feats are achieved by using its thoracic musculature, which progressively compresses an elastic protein called resilin held in the pleural arches. Simultaneously, cuticular catches are engaged that ensure that the resilin remains compressed when the thoracic musculature relaxes. Jumping occurs when the flea releases the catches and the energy stored in the compressed resilin is released with explosive force. The energy is transmitted to the substrate through the flea's hind legs (Bossard, 2002; Rothschild *et al.*, 1972). At room temperature a flea can jump about every 5 s and can do so repeatedly almost indefinitely.

Resilin is a remarkable rubber-like substance. It is the most efficient material known, natural or man-made, for the storage and sudden release of energy. It is little affected by temperature, and therefore the flea can manage to jump at temperatures that would certainly ground a flying insect. How is this achieved? Although the efficiency of the muscles is reduced at lower temperatures, as long as they can compress the resilin, no matter how long it takes, the fleas can still jump. This is unlike flying insects, in which decreased muscular efficiency at lower temperatures rapidly limits the ability to fly. Indeed, flight becomes impossible below a certain threshold temperature.

Fleas use their prodigious jumping feats to contact hosts passing in their vicinity. A flea in the correct physiological state is triggered to jump by host-related stimuli. The jump is aimed towards the source of the stimuli, and the legs, especially the middle pair, are held out ready to grasp onto a host if the jump is successful. The success of fleas is a testimony to the efficiency of this means of moving onto a host. Jumping seems to combine the ability to move rapidly, normally restricted to flying insects, with the capacity to do away with wings, which are such a hindrance to movement within the host's covering. Jumping is also compatible with the production of a flattened body, which further enhances the insect's mobility on the host, but the mechanism is clearly not as efficient as flight in host location as we can see from the distribution of fleas. Fleas are mainly limited to hosts that rear their young in some sort of nest, and so fleas are generally not found on grazing and browsing animals (Traub, 1985). Why should this be so? Fleas, in common with the majority of temporary ectoparasitic Diptera, have larval stages that live independently of the host. When the adult flea emerges it must locate a host and jumping is clearly going to be efficient over a much smaller range than flight. So although the newly emerged dipteran can fly considerable distances to find a host, the flea must be confident that it will find a host in its immediate vicinity.

5

Ingestion of the blood meal

5.1 Probing stimulants

Probing occurs in response to the quality and quantity of host-related stimuli (Friend and Smith, 1977). As in the other phases of host location, the response is not performed in a completely stereotyped way. Anyone who has slept under a mosquito net is likely to have had firsthand experience of this flexibility of responsiveness. If the skin becomes pressed against the net, mosquitoes are quite happy to probe and feed through it. The set of stimuli received in these circumstances must be quite different from the range of host-related stimuli that an insect landing directly on the skin would normally receive. The new set of stimuli received after landing can still influence host choice even at this very late stage, with insects choosing to leave rather than feed (Gikonyo et al., 2000). Post-landing responses can also vary with internal changes of circumstance such as the insect's degree of hunger (Brady, 1972; Brady, 1973; Friend and Smith, 1975) or water deprivation (Khan and Maibach, 1970; Khan and Maibach, 1971), feeding experience (Mitchell and Reinouts van Haga, 1976) or reproductive state (Tobe and Davey, 1972). Even if internal and external factors are carefully controlled, different individual insects still show a considerable degree of innate variation in their response to a host (Gatehouse, 1970). This flexibility of close-range responsiveness allows the insect to make the most of the differing circumstances in which it contacts hosts.

External factors affecting the readiness of insects to probe include vibration; surface texture; skin, hair and feather thickness; carbon dioxide and other odour levels; visual stimuli; contact-chemical stimuli; and heat and moisture levels. Of these, heat is an important probing stimulant in many insects (Friend and Smith, 1977). It can be sufficient on its own to stimulate probing in hungry R. prolixus, which will attempt to probe the inside of glass containers warmed on the outside by a hand. In these insects the heat receptors are restricted to the antennae (Flores and Lazzari, 1996; Schmitz et al., 2000). In the tsetse fly the heat receptors are found on both the antennae and the prothoracic leg tarsi (Dethier, 1954; Reinouts van Haga and Mitchell, 1975). Using these receptors the tsetse fly can monitor substrate temperature and, providing there is a temperature differential of

14 °C between the substrate and the air, probing may be initiated (Dethier, 1954; Van Naters *et al.*, 1998). Similarly, a rapid increase in substrate temperature induces probing in the stablefly *Stomoxys calcitrans* (Gatehouse, 1970). In contrast the bedbug *Cimex lectularius* may be induced to probe by a temperature differential of only 1–2 °C (Aboul-Nasr, 1967). Heat is also required to stimulate feeding in simuliids (Sutcliffe and McIver, 1975) and tabanids (Lall, 1969). The situation is not so clear-cut in mosquitoes, in which heat may be an important factor inducing probing (Davis and Sokolove, 1975; Eiras and Jepson, 1994; Moskalyk and Friend, 1994), but is not always a necessary stimulus (Jones and Pillitt, 1973).

5.2   Mouthparts

Vertebrate skin is divided into an outer epidermis and an inner dermis. Only the dermis is vascularized and any insect in search of a blood meal must first find a means of getting to the blood. The mouthparts of blood-sucking insects are often highly specialized for this purpose (Bergman, 1996). Some insects are opportunists and will feed on blood when it leaks to the surface through a wound. This is fairly common in the Muscidae; for example, the non-biting flies *Fannia benjamini* and *Hydrotaea armipes* take blood from wounds caused by tabanids (Garcia and Radovsky, 1962). *Hydrotaea* spp. may even become impatient and crowd around the mouthparts of the biting fly, feeding concurrently with it (Tashiro and Schwardt, 1953), sometimes interrupting the feeding of the biting fly and forcing it away from the wound in the process.

Other insects are more 'professional' in their blood-feeding activities and do not depend on chance for their blood meal. Species such as *Philaematomyia lineata* have prestomal teeth which may be sufficiently well developed to break through the partially dried and clotted surface of a wound (Patton and Craig, 1913). Other species have gone a step further and have mouthparts that can penetrate intact skin. Mouthparts designed for this purpose have arisen independently in several different insect groups. They can be crudely divided into two categories. Piercing and sucking mouthparts are seen in the bugs, lice, fleas and mosquitoes. In all of these the mandibles and/or maxillae have been modified to form long, thin, piercing stylets which are also interconnected to form a long tube through which blood can be sucked. Commonly, insects with these mouthparts take their food directly from a blood vessel that has been lanced by the mouthparts. The second category includes mouthparts that are used to rip, tear or otherwise cut the skin, and then to lap or suck blood from the haemorrhagic pool that forms. These mouthparts are seen in the tabanids, blackflies and biting flies. A guide to the different mouthpart components of the different insect groups, and an outline of the ways in which they have been modified

Table 5.1 *Adaptations of mouthpart components for different purposes in various haematophagous insect groups.*

	Sheath	Salivary canal	Food canal	Anchorage	Puncture	Penetration	Site of feeding
Hemiptera	Labium	Maxillae	Maxillae	Mandibular teeth	Mandibular stylets	Maxillae (and mandibles in *Cimex*)	Vessel
Anoplura	Investigation of the labium	Median stylet (labial?)	Hypopharynx	Haustellar teeth (labral?)	Haustellum	Dorsal, median and ventral stylets	Vessel
Siphonaptera	Labium	Maxillae	Epipharynx	?	Maxillae	Vessel or pool	
Diptera, Culicidae	Labium	Hypopharynx	Labrum (and mandibles?)	Maxillary teeth	Maxillae	Maxillae	Vessel or pool
Simuliidae	Labium	Hypopharynx	Labrum	Maxillary teeth	Mandibles	Mandibles and maxillae	Pool
Tabanidae	Labium	Hypopharynx	Labrum	Maxillary teeth	Mandibles (and maxillae?)	Mandibles and maxillae	Pool
Muscidae	Maxillary palps	Hypopharynx	Labrum	–	Labellar teeth	Labellar teeth	Pool
Hippoboscidae	Labium	Hypopharynx	Labrum	–	Labellar teeth	Labellar teeth	Pool or vessel

Information largely from Marshall (1981) and Smith (1984).

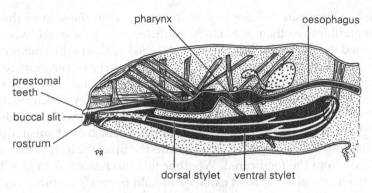

Figure 5.1 A longitudinal section of the head showing the arrangements of the mouthparts of a blood-sucking louse. (James and Harwood, 1969).

for each feeding task, is given in Table 5.1. The way that these mouthparts are used to cut or pierce the skin is still not clear for many species. Some examples, of the better studied and understood cases, are described below.

5.2.1 Lice

The labrum forms a short snout at the front of the louse. This is everted during feeding to expose a series of teeth that grip the host's skin. Within the head is the trophic sac which contains the three stylets that are the penetrative elements of the mouthparts (together these are known as the syntrophium or fascicle) (Jobling 1976). The fascicle is formed of a dorsal (hypopharyngeal) stylet, a median stylet (the homologies of which are uncertain but it carries the salivary canal within it), and a ventral (labial) stylet (Fig. 5.1). The ventral stylet is armed with teeth and is used for piercing the skin (Smith and Titchener, 1980; Stojanovich, 1945).

5.2.2 Bugs

The structure of the mouthparts, and the way in which they are used in feeding, has been extensively studied in the reduviid bugs *Triatoma* and *Rhodnius* (Friend and Smith, 1971; Guarneri *et al.*, 2000; Lavoipierre *et al.*, 1959; Snodgrass, 1944). The mouthparts are in the form of a long, thin beak which, when the insect is not feeding, is folded back under the head. The food canal and salivary canal are both formed by the maxillae, which are joined together by hooked structures that have been named coaptations. The maxillae are flanked by the mandibles, which have backward-pointing teeth on their lateral edge. The maxillae and mandibles are in the form of long, thin stylets. The labium is the largest structure in the mouthparts of these bugs. The dorsal surface of the labium is deeply indented to form a sunken groove in which the stylet-like maxillae and mandibles lie. When the bug is about to feed the mouthparts are swung forwards in front of the rest of the head and are placed in contact with the skin of the host. The mandibles are then alternately retracted and protracted, when it is presumed that the mandibular teeth cut the skin in a saw-like manner. The mandibles do not penetrate far into the skin, but act as an anchorage point through which the maxillae are projected. The maxillae are flexible and capable of changing direction within the skin which they penetrate deeply. Once a blood vessel is penetrated, the left maxilla slides backwards over the right for some distance, thus disconnecting a catch mechanism that has been holding the two together, and permits the tip of the left maxilla to fold outwards from the food canal. Whether this mechanism is to give blood cells easier access to the food canal or to hold open a punctured capillary is a matter of conjecture.

5.2.3   Blackflies

These are pool-feeding flies and their mouthparts and feeding method are thought to represent the primitive condition in the Nematocera. The food canal is a chamber lying between the labrum and the hypopharynx. The salivary canal is contained in the hypopharynx. The skin is cut by the mandibles and laciniae of the maxillae. Both elements have backward-pointing teeth at their distal ends. The prominent snout of the blackfly is formed by the labrum, which has a deep groove on its ventral surface that holds the other elements of the mouthparts when they are not in use. The cutting of the skin is a serial process (Wenk, 1962). The mouthparts are opened and applied to the surface of the skin. The mandibles are repeatedly protracted and retracted, making the initial cut in the skin on the retracting stroke. The wound is opened by the insertion of the labrum and the laciniae. The laciniae appear to be used as anchorage points while all the mouthparts except the labium are thrust into the wound. This series of events is repeated and the wound is deepened until blood vessels are lacerated and a pool of blood forms. The labella and ligula of the labium form a seal around the entrance to the wound, which aids the fly in efficiently sucking blood from the forming pool.

5.2.4   Mosquitoes

Mosquitoes have piercing mouthparts which they use for feeding directly from blood vessels or from a pool of blood in the skin. These mouthparts, and this feeding method, are thought to be the most highly evolved for blood feeding in the Nematocera. During feeding the mouthparts divide into the fascicle (or stylet bundle) and the labium that encloses it. When feeding, only the fascicle passes into the skin, while the labium becomes progressively bent as the mosquito probes increasingly deeper into the tissues (Fig. 5.2). The fascicle is formed of the labrum, maxillae, hypopharynx and mandibles, which are all drawn out into long, thin stylets. These

MOUTHPARTS

Figure 5.2 As the mosquito fascicle penetrates the tissues of the host it springs out of the channel in the labium, which is progressively bent as the fascicle passes deeper into the host.

individual components are usually held tightly together. It has been suggested that adhesion of the stylets is aided by the surface tension of the saliva (Lee, 1974). While this may be the case in the non-feeding mosquito, it is not true during feeding as the mouthparts are immersed in liquid. The maxillae have a pointed tip and recurved teeth at their distal ends; they are believed to be the main penetrative elements of the mouthparts. The maxillae are thought to thrust alternately and, by using the teeth to anchor themselves in the tissues, to pull the labrum (food channel) and hypopharynx (salivary channel) with them as they penetrate the tissues (Robinson, 1939). The function of the mandibular stylets is unclear, but may involve protection of the bevelled end of the labral food canal during penetration, or the separation of the salivary canal from the food canal in the feeding process (Lee, 1974; Robinson, 1939; Wahid *et al.*, 2003). Retraction of the mouthparts from the wound is also thought to involve the active participation of the maxillae. When the mouthparts are not in use, the fascicle is enclosed in a dorsal or ventral channel in the labium.

5.2.5 Tabanids

Close personal contact with some of the larger tabanids can be an unforgettable experience. They are pool feeders, drawing blood from a considerable wound. The maxillae and mandibles bear teeth which are used to cut the skin in a scissor-like fashion on the inward sweep (Dickerson and Lavoipierre, 1959). The toothed maxillae are probably used for anchoring the mouthparts, giving the mandibles the purchase necessary to work on the skin. The labellar lobes, at the tip of the labium, are muscoid in type and bear the characteristic pseudotracheal (sponging) channels of these flies. The labium, however, does not enter the wound and is not used in blood feeding; instead the food channel is formed by a deep gutter in the labrum.

5.2.6   Tsetse flies and stableflies

These pool-feeding flies pierce the skin in an essentially different manner from other blood-sucking Diptera. Instead of using the mandibles and/or maxillae as cutting or penetrating stylets, the biting flies use the labium. This forms the major part of a semi-rigid set of mouthparts normally called the proboscis, or less often the haustellum. The labium bears two labellar lobes at its distal end. These lobes are everted during feeding, exposing an array of highly sclerotized teeth which are hidden from view in the non-feeding fly (Fig. 5.3). To cut the skin the lobes are repeatedly and rapidly moved outwards and backwards (the cutting stroke) (Gordon et al., 1956). It is unclear how the labium is forced into the skin, but it has been suggested that the labellar teeth anchor the mouthparts in the tissues so that the backward stroke of the labellar lobes forces the tip of the labium further into the tissues. Once the skin is pierced the proboscis is often partly withdrawn before being thrust in again at a slightly different angle, possibly to locate suitable blood vessels or to increase the size and rate of formation of the blood pool. The whole tip of the proboscis is also capable of considerable rotation about the long axis of the proboscis. In the stablefly this rotation occurs when the proboscis is fully inserted into the skin. This may also be a means of increasing the damage to the blood vessels and the rate at which the blood pool will form. From personal experience, it also seems to be the most painful stage in the bite of the stablefly. The dorsal surface of the labium has a deep gutter which forms the floor of the food canal. It also holds the hypopharynx (through which the salivary canal runs) and the labrum (which forms the roof of the food canal).

5.2.7   Fleas

The mouthparts of the flea are shown in Figure 5.4. The food canal is formed by the apposition of the epipharyngeal stylet dorsally and laterally, with the laciniae of the maxillae ventrally. Both the epipharyngeal stylet and the maxillae have an extended stylet-like form, and the sides of the maxillae have backward-pointing teeth at their distal end (Snodgrass, 1944). The teeth on the maxillae are used to cut the skin of the host. When a blood vessel has been located and feeding is taking place, only the epipharyngeal stylet is within the blood vessel (Lavoipierre and Hamachi, 1961).

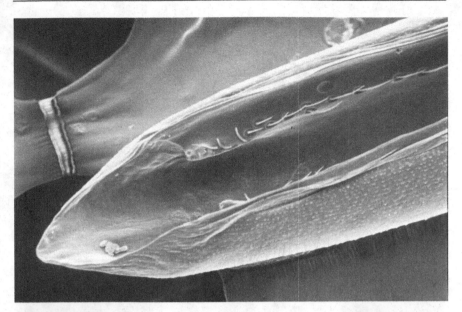

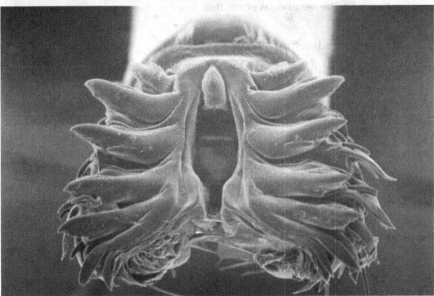

Figure 5.3 Top: the labium of the stablefly, *Stomoxys calcitrans*, is shown after the labrum and hypopharynx have been removed. The labium forms a deep gutter and the sensory organs that monitor blood flow can be seen. Bottom: the labellar lobes at the tip of the proboscis of *S. calcitrans* are everted during feeding to expose these highly sclerotized teeth.

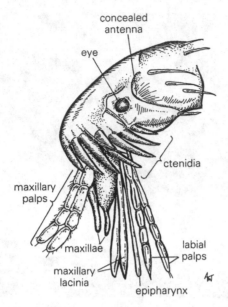

Figure 5.4 The mouthparts of the flea.

5.3   Vertebrate haemostasis

Bleeding occurs at sites of tissue damage in the vertebrate host. The bleeding is stopped by a series of interrelated mechanisms involving the blood vessels themselves, coagulation and platelet activity. This process is collectively known as haemostasis. Haemostasis is clearly of great importance to the blood-feeding insect, which has developed a series of mechanisms to overcome the problems posed by it. Before discussing these adaptive features of blood-sucking insects, a brief outline will be given of the major mechanisms involved in vertebrate haemostasis. I emphasize that this is only a brief outline because haemostasis, with its central importance in human health, has been studied in considerable depth, and complete books are produced on this subject alone (Colman, 2001).

5.3.1   Vasoconstriction and platelet plugs

The amount of blood lost from a damaged vessel is directly proportional to the quantity of blood that flows through it. As a consequence venules and arterioles contract after injury, reducing the local flow of blood and in turn blood loss. The capillaries lack the necessary muscle layer to contract in this way, but it is possible that following injury blood flow in the affected capillary bed is reduced by contraction of the precapillary sphincter.

The blood vascular system is lined by endothelium. The cells of the endothelium have a regulatory role, inhibiting inappropriate episodes of coagulation or platelet aggregation, which could lead to thrombus formation in the closed vascular system. The rupture of the wall of a blood vessel leads to the exposure of blood to non-endothelial tissues. Many of the components of these tissues, collagen in particular, carry negatively charged groups on their surface. Platelets adhere strongly to these negatively charged surfaces and are induced to synthesize and secrete a number of factors. One of the most powerful of these substances is thromboxane A_2 which, along with serotonin (5HT) (which is also secreted by platelets), causes local vasoconstriction. The platelets also secrete adenosine diphosphate (ADP), which supplements the ADP released during cell damage at the site of injury. The ADP, in combination with thromboxane A_2 and thrombin, induces aggregation of platelets at the site of injury, and the forming platelet plug begins to block the hole in the vascular system (Fig. 5.5). The platelet plug is sufficient on its own rapidly (a few seconds) to block small injuries to capillaries. To block more extensive injuries the platelet plug needs to be stabilized and strengthened. This is achieved by the introduction of a network of fibrin into the plug during blood coagulation.

5.3.2 Coagulation

Damage to the blood vessels, and the subsequent exposure of the subendothelial tissues, not only induces platelet aggregation at the site of injury, but also blood coagulation. The coagulation system is organized as a biological amplifier. A small initial stimulus is turned into a major response by a chain of enzyme reactions (the coagulation cascade) interlinked by a series of positive feedback loops. The coagulation cascade can be initiated in two ways. First, damage to the tissues leads to the release of tissue thromboplastin (tissue factor); this begins coagulation by the extrinsic pathway. Alternatively, the exposure of factor XII (which is present in blood plasma) to negatively charged surfaces such as collagen (as happens when the endothelial lining of the vascular system is broken) causes its conversion into activated factor XII (i.e. XIIa). This begins coagulation by the second route, the intrinsic pathway. It is now clear that the extrinsic pathway, dependent upon tissue factor, is the predominant pathway physiologically. Tissue factor is not present on normal endothelium and is absent from red cells. All other tissues express tissue factor and it is crucial in the haemostatic response to injury. The extrinsic and intrinsic pathways are connected by feedback loops and both lead into a common final portion of the coagulation cascade (Fig. 5.5).

Activated factor XII activates prekallikrein to kallikrein which, in the presence of high molecular weight (HMW) kininogen, causes the conversion of more XII to XIIa in a positive feedback loop. Factor XIIa in the presence of HMW kininogen causes the conversion of factor XI to XIa. Factor XIa in the presence of calcium ions in turn converts factor IX to IXa.

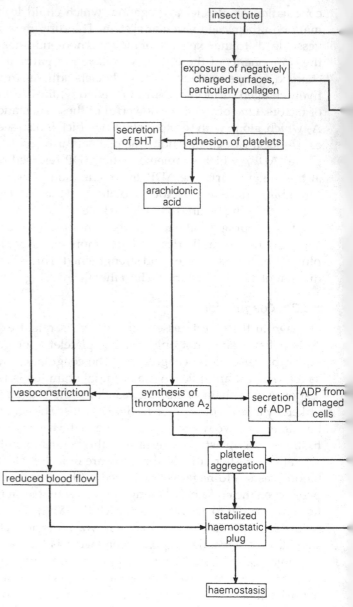

Figure 5.5 An outline of haemostasis. Mechanisms leading to vasoconstrict
the diagram. The main elements of the coagulation system are shown on t

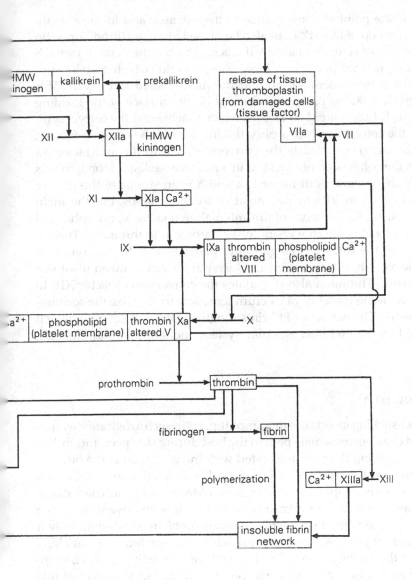

plugging of the puncture by platelet activity are shown on the left side of
t-hand page.

This step is the point of convergence of the extrinsic and intrinsic pathways. Conversion of IX to IXa can also be caused by tissue thromboplastin in the presence of activated factor VII. Factor IXa can then cleave factor X to Xa in the presence of calcium ions, phospholipid (which may be supplied by the platelet membranes) and thrombin-altered factor VIII. This conversion (X to Xa) is probably achieved on the surface of the forming platelet plug (phospholipid co-factor), whereas achieving the correct symmetry for the conversion is probably the function of the other co-factors. Factor VIIa can also stimulate the conversion of X to Xa, and factor Xa stimulates the conversion of IX to IXa in a positive feedback loop that acts to amplify the system. Both factor IXa and Xa can stimulate the further conversion of VII to VIIa in two positive feedback loops, but the main function of Xa (in the presence of thrombin-altered factor V, phospholipid and calcium ions) is the conversion of prothrombin to thrombin. Thrombin stimulates the further aggregation of platelets and their secretion of ADP. It cleaves fibrinogen to fibrin, which polymerizes into an insoluble fibrin network. Thrombin also stimulates the conversion of factor XIII to XIIIa which, in the presence of calcium ions, acts to stabilize the forming fibrin network. The network of fibrin strengthens the platelet plug, which blocks the hole in the blood vascular system and stops further bleeding (Fig. 5.5).

5.4   Host pain

Most blood-sucking insects use their mouthparts in such a delicate way that they avoid causing immediate pain to the host during skin penetration, but avoiding triggering the pain associated with inflammation at the bite site is more difficult. Tissue injury in the host will lead to inflammation, which in simple terms is a triple response of pain, redness and heat. The redness and heat are associated with vasodilation, which is an advantage to the insect. The pain potentially presents a serious problem. Most commonly it is the irritability promoted by pain and itchiness at the bite site that alerts the host to the feeding insect. For the first insect feeding at least, there is a delay period between the initiation of feeding and the onset of this irritability, generally referred to as the safe feeding period (see Chapter 7). The insect tries to complete its meal inside the safe feeding period or to extend the safe feeding period by interrupting host processes leading to pain and irritation. The major host factors leading to pain are the release of adenosine triphosphate (ATP) from damaged cells, release of serotonin and histamine from platelets and mast cells, and bradykinin production following the activation of factor XII by tissue-exposed collagen (Cook and McCleskey, 2002; Julius and Basbaum, 2001).

5.5   Insect anti-haemostatic and anti-pain factors in saliva

The introduction of saliva into the host stimulates immune and irritant responses that alert the host to the presence of the insect. This creates an immediate danger for the insect. In addition, in the longer term, immune responses to saliva may increase mortality in the host (Ghosh and Mukhopadhyay, 1998) and may deny the insect access to a meal altogether (see Chapter 7). Clearly, in an evolutionary sense, these are all things the insect would wish to avoid. So it is clear, given these negative aspects of saliva injection, that there must also be considerable benefits. Saliva is produced by most, possibly all, terrestrial animals. Its original function was probably as a lubricant for the mouthparts as they worked against one another. It is proposed that the specific components of the saliva, which vary according to the diet of the animal, appeared later in evolution to provide secondary functions to the lubricatory role. Perhaps the most obvious advantage that saliva can bring is the presence of enzymes which are used in the digestion of the meal, but in blood-sucking insects this is not the case. Few digestive enzymes have been found in these insects, and those that are present are at concentrations too small to make it likely that they are important factors in blood meal digestion (Gooding, 1972). Clearly the introduction into the host of macromolecules for the purpose of digestion is likely to stimulate a rapid and pronounced response alerting the host to the insect's presence and placing the insect in great danger. So their almost complete absence in blood-sucking insects is probably an adaptation designed to eliminate this response. Exceptional examples such as the salivary hyaluronidase in sandflies and blackflies are unlikely to be digestive. The probable function of this hyaluronidase, for example, is in increasing the size of the lesion for these pool feeders and spreading the pharmacologically active salivary agents beyond the wound (Ribeiro *et al.*, 2000).

So what do blood-sucking insects introduce into the wound in their saliva? Well virtually all of the relatively few species studied produce an anti-coagulant, an anti-platelet compound and a vasodilator (sometimes more than one of each), and these in combination allow the insect to overcome the haemostatic defences of the host and to feed successfully. (The only exception to this found to date is the face fly *Haematobia irritans* in which only an anti-clotting agent is present (Cupp *et al.*, 1998a), and the suggestion is that this fly is at a very early stage of adaptation to blood-feeding (Ribeiro and Francischetti, 2003).) These compounds also often ameliorate the pain caused by the feeding process, thus ensuring feeding is a much safer process for the insect. It is clear that in evolutionary terms these compounds have been appropriated for use by blood-sucking insects on many different occasions because closely related species often use quite separate molecular mechanisms. For example, the sandflies

Lutzomyia and *Phlebotomus* have completely different molecular species leading to vasodilation, as do the mosquitoes *Aedes* and *Anopheles* (Ribeiro and Francischetti, 2003). Another example is that of the anti-clotting mechanisms used by mosquitoes. *Anopheles* has a small, unique anti-thrombin molecule (Francischetti *et al.*, 1999; Valenzuela *et al.*, 1999) while *Aedes* uses a serpin inhibitor of factor Xa (Stark and James, 1995). Mass-sequencing approaches are going to lead to further rapid progress in our understanding of the pharmacology of insect salivary gland secretions (Valenzuela *et al.*, 2002). Let us look in more detail at the range of compounds we already know are found in blood-sucking insect saliva.

Blood-sucking insects release blood from the circulatory system of their hosts by use of their mouthparts. They then take their blood meal either from a pool that forms on the surface of the skin, from a haematoma formed beneath the surface of the skin, or sometimes directly from the blood vessel. It is vitally important for the insect that the blood remains in a liquid form until feeding is complete. Should the blood coagulate, not only will the insect be unable to complete the blood meal, but its mouthparts will be blocked by the forming clot. Given the possibility of this unpleasant and probably fatal event, it is not surprising that the saliva of most blood-sucking insects contains anti-coagulants (Table 5.2). Consistent with the polyphyletic origins of blood-feeding in insects, it has been shown that different insects produce different anti-coagulins that act at various points in the coagulation cascade. For example, the saliva of the tsetse fly, *Glossina morsitans*, contains an anti-thrombin (Cappello *et al.*, 1998). The salivary anti-coagulin of *Rhodnius prolixus*, which has been named Prolixin S, disrupts the coagulation cascade by preventing factor VIII sitting in the Xase complex (Isawa *et al.*, 2000; Ribeiro *et al.*, 1995; Zhang *et al.*, 1998). *Anopheles stephensi* produces hamadarin, which acts against factor XII (Isawa *et al.*, 2002). From the widespread occurrence of anti-coagulins in the saliva it seems clear that they play an important role in feeding by many blood-sucking insects, but it is interesting to note that insects can feed without them. Experiments in which the salivary glands are surgically removed from insects known to possess anti-coagulins have shown that the operated insects are still capable of taking blood meals for a limited time. Operated *R. prolixus* can successfully feed from a live rabbit (Ribeiro and Garcia, 1981a), and tsetse flies with their salivary glands removed can still successfully complete a number of feeds on a host (although eventually lethal clots did form in the mouthparts of these flies) (Lester and Lloyd, 1929). Given this, it is not surprising to find that saliva performs other functions in addition to preventing blood coagulation.

For temporary and periodic ectoparasites, visiting the host to obtain blood is a very hazardous part of their lives. For most there will be strong evolutionary pressures to minimize host contact time. Mechanical injuries

Table 5.2 *Blood-sucking insects produce a wide range of anti-haemostatic factors in their salivary secretions. This table gives some examples with a range of different activities*

Insect	Anti-haemostatic factor	Function	Reference
Haematobia irritans	Thrombostatin	Anti-thrombin	(Zhang *et al.*, 2002)
Anopheles stephensi	Hamadarin	Anti-factor XII	(Isawa *et al.*, 2002)
Rhodnius prolixus	Prolixin S	Anti-factor IXa / Xa	(Isawa *et al.*, 2000)
Phlebotomus papatasi	Apyrase	Hydrolysis of ADP and ATP	(Valenzuela *et al.*, 2001)
Rhodnius prolixus	RPAI	Anti-platelet aggregation	(Francischetti *et al.*, 2000)
Chrysops	Chrysoptin	Fibrinogen receptor antagonist	(Reddy *et al.*, 2000)
Rhodnius prolixus	Nitrophorin	Anti-histaminic	(Ribeiro, 1998)
Triatoma infestans	Unnamed	Local anaesthetic: Sodium channel blockage	(Dan *et al.*, 1999)
Lutzomyia longipalpis	Maxadilan	Peptide vasodilator	(Moro and Lerner, 1997)
Anopheles albimanus	Peroxidase	Destroying vasoconstrictors	(Ribeiro and Valenzuela, 1999)
Aedes	Sialokinin	Peptide vasodilator	(Champagne and Ribeiro, 1994)
Cimex lectularius	Nitrophorin	Nitrovasodilator	(Valenzuela and Ribeiro, 1998)

to small blood vessels, such as those that occur during the probing of insect mouthparts, are plugged within seconds by aggregating platelets (Vargaftig *et al.*, 1981) (the process is outlined above and in Fig. 5.5). This method of controlling blood loss from small blood vessels is very important for the animal. If platelet activity is inhibited, even minor damage to small blood vessels can lead to severe bleeding. The more efficiently blood-sucking insects can overcome this platelet blocking, the more rapidly they can feed and get away from the host. The saliva of at least 23 species in 8 families (Champagne *et al.*, 1995a) contains an enzyme, apyrase, that acts on the platelet-aggregating factor ADP, converting it to non-active adenosine monophosphate (AMP) and orthophosphate (Ribeiro *et al.*, 1985). This inhibits platelet aggregation (Fig. 5.6). There are at least two distinct molecular families of apyrases in blood-sucking insects, the mosquitoes

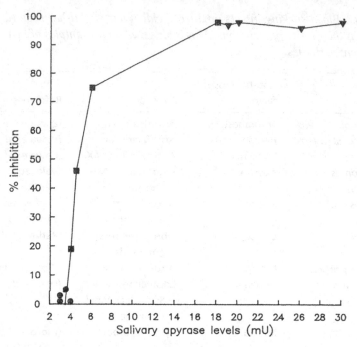

Figure 5.6 Platelet aggregation is inhibited by apyrase from the salivary glands of anopheline mosquitoes: *Anopheles* sp. nr. *salbaii* (•); *An. stephensi* (■); *An. freeborni* (▼) (Ribeiro *et al.*, 1985).

using apyrases from the 5′ nucleotidase family (Champagne *et al.*, 1995b) while sandflies and bedbugs have apyrases belonging to a novel protein family (Valenzuela *et al.*, 2001). The time taken by a mosquito to feed is directly dependent upon the amount of apyrase it can inject into the wound (Fig. 5.7). An apyrase that inhibits platelet aggregation is also contained in the saliva of *Rhodnius prolixus* (Sarkis *et al.*, 1986), together with RPAI-1, which also inhibits platelet aggregation but by scavenging low concentrations of ADP (Francischetti *et al.*, 2000). Deerflies use another method of inhibiting platelet aggregation. They use a protein called chrysoptin which inhibits the binding of fibrinogen to the fibrinogen/glycoprotein IIb/IIIa receptor on platelets (Reddy *et al.*, 2000).

   Damage to host blood vessels usually results in vasoconstriction, which, if it is not countered, will increase feeding time for blood-sucking insects because blood flow to their mouthparts is restricted. Blood-sucking insect saliva usually contains components promoting vasodilation. *Cimex lectularius* and *Rhodnius prolixus* contain nitric oxide (NO) which promotes vasodilation (and inhibits platelet aggregation). Nitric oxide is unstable and in insects proteins have developed that carry it to its site of action. In bugs these are called nitrophorins (Champagne *et al.*, 1995a; Montfort *et al.*, 2000; Valenzuela and Ribeiro, 1998). In some insects peptide vasodilators

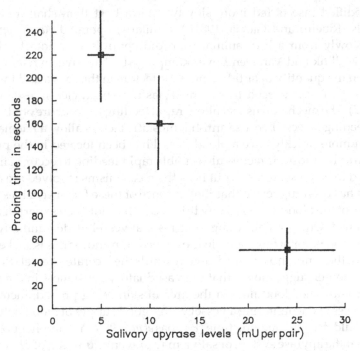

Figure 5.7 Feeding time in mosquitoes is directly related to the quantity of apyrase in their salivary glands. This is illustrated here, where median probing time for *Anopheles freeborni* (lowest probing time), *An. stephensi* and *An. sp.* nr. *salbaii* (highest probing time) are plotted against the mean quantity of salivary apyrase present in the pair of salivary glands of each species. Bars = standard errors. (Ribeiro *et al.*, 1985).

are present, for example maxadilan (the most potent vasodilator known) in the sandfly *Lutzomyia longipalpis* (Champagne and Ribeiro, 1994; Cupp *et al.*, 1998b; Moro and Lerner, 1997). Another system operates in the saliva of the mosquito *Anopheles albimanus*, which contains a peroxidase/catechol oxidase that promotes vasodilation presumably by destroying catecholamines in the host skin (Ribeiro and Valenzuela, 1999). Adenosine is produced by the sandfly *Phlebotomus papatasi*, and this is its main vasodilatory substance (Katz *et al.*, 2000; Ribeiro *et al.*, 1999).

It seems probable that this minimization of contact time between the host and the feeding insect is the key function of the saliva of blood-sucking insects. As stated above, this minimization of contact time has a strong selective advantage, particularly for the temporary ectoparasite, increasing its chances of surviving these most dangerous episodes in its life. Strong evidence that salivary agents permit the minimization of probing and feeding times was obtained from experiments in which *R. prolixus* had their salivary glands surgically removed. While these insects could feed equally rapidly from blood presented in an artificial feeding apparatus, the modified insects fed more slowly from a host than sham operated controls (Ribeiro and Garcia, 1981b). Similarly, operated *Ae. aegypti* fed more slowly from a host animal than did controls with intact salivary ducts (Mellink and Van Den Bovenkamp, 1981). The inverse relationship between the quantity of anti-haemostatic factors in the saliva and the time it takes the insect to feed from a host has now been clearly established (Fig. 5.7). The mechanisms that allow rapid feeding to occur are obvious for pool-feeding insects, because anti-haemostatic factors allow a haematoma to form more quickly once a blood vessel has been located in the probed skin. Anti-haemostatic factors also enable rapid feeding to occur in insects believed to be vessel feeders, but here the mechanisms involved are not so clear. It has been suggested that the presence of these factors enables rapid location of the blood vessels. How this occurs has not been experimentally determined. One possibility suggested is that vessel feeders initially feed from the haematoma they cause in the probed skin and, as this is sucked dry of blood, their mouthparts are drawn towards the lacerated vessel (Ribeiro, 1987). Another suggestion is that the vasodilatory substances in the saliva may increase the blood flow in the area of skin being probed, increasing the chances of encountering a vessel with a good supply of blood, which in turn would decrease feeding time (Pappas *et al.*, 1986). Yet another possibility is that the apyrase system of saliva may prevent clumping of platelets at the tip or inside the feeding canal of the feeding insect, which would also expedite feeding. None of the above hypotheses are mutually exclusive and they could all play a role in the feeding process.

The saliva of some blood-sucking insects (*R. prolixus*, for example) is known to contain anti-histamine (Ribeiro, 1982), which in addition to acting as an antagonist to vasoconstriction also acts as an anti-inflammatory agent. Such an agent may extend the 'safe feeding period', that is, the period before the inflammatory reaction and the itchiness associated with it draw the attention of the host to the biting site. Even relatively modest increases in this period could have considerable advantages for the biting insect and particularly for temporary ectoparasites (see Chapter 7). Other agents acting on pain agonists, in addition to anti-histaminics, are also present in saliva. Bradykinin is an inducer of pain and so hamadarin, which can prevent bradykinin formation, would decrease pain at the site of the bite (Isawa *et al.*, 2002). Serotonin, which also induces pain, is removed from the site of the bite by some insects (Ribeiro, 1982), and ATP is also a pain-inducing substance that is removed from the site of the bite by salivary apyrases in many insects. Some insects, such as the yellow fever mosquito *Aedes aegypti* and the sandfly *Lutzomyia longipalpis*, secrete adenosine deaminase in their saliva (Charlab *et al.*, 2000; Ribeiro *et al.*, 2001). This enzyme will remove from the site of the bite adenosine, a molecule associated with both the initiation of pain perception and the induction of mast cell degranulation. In the process it will produce inosine, a molecule that potently inhibits the production of inflammatory cytokines (Ribeiro *et al.*, 2001). However, the complex and varied nature of the evolution of salivary pharmacological agents is well illustrated here because other blood-feeding insects adopt the opposite strategy and actually secrete pharmacologically significant quantities of adenosine in their saliva. In these insects it is suggested the adenosine plays a vital role as a vasodilator, a role that has been successfully filled by appropriating other molecules in insects destroying adenosine (Ribeiro and Francischetti, 2003; Ribeiro *et al.*, 1999). It might also be speculated that these other insects also have as yet undiscovered means of dealing with the mast cell degranulation that these copious amounts of adenosine would induce.

As we saw above, pain can be reduced by the destruction of host agonists that stimulate nerves. An anaesthetic, a substance acting directly on nerve conduction, would also help increase the safe feeding period. It has been suggested that the saliva of *Ae. aegypti* contains such an anaesthetic, serving to deaden the response of local nerve endings when the insect is probing. The benefits of such a component in the saliva are clear, but its presence has not been substantiated (Mellink and Van Den Bovenkamp, 1981). There is more compelling evidence for an as yet unidentified anaesthetic compound in the saliva of *Triatoma infestans* (Dan *et al.*, 1999). This compound works through the blocking of sodium channels in the nerves. The presence of such a compound in *Triatoma infestans* might be expected as these large insects take a particularly long time to feed, up to 15 minutes for an adult insect.

5.5.1 Other salivary functions

Secondary roles for saliva have been suggested, including playing a part in skin penetration. It is common for small parasites (the infective third stage larva of hookworms and the cercaria of schistosomes, for example) to use softening agents and enzymes to facilitate their passage across the skin. But blood-sucking insects seem to rely almost entirely on mechanical means of penetration. This would seem sensible for temporary ectoparasites at least because of the selective advantage to be gained from feeding quickly. Chemical-based systems to soften the skin are likely to slow down feeding because they are relatively slow-acting. But in some instances chemicals may be useful. Thus pool-feeding sandflies and blackflies have a salivary hyaluronidase, and it is suggested this may help the spread of salivary pharmacologically active agents in the vicinity of the feeding lesion, perhaps to increase the size of the feeding lesion itself (Ribeiro *et al.*, 2000). Also it has been reported that the saliva of the cat flea *Ctenocephalides felis* has a skin-softening agent that makes it easier for the mouthparts to penetrate (Feingold and Benjamini, 1961).

Another interesting role suggested for saliva in mosquitoes is in holding the stylet bundle together by surface tension (Lee, 1974).

5.6   Phagostimulants

Having successfully cut the skin and dealt with haemostasis, the insect is now in contact with the blood meal. The insect still makes no assumptions, but carefully looks for a set of blood-associated cues before it begins ingestion. In line with their polyphyletic origin, there are several cues that different species of blood-sucking insect may look for, but each particular species tends to use a very small range of clues. The fact that relatively few signals are acceptable to a particular insect is reflected by the relatively small numbers of receptors found on the mouthparts of most blood-sucking insects (Chapman, 1982).

Blood-sucking insects can be classified into three groups dependent on the blood-associated clues they use as phagostimulants (Galun, 1986). This classification (outlined below) is interesting in that it shows that, even within a closely related evolutionary group, different clues have emerged for the recognition of blood. The mosquitoes are an example: culicines respond maximally to ADP while aedines prefer ATP (Galun, 1987; Galun et al., 1988; Galun et al., 1993) and anophelines will feed in the absence of nucleotides.

(a)   *Insects using clues associated with the cellular fraction of the blood meal.* Extrapolating from the few species studied to date, it seems probable that most blood-sucking insect species fall into this group. Tsetse flies (Galun and Kabayo, 1988), tabanids (Friend and Stoffolano, 1984), blackflies (Smith and Friend, 1982) and culicine mosquitoes (Friend, 1978; Galun et al., 1963; Hosoi, 1958; Hosoi, 1959) all feed readily on whole blood, but are very reluctant to feed on plasma presented alone. It is clear that the convergent evolution that has occurred among the different members of this group has produced a range of slightly different mechanisms allowing them to discriminate between plasma and whole blood. Most rely on nucleotides derived from platelets (Galun et al., 1993) modulated by other factors in the blood (Werner-Reiss et al., 1999). The major stimulant for most of the insects in this group is ATP, but some, like *Simulium venustum*, are maximally responsive to ADP (Smith and Friend, 1982).

Work with ATP analogues suggests that the range of responsiveness seen reflects differences in the binding of the stimulants to the membrane of receptors of the gustatory sensillae in the insect's foregut. Stimulation is not dependent on ATP acting as an energy source for the receptor. Rather, it has been suggested that stimulation occurs in response to conformational changes in the membrane brought about by nucleotide binding, which in turn opens membrane channels permitting an influx of sodium ions to the receptor (Galun, 1986). Like so many of the responses occurring during host location, it is notable that sensitivity to ATP is not a fixed phenomenon (see Chapter 4). As the insect becomes hungrier so it becomes more sensitive to ATP (Friend and Smith, 1975).

(b)   *Insects using clues present in the blood plasma.* The sandfly *Lutzomyia longipalpis* and the anopheline mosquitoes *Anopheles freeborni*, *An. stephensi*, *An. gambiae* and *An. dirus* all feed happily on cell-free plasma (Galun et al., 1985; Ready, 1978). In fact the sandfly will even feed readily on isotonic saline while *An. freeborni*, *An. stephensi* and *An. gambiae* will feed on isotonic saline when 10^{-3} M sodium hydrogen carbonate is added. *Anopheles dirus* feeds on this solution if albumin is added. It seems that the major consideration in this group is the tonicity of the solution, and that ATP has no impact on feeding.

(c)   *Insects intermediate between (a) and (b) above.* These insects rely on more than one stimulus for full engorgement. The flea *Xenopsylla cheopis* is a straightforward example. It responds to the tonicity of the meal and feeds on solutions isotonic with plasma; a full feeding response can be elicited to such solutions if they are supplemented with ATP (Galun, 1966). *Rhodnius prolixus* also responds to a range of other nucleotides and phosphate derivatives from the cellular fraction of blood, but it will, despite being an obligate haematophage, take isotonic saline in the absence of nucleotides (Guerenstein and Nunez, 1994).

Whether they use nucleotides as a stimulant or not, all blood-feeding insects require isotonic saline before they will display optimal feeding in an artificial system. Tonicity and the presence of adenine nucleotides are not the only engorgement stimuli used by blood-sucking insects. This can be seen in the body lice which will feed, but not fully engorge, on plasma and will not feed on isotonic saline. Nor can they be stimulated to take a full meal by the addition of ATP to plasma. It seems that other, small molecular weight components of the cellular fraction are important here (Mumcuoglu and Galun, 1987).

Thickness of the substrate being probed can also be an important part of the series of events leading up to blood feeding. Tsetse flies feed far more readily when presented with blood under thicker, agar membranes than under thinner, parafilm membranes (Burg et al., 1993; Langley and Maly, 1969; Margalit et al., 1972). It has been suggested that the thicker membrane may be more effective because of the stimulation of increased numbers of external receptors on the inserted mouthparts (Rice et al., 1973). The inclusion of ATP in the thicker membranes further increases the probing response.

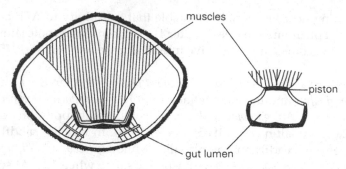

Figure 5.8 The cibarial pump of *Rhodnius prolixus*, showing the relaxed pump with a closed lumen (left) and the cavitated lumen (right), caused by the contraction of the pump's well-developed musculature (Bennet-Clark, 1963).

5.7   Blood intake

Having located the blood, the insect has to transfer it from the host to its gut. To do this it uses pumps located in the head capsule. There may be only one, the cibarial pump, as in the bugs and the muscoid flies, or this may be supplemented by a second, pharyngeal pump, as in the mosquitoes, blackflies and tabanids. Essentially the pumps work by creating a negative pressure difference between the tip of the mouthparts and the pump by muscular cavitation of the pump's lumen. The action of the cibarial pump of *Rhodnius prolixus* has been studied in some detail (Bennet-Clark, 1963; Smith, 1979; Smith and Friend, 1970) and I will use it as an example.

The head of a fifth instar nymph of *R. prolixus* is about 5 mm long and 0.8 mm in diameter. The pump and its associated musculature nearly fill the head capsule. The pump is formed from a rigid, ventrally positioned V-shaped girder 3.5 mm long and 0.28 mm wide. Attached to this, by 'rubbery' ligaments on either side, is a more flexible overlying element which acts as a piston. The piston is attached to the well-developed muscles of the pump (Fig. 5.8). The fifth instar nymph is capable of ingesting a meal of 300 mg in 15 minutes. The terminal diameter of the mouthparts is about 8 μm (or 10 μm if the left maxilla is fully retracted). This gives an astonishing flow rate at the opening of the mouthparts of between approximately 4.4 and 6.6 m s^{-1}. The cibarial pump supplies the force to move the blood. Given that the viscosity of blood at 37 °C is probably about 3 centipoises (cP) (Altman and Dittmer, 1971), the pressure differential the cibarial pump must generate to move the blood can be calculated from the Hagen–Poiseuille equation to be about 576 kPa (5.76 atm) at a terminal diameter of 10 μm, and a massive 1405 kPa (14.05 atm) at a terminal diameter of 8 μm. How the blood avoids cavitation at these sorts of pressures is hard to understand and it is possible that we are misunderstanding the mechanics of this system.

Many of the calculations carried out on the movement of blood in insects have used the Hagen–Poiseuille equation, which describes the mechanics of Newtonian fluids. There is some uncertainty as to the validity of these calculations because blood is a non-Newtonian fluid, that is, its apparent viscosity will depend upon the shear rate to which it is subjected. However, at the tube radii and feeding pressures seen in piercing and sucking insects it is believed that blood behaves essentially as a Newtonian fluid and so the Hagen–Poiseuille equation is still used in calculations on blood movements (Loudon and McCulloh, 1999). Another problem encountered when performing calculations on blood flow in insect mouthparts is that there is still considerable doubt about the viscosity of blood at the tube diameters and pressures seen in insect mouthparts. Estimates for the viscosity of blood in vivo range from about 1.5 cP, which is the measured value for plasma alone, to about 3 cP, which has been measured for whole blood (but in tubes with much larger diameters than insect mouthparts).

Despite these uncertainties, models describing the flow of blood in narrow tubes have been developed and used to describe the mechanics of feeding in blood-sucking insects (Daniel and Kingsolver, 1983; Tawfik, 1968). Analysis of these models suggests that the biomechanics of feeding are reasonably accurately described by the Hagen–Poiseuille equation:

$$Q - \frac{\pi r^4 pt}{8\eta l}$$

where Q is the meal size, p is the pressure difference between the tip of the mouthparts and the lumen of the pump, r is the radius of the feeding tube, l is the length of the feeding tube, t is the feeding time and η is the dynamic viscosity of the blood. The feeding models developed have demonstrated several interesting relationships. First, the time required to complete a blood meal is most sensitive to the radius of the food canal. As a rule of thumb, this means that for a food canal of fixed length, and for a fixed pressure drop applied by the cibarial pump, feeding time will increase 16-fold for a halving in the radius of the canal. To a lesser extent feeding time is also directly proportional to the length of the food canal, and is in inverse proportion to the pressure difference between the cibarial pump and the blood source. These relationships are illustrated in Figure 5.9.

Because of the dangers inherent in obtaining a blood meal, short feeding time is likely to be a character that is strongly selected for in temporary ectoparasites. As we have seen, small increases in the radius of the feeding canal can bring large decreases in the time taken for the insect to achieve a full blood meal. Therefore we might expect that natural selection is acting to maximize the diameter of the feeding canal. The fact that insects such as the triatomine bugs retain a food canal with a terminal diameter (8–10 μm) little larger than that of blood cells is strong evidence that other equally important selection factors must be in operation to prevent a larger terminal

Table 5.3 *The size of red corpuscles varies widely in different animals. Given that many blood-sucking insects have mouthparts with a terminal diameter of around 10 μm this may be a factor affecting the feeding efficiency of blood-sucking insects feeding on different host species.*

Host	Red blood corpuscle dimensions (μm) estimated from dry films	Host	Red blood corpuscle dimensions (μm) estimated from dry films
Mammals		Reptiles	
Man	7.5	Alligator	23.2
Horse	5.5	Tortoise	18.0
Cow	5.9		
Sheep	4.8	Amphibians	
		Congo 'snake'	62.5
Birds		(*Amphiuma means*)	
Turkey	15.5	Frog	24.8
Chicken	11.2	(*Rana catesbeiana*)	

Data from Altman and Dittmer (1971).

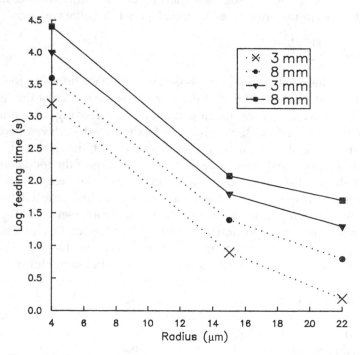

Figure 5.9 The time needed to complete a blood meal of 4.2 mm³ is plotted against the radius of the food canal for two lengths of the feeding canal and for two pressure differences, 100 kPa (. . . . .) and 10 kPa (——) (Daniel and Kingsolver, 1983).

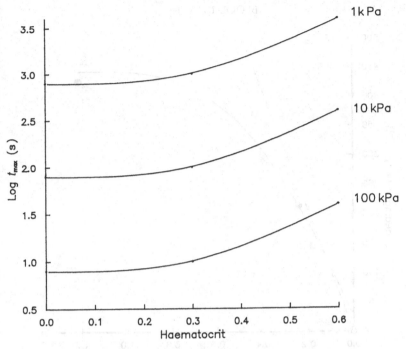

Figure 5.10 The feeding time (expressed as log t_{max} s) for *Aedes* decreases with decreasing haematocrit, irrespective of the pressure differential exerted by the insect. Because many parasitic infections cause a fall in the haematocrit of the host, this would favour those insects feeding on infected hosts (but other factors must also be taken into consideration) (Daniel and Kingsolver, 1983).

aperture from emerging. The most obvious of these selection pressures is that a small terminal diameter is required in order to efficiently (and painlessly?) penetrate the tissues of the host. If we look at the terminal diameter of the food canal of other species of piercing insects we see a convergence in size that supports this argument. For example, the terminal diameters of the food canal of *Cimex*, *Aedes* and *Pediculus* are 8 μm, 11 μm and 10 μm, respectively (Tawfik, 1968). However, the factors defining the size of the terminal aperture in *Rhodnius prolixus* must be in very fine balance because even an increase in terminal diameter from 8 to 11 μm (which is the diameter successfully used by *Aedes*) has not arisen, although this would theoretically decrease the feeding time by a factor of approximately 3.6, that is from about 15 minutes to under 5 minutes! This would seem to be an enormous potential advantage to the insect, yet it has not been adopted. It is possible that the morphological adaptation seen in the triatomine bugs, in which the left maxilla is redrawn on the right to increase

BLOOD INTAKE

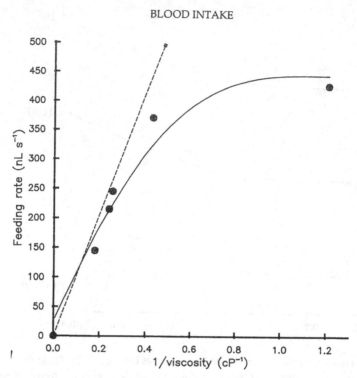

Figure 5.11 The measured feeding rate for *Rhodnius prolixus* is plotted on the reciprocal of the viscosity of the fluid ingested (•, solid line). The dashed line shows the feeding time predicted from the Hagen–Poiseuille equation. Blood viscosity at 37 °C is estimated to be about 3 cP. Increasing viscosity will lead to a decrease in feeding rate which can be predicted by the Hagen–Poiseuille equation. Decreasing viscosity due to, for example, decreasing haematocrit, will cause an increase in feeding rate, but this will not necessarily be predicted by the Hagen–Poiseuille equation (Smith, 1979).

the effective opening of the terminal aperture from 8 to 10 µm, is an evolutionary attempt to have the best of both worlds, a fine diameter stylet during probing and a wider food canal during feeding. In many host species the diameter of red blood corpuscles is considerably greater than 10 µm (Table 5.3) and this may have interesting consequences in terms of feeding time and/or host choice.

Another notable relationship arising from these models is that between the haematocrit of the blood (that is, the proportion of red cells present) and its viscosity. A decrease in the haematocrit leads to a decrease in the apparent viscosity of blood. In the original interpretation of the model it was suggested that a decrease of 30 per cent in the haematocrit would lead to about a 30 per cent decrease in the time needed to complete the meal (Fig. 5.10) (Daniel and Kingsolver, 1983). This is predicted from the Hagen–Poiseuille equation, but experimental work on *R. prolixus* suggests this is an overestimate and that a 15 per cent decrease in ingestion time would be expected (Fig. 5.11) (Smith, 1979) – still a significant decrease. The insect is quite likely to encounter lowered haematocrits in the field as it is known that blood-feeding insects may directly cause anaemia in their hosts (Schofield, 1981) and the parasites that they transmit may also decrease the host's haematocrit (Taylor and Hurd, 2001). It would be advantageous for the insects to feed on these hosts, in the sense that this would decrease the time necessary to complete feeding.

One fact that seems to have been ignored in the exploration of these models is that blood viscosity is inversely related to temperature. The measured absolute viscosity of blood from female humans rises from about 3 cP at 37 °C to about 4.46 cP at 20 °C (Altman and Dittmer, 1971). The models have been interpreted with the assumption that the blood being ingested is always at 37 °C. This is not necessarily the case. The temperature of the skin at many points on the host's body will be significantly lower than core body temperature. The Hagen–Poiseuille equation predicts that because of the higher viscosity of blood, at 20 °C it will take about 1.49 times longer to complete a full blood meal. The effect of viscosity of the meal on feeding time has been verified experimentally (Fig. 5.11) (Smith, 1979). Because of this effect on the rate of ingestion, increased blood viscosity at lower temperatures may well be a factor that influences the choice of insect feeding sites on the host.

6

Managing the blood meal

6.1   Midgut anatomy

Blood-sucking insects can be divided into two groups depending on the design of the alimentary canal for the storage of the blood meal. In one group, typified by Hemiptera and fleas, the alimentary canal is a simple tube with no diverticulae and the blood is stored in the midgut. In the second group, typified by Diptera, the gut has between one and three diverticulae which may be used, in addition to the midgut, for the storage of the blood meal (Fig. 6.1).

The midgut is the site of blood meal digestion and absorption. Two basic patterns of digestion are seen in blood-sucking insects: a batch system and a continuous system (Fig. 6.1). In the batch system, which is well illustrated by mosquitoes, sandflies and fleas, digestion proceeds almost simultaneously over the entire surface of the food bolus. The continuous system is typical of higher Diptera and Hemiptera, the blood meal being held in a specialized portion of the anterior midgut where no digestion takes place. Portions of the blood meal are then gradually passed down through the digestive and absorptive mid and posterior regions of the midgut. In this continuous system much of the meal will have been completely processed and defecated before some has even entered the digestive section of the midgut.

The blood meal is normally separated from the midgut epithelium by an extracellular layer known as the peritrophic matrix (previously known as the peritrophic membrane). The major constituents of this layer appear to be glycosaminoglycans overlaying a chitin scattold. Two types of peritrophic matrix are recognized based on their method of production (Lehane, 1997; Waterhouse, 1953). The most common production method in insects is by secretion from cells along the complete length of the midgut. This type of peritrophic matrix (type I) is absent in the unfed insect, and is only produced when the blood meal has been taken. Type I peritrophic matrix is found in adult mosquitoes (Fig. 6.2), blackflies, sandflies and tabanids.

Type II peritrophic matrix is found in adult muscids, tsetse flies and the hippoboscids. Type II peritrophic matrix is produced by a special organ,

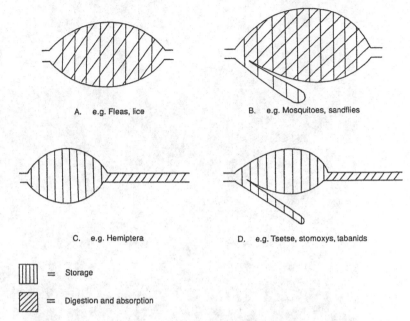

A.   e.g. Fleas, lice

B.   e.g. Mosquitoes, sandflies

C.   e.g. Hemiptera

D.   e.g. Tsetse, stomoxys, tabanids

= Storage

= Digestion and absorption

Figure 6.1 Divided on the basis of structure, there are two basic designs of alimentary canal, either the straight tube (A and C) or an alimentary canal with diverticula (B and D). Blood-sucking insects can also be divided into two categories based on their method of digesting and absorbing the blood meal. Batch processors (A and B) commence digestion over the entire surface of the food bolus; continuous processors (C and D) segregate the intestine into a 'production line', gradually passing a stream of food down the 'line' from the storage region to the digestive and absorptive regions.

the proventriculus (= cardia), which is sited at the junction of the fore- and midgut (Lehane, 1976b; Lehane *et al.*, 1996a). Type II peritrophic matrix is secreted continuously and forms a complete cylinder lining the midgut in both fed and unfed flies. Consequently it is always present to separate the blood meal from the midgut epithelium. Contrast that to type I peritrophic matrix, which is only produced after the blood meal is taken thus ensuring there is a period when the blood meal is in direct contact with the midgut epithelial cells. This may be a key factor in parasite and pathogen interactions with blood-sucking insects (see Section 8.7).

The peritrophic matrix is a semipermeable filter with a pore size of about 9 nm in tsetse flies (Miller and Lehane, 1990). This pore size permits the passage of globular proteins of about 140 kDa and so presents no barrier to either trypsins (about 25 kDa) or to the passage of the peptides produced during digestion. A peritrophic matrix is not produced in hemipteran bugs, but they do produce an extracellular coating in response to the blood meal

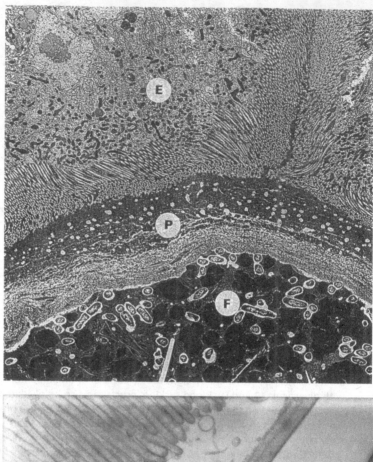

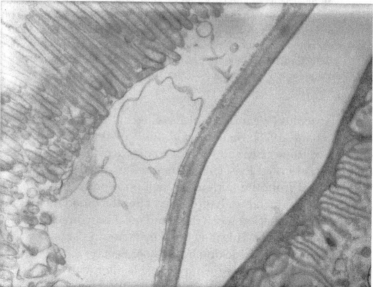

Figure 6.2 Top: an example of a type I peritrophic matrix from the posterior midgut of *Anopheles stephensi* at about 60 hours after the blood meal. The peritrophic matrix (P) can be seen lying between the midgut epithelium (E) and the food bolus (F). (Micrograph kindly provided by W. Rudin.) Bottom: The five-layered, type II peritrophic matrix of *Stomoxys calcitrans*.

(Billingsley and Downe, 1983; Billingsley and Downe, 1986; Lane and Harrison, 1979; Silva *et al.*, 1995). This surface coating is known as the extracellular membrane layer (ECML), and possibly fulfils the same function as a peritrophic matrix. It is interesting to note that structures similar to the ECML are beginning to be described in mosquitoes (Zieler *et al.*, 2000).

The function of the peritrophic matrix is still an enigma, although several suggestions have been made. One function that seems clear is that the peritrophic matrix is the first line of anti-microbial protection for the cells of the midgut, and the presence of lectins capable of binding bacteria has been shown in blowflies and tsetse flies (Grubhoffer *et al.*, 1997; Lehane, 1997; Peters *et al.*, 1983). Other suggested functions are based on the division of the midgut lumen into an endoperitrophic and ectoperitrophic space by the peritrophic matrix. It is suggested that this division increases the efficiency of digestion by spatially separating the various digestive processes, or that it permits the conservation of digestive enzymes which can be recirculated back up the midgut in the ectoperitrophic space (Terra, 1988a; Terra, 2001), or that the production of an unstirred layer next to the midgut cells may increase absorption efficiency. These various suggestions are at least partly true, but there is still a nagging feeling that there is some fundamental and still unrecognized role for the peritrophic matrix in digestive physiology and some speculations have been made (Lehane, 1997). (See Section 8.7 for further discussion of the peritrophic matrix.)

6.2   The blood meal

The size of the blood meal is affected by a range of factors including ambient temperature, insect age, mating status, stage of the gonotrophic cycle, previous feeding history, and source of the blood meal. Nevertheless, it is fair to say that temporary ectoparasites often take very large blood meals. Adult insects will commonly take meals that are twice their unfed body weight, while nymphal stages of the blood-sucking Hemiptera may take meals of ten times their unfed weight! These large blood meals certainly impair the mobility of the insect, increasing the short-term chances of its being swatted by the host or eaten by a predator. For example, the flight speed of *Glossina swynnertoni* is reduced from 15 to 3–4 miles per hour after feeding (Glasgow, 1961), with the fly often only capable of a downward 'glide' away from the host. However, as can be seen by its widespread adoption by so many different groups of blood-sucking insect, the disadvantages of taking such large blood meals must clearly be outweighed by the advantages. A cost–benefit analysis of feeding frequency, which clearly affects blood meal size, has been made for tsetse (Hargrove and Williams, 1995).

Table 6.1 *The size of the red blood meal and the time taken in its digestion are affected by a range of factors including ambient temperature, age of the insect, mating status, stage of the gonotrophic cycle, previous feeding history, source of the blood meal, etc. The figures given here are a rough guideline to the 'average' meal size and time for digestion in a variety of haematophagous insects.*

Species	'Average' time for the digestion of a blood meal by the adult female insect (hours)	'Average' meal size for an adult female insect (in an early phase of the reproductive cycle) (% unfed body weight)
Pediculus humanus	4	30
Cimex lectularius	168	130
Triatoma infestans	336	210
Aedes aegypti	60	109
Anopheles quadrimaculatus	60	122
Culex quinquefasciatus	70	140
Culicoides impunctatus	130	–
Stomoxys calcitrans	48	110
Glossina morsitans morsitans	48	170
Pulex irritans	36	–

The main benefit of taking very large blood meals is probably the minimization of the number of visits that the temporary ectoparasite must pay to the host; this could be advantageous in several ways. If hosts are difficult to find, then the insect must make the most of each encounter. It would also be selectively advantageous to take the largest meal possible if the extra time and energy required for more frequent host visits could be better spent in reproductive activity. But the main advantage to the insect is probably the reduced risk of being swatted which goes with visiting the host as few times as possible. Taking such large and discomforting meals is likely to be a disadvantage to permanent ectoparasites because it would restrict their mobility in the host's covering and none of the benefits outlined above would be gained. In consequence many of the permanent ectoparasites take smaller, more frequent meals. For example, the anopluran lice feed every few hours and take meals that are only 20 per cent to 30 per cent of their unfed weights (Buxton, 1947; Murray and Nicholls, 1965).

Temporary ectoparasites have adapted their morphology and physiology to minimize the risks involved in taking such large meals. When a blood meal is taken it imposes considerable mechanical stresses on the

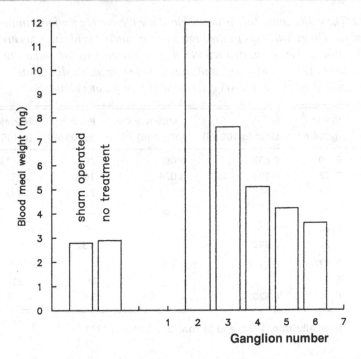

Figure 6.3 Overdistension of the abdomen during blood feeding is prevented by abdominal stretch receptors. This can be illustrated by the overdistension of the abdomen that occurs when the abdominal cord of adult female *Aedes aegypti* is cut at various points along its length. (Redrawn from Gwadz (1969).)

storage zone of the gut and the abdominal wall. The crop of tsetse flies and the midgut storage regions of other blood feeders are capable of considerable stretching to accommodate the blood meal. Undoubtedly the nature of the intercellular junctions (Billingsley, 1990) and underlying muscular coat of the midgut cells helps blood feeders to withstand these mechanical stresses. The abdominal wall is also very elastic or has folds permitting the considerable distension during feeding which is so characteristic of these blood-sucking insects. There is some evidence that the abdominal wall of *Rhodnius prolixus* may be plasticized in response to feeding (Bennet-Clark, 1963), the elasticity of the abdominal wall being switched on and off in response to the blood meal.

The abdominal wall is provided with stretch receptors to prevent overdistension (Fig. 6.3) (Gwadz, 1969; Maddrell, 1963). This is neatly shown in female tsetse flies, which retain the developing larva inside the abdomen until the larva is fully mature. As the larva grows, the size of each blood meal diminishes so that the abdomen never exceeds a certain volume (Tobe and Davey, 1972). The disadvantages of taking very

Table 6.2 *The major constituents of the blood are reasonably uniform in most host animals. The exception is in the high levels of nucleic acids in birds and reptiles because of their nucleated red blood cells. Proteins are far and away the most abundant nutrients in blood, and nutrients are unevenly distributed between whole blood (B), red blood cells alone (E), or plasma alone (P).*

	Water (g/100 ml)	Lipid (g/100 ml)	Carbohydrate (g/100 ml)	Protein (g/100 ml)	Nucleic acid (g/100 ml)
Man	B 80	0.652	0.088	20.5	–
	E 72	0.596	0.074	36.8	0.136
	P 94	0.60	0.097	7.41	0.057
Cow	B 81	–	0.046	–	–
	E 62	–	–	29	–
	P 91	0.348	–	8.32	–
Chicken	B 87	–	0.170	–	–
	E 72	–	–	29	4.216
	P 94	0.520	–	3.6	–

Data drawn from Albritton, 1952, and Altman and Dittmer, 1971.

large blood meals are also overcome by making use of the fact that about 80 per cent of the blood meal is water (Table 6.2). Most of this water is not required by the insects, and they possess very efficient physiological systems for its rapid excretion, thereby reducing their weight and restoring their mobility. To achieve this the meal is held in a distinct region of the midgut where the epithelium is adapted for rapid water transfer (Fig. 6.4). In tsetse flies, stableflies and triatomine bugs, water movement across this epithelium is linked to a ouabain-sensitive, $Na^+–K^+$-ATPase located in the basal membranes of the epithelium with chloride as the counter-ion (Farmer *et al.*, 1981; Gooding, 1975; Macvicker *et al.*, 1994; Peacock, 1981; Peacock, 1982). This very efficient pump works by generating an osmotic gradient across the epithelium which drags water passively after it. It is possible that the pump in *Rhodnius* is switched on by the same diuretic hormone, released from the mesothoracic ganglion in response to the blood meal, which stimulates a 1000-fold increase in fluid secretion from the Malpighian tubules. These systems are so efficient that tsetse flies can shed about 40 per cent of the weight of the meal in the first 30 minutes following feeding (Gee, 1975; Moloo and Kutuza, 1970), and most of the fluid in the very large blood meals of triatomine bugs is discarded within four hours of ingestion (Pereira *et al.*, 1998). A possible danger in this massive flux of water through the haemolymph of the insect is that the haemolymph balance will be disturbed. This does not occur, probably

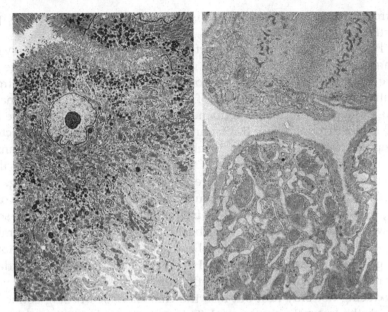

Figure 6.4 Left: an electron micrograph of the epithelium of the reservoir region of the anterior midgut of *Stomoxys calcitrans*. The blood meal is stored in this region. Right: the extensive basal infoldings of the reservoir region and the accumulations of mitochondria around them, which are typical of tissues designed for the rapid transport of water.

because any flux in the haemolymph volume alters the concentration of the diuretic hormone. This in turn causes a feedback response in the rate of absorption by the midgut and/or secretion by the Malpighian tubule epithelium to correct the situation (Maddrell, 1980).

Most of the osmotic pressure in the blood meal is exerted by the salts it contains. However, the rapid and energy-efficient removal of water from the blood meal would be assisted slightly by the precipitation of the soluble proteins in the plasma, because this would reduce the osmotic pressure of the gut contents to some extent. Given the rubbery consistency of the blood meal soon after its ingestion in many insects, it is possible that they have mechanisms to do precisely this, but the removal of water from the meal is also likely to lead to this rubbery consistency. The 'coagulins' secreted into the ingested meal by many blood-sucking Diptera (Gooding, 1972) precipitate at least a part of the soluble components of the plasma, thereby reducing its osmotic pressure. However, coagulation cannot explain the consistency of the blood meal in all blood feeders because blood-sucking Hemiptera actually produce anti-coagulants in the midgut (Gooding, 1972). Whether the jelly-like consistency of the blood is brought about by water removal, haemagglutination, the introduction of precipitating agents, or a combination of factors requires further investigation.

Type II peritrophic matrix, by virtue of its filtration properties, may make water removal from blood meals more energy efficient. The large meal size in tsetse flies produces hydrostatic pressure within the peritrophic matrix, leading to bulk filtration of the blood meal such that plasma, and those blood solutes having a diameter of less than 9 nm, are exuded into the ectoperitrophic space (Miller and Lehane, 1990). Once in the ectoperitrophic space this filtrate can be efficiently absorbed by active transport through cells in the anterior part of the midgut. Bulk filtration saves the fly energy by reducing the osmotic pressure of the fluid it is absorbing.

There are other ways in which insects can minimize the dangers of taking such large blood meals. Lift in tsetse flies is proportional to the square of wing beat frequency, and wing beat frequency increases with temperature up to 32 °C (Hargrove, 1980). Fed tsetse flies maximize their mobility by generating heat in the thorax (Howe and Lehane, 1986). They do this by rapidly vibrating the thoracic flight-box after uncoupling the wings, producing not only heat but also the characteristic buzzing sound after which the tsetse flies are named. This endogenously generated heat can have an important impact on the fly. A fly at an ambient temperature of 20 °C can increase its lift potential by about 17 per cent by raising thoracic temperature to the optimum of 32 °C. Buzzing also increases abdominal temperature, which allows more rapid excretion of water from the meal, further improving the mobility of the fed fly.

The time taken for digestion of the blood meal varies widely both intra- and interspecifically, and is strongly influenced by ambient temperature, blood source, meal size, and several other factors (approximate times are given in Table 6.1). This can be of importance when using serology or DNA-based techniques in epidemiological studies to determine host choice rates (see Section 3.1) because it will determine the time period over which the blood meal is still identifiable. Determining digestion time can be complicated in some insects by their habit of jettisoning the semi-digested remains of one meal if another is offered. This potentially wasted resource is put to good use in some fleas: the larval forms feed on this supply of semi-digested blood from the adults.

One of the most important events in blood digestion is the lysis of the red blood cells because these contain much of the protein in the meal (Table 6.2). Some mechanical haemolysis occurs in a few blood-sucking insects. In fleas this is achieved by repeatedly pushing the food bolus forwards against a series of backwardly projecting proventricular spines, and some mosquitoes possess a cibarial armature which ruptures some (15 per cent to 50 per cent has been reported) of the blood cells as they flow past (Chadee *et al.*, 1996; Coluzzi *et al.*, 1982). Cibarial armatures are found in other blood-sucking Diptera as well as mosquitoes, although it is not known if they cause haemolysis of their blood meals. It is known that cibarial armatures damage ingested filarial worms, and it has been suggested that they have developed as a defence against these parasites. It seems unlikely that the primary role of a cibarial armature is in haemolysis given their relative inefficiency. In most blood-sucking insects it is clear that blood meal haemolysis is achieved by chemical means.

Chemical haemolysins may be produced in the salivary gland, as in *Cimex lectularius* (Sangiorgi and Frosini, 1940), but are more normally produced in the midgut. In *Rhodnius prolixus* haemolysin is produced in the anterior, storage region of the midgut where no proteolytic digestion of the blood takes place (De Azambuja *et al.*, 1983); the molecule involved is a small basic peptide. In the tsetse fly and stablefly the haemolytic agent is secreted in the posterior, digestive region of the midgut, in response to some component of the cellular fraction of the blood meal (Gooding, 1977; Kirch *et al.*, 1991a; Kirch *et al.*, 1991b; Spates, 1981; Spates *et al.*, 1982). In the tsetse fly the haemolysin is proteinaceous, but in the stablefly particular species of free fatty acid may bring about haemolysis.

The salivary glands of blood-sucking insects are unusual in producing virtually no digestive enzymes (Gooding, 1972). The loss of hydrolytic enzymes in the saliva is probably related to the need for the insect to cause the least possible disturbance to the host during feeding. Most digestive enzymes in blood-sucking insects arise from the midgut cells. Because the blood meal is predominantly protein (about 95 per cent), it is not surprising that the major digestive enzymes are proteinases, and molecular information is beginning to be gathered on these (Lehane *et al.*, 1998; Muller *et al.*, 1995; Muller *et al.*, 1993). Trypsins (proteinases active at alkaline pH) are predominant, with carboxypeptidases and chymotrypsins playing a subsidiary role in blood meal digestion in most blood-sucking insects. Hemipteran bugs are exceptional in using cathepsin-like proteinases (active at acid pH). This is consistent with a proposed evolutionary path for blood-sucking hemipteran bugs from sap-sucking ancestors or from bugs feeding on seeds containing serine proteinase inhibitors. Hemiptera with such an ancestry either did not need trypsins (sap feeders) or were forced to use other proteolytic means because of the plants' defences (seed feeders). When they became blood feeders they required digestive proteinases once more. All cells produce cathepsins for use in lysosomes, and hemipterans may possibly have re-routed these enzymes for extracellular digestion (Billingsley and Downe, 1988; Houseman *et al.*, 1985b; Terra *et al.*, 1988b).

Some blood-sucking insects, such as *Stomoxys calcitrans*, secrete digestive enzymes by the regulated route; that is, digestive enzymes are stored in secretory granules in midgut cells and are secreted in response to the meal.

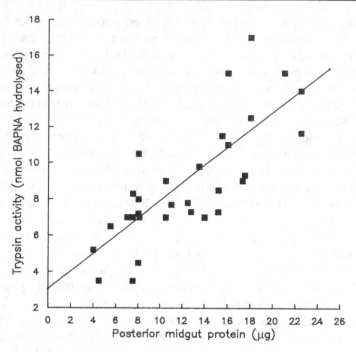

Figure 6.5 The level of trypsin secreted into the intestine is directly controlled by the quantity of protein present in the digestive region of the midgut. (Redrawn from Houseman et al. (1985a).)

In these insects digestion starts immediately the meal is taken, with more enzymes being rapidly synthesized and secreted (Lehane, 1976a; Lehane, 1987; Lehane, 1988; Lehane, 1989; Moffatt et al., 1995). In other insects, such as the mosquitoes, no significant store of enzymes is held in the unfed midgut; enzymes are only produced in significant quantities some hours after the meal is taken (Graf et al., 1986; Hecker and Rudin, 1981; Noriega and Wells, 1999). In all blood-sucking insects, however, the levels of diges-tive enzymes increase after the blood meal, reaching a peak before declining to resting levels as digestion is completed. Secretion of digestive enzymes is regulated by the quantity of protein found in the midgut (Fig. 6.5) (Houseman et al., 1985a; Lehane, 1977a; Lehane et al., 1996b; Noriega et al., 2002; Noriega and Wells, 1999). It is not known if control is exercised directly on the enzyme-secreting cells or if it involves hormones released from the large number of endocrine cells that are interspersed among the digestive cells of the midgut. Some of these cells have been shown to contain FMRFamide and pancreatic polypeptide-like immunoreactivity, indicating a role for these hormones in the regulation of digestive events (Billingsley, 1990; Lehane and Billingsley, 1996). It seems that different ratios of the various digestive enzymes can be induced by blood meals from different hosts, suggesting a fine level of control. For example, blood from bird hosts, which contains nucleated red cells, may induce high levels of digestive DNAase. Variation in the digestive response to blood of different origins may have an important bearing on the susceptibility of the insect to parasites. Thus sandfly susceptibility to *Leishmania* is dependent upon the sources of its blood meals (Nieves and Pimenta, 2002; Schlein and Jacobson, 1998; Schlein *et al.*, 1983) and success of trypanosome infections in tsetse flies depends on which hosts the flies feed on after the infectious blood meal (Masaninga and Mihok, 1999; Olubayo *et al.*, 1994).

Vertebrate serum contains proteinase inhibitors. These are designed to control tissue destruction by endogenously produced serine proteinases, such as those produced by granulocytes. Insects must overcome the effect of these inhibitors if they are to digest the blood meal. These inhibitors slow the rate of blood meal proteolysis, especially in the early phases of digestion when relatively small amounts of insect trypsin are present (Huang, 1971). Blood from different vertebrates displays different levels of inhibition of insect trypsins. Pig and human serum possess only about a third of the inhibitory effect of cow plasma on tsetse fly trypsins (Gooding, 1974). It is possible that these inhibitors may play a part in causing the different rates of digestion seen in an insect dealing with blood from different sources, but this is certainly not always the case, as we can see if we look at the tsetse again. Despite a 300 per cent difference in trypsin inhibition levels in the sera of the two blood meals, tsetse flies can digest a cow meal and a pig meal at the same rate. There is some evidence that insects compensate for the effects of these inhibitors. For example, blood possessing high levels of proteinase inhibitor may stimulate tsetse flies to secrete increased levels of trypsin (Gooding, 1977).

Following extracellular digestion, further digestion of the meal takes place on and in the midgut epithelial cells. We know that aminopeptidase is present (Lemos *et al.*, 1996; Noriega *et al.*, 2002; Rosenfeld and Vanderberg, 1998), probably bound to the microvillar membranes of midgut epithelial cells, and it is possible that intracellular digestion of absorbed peptides occurs in lysosomes (Billingsley and Downe, 1985; Terra and Ferreira, 1994). Other digestive enzymes such as invertase, amylase, and esterases can also be found in the gut (Gooding, 1972); these probably play a subsidiary role in blood digestion while some may be necessary for dealing with the insect's non-blood meals.

Absorption of the products of digestion also occurs in the midgut, but relatively little is known of the processes involved. In those insects showing continuous digestion, absorption of the digestive products occurs in a specialized region of the posterior midgut. The region involved is often characterized by the gradual appearance and accumulation of large lipid globules in the apices of its cells, followed by their gradual disappearance as the insect is starved. This reflects the metabolic fate of the blood meal in adult insects in which the absorptive products are largely converted to and stored as fats, even though the blood meal consists of about 95 per cent protein and only about 4 per cent lipid. These absorptive cells are probably involved in the conversion of the protein products of digestion to lipids (Lehane, 1977b). Most of these lipids will eventually be transferred to other organs such as the fat body or ovaries (Arrese *et al.*, 2001).

It is also possible, but less likely, that all conversions occur in the fat body and that the midgut wall is merely a storage organ, as seems to be the case in one part of the midgut of *Rhodnius prolixus* (Billingsley and Downe, 1989). In insects where the immature stages are also blood feeders, there may be less emphasis on conversion of digestive products to lipid because a considerable amount of protein is used in growth and development. In insects in which digestion is a batch process there is less evidence for cell specialization in the midgut; most cells in these forms are probably responsible for secretion as well as absorption (Rudin and Hecker, 1979; Rudin and Hecker, 1982). Lipid accumulations still occur in the midgut cells and the blood is still largely converted to lipid for storage.

6.3   Gonotrophic concordance

In adult insects the blood meal is largely used to provide resources for the reproductive effort. Thus in females digestion and ovarian development are physiologically integrated. The coordination of these events has been well studied in anautogenous mosquitoes in which the blood meal triggers egg development. In these mosquitoes there is a previtellogenic period before the blood meal is taken when the fat body becomes capable of intense synthesis of yolk protein precursors. This process is thought to be under the control of juvenile hormone III. There is then a period of arrest until a blood meal is taken. Following the blood meal the mosquito enters an intense phase of production of yolk proteins (the vitellogenic period). These are produced by the fat body and transferred to and accumulate in the yolk bodies of the oocytes. Hormonal controls over this process, which centre on 20-hydroxyecdysone, are complex and are reviewed by Raikhel *et al.* (2002).

Many female blood-sucking insects will develop and lay a batch of eggs each time a blood meal is taken, providing the quantity of blood ingested exceeds a minimum threshold level. This process is called gonotrophic concordance (Swellengrebel, 1929) and is exhibited by many mosquitoes, blackflies, sandflies, tabanids and, in a rather poorly defined way, tri-atomine bugs. The number of eggs produced by these insects is directly proportional to the size of the blood meal (Goodchild, 1955; Roy, 1936) up to a maximum that is determined by the number of ovarioles in the

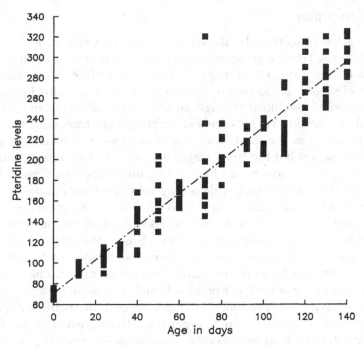

Figure 6.6 The graph shows the regular accumulation of fluorescent pigments in the head of an adult female tsetse fly (*Glossina morsitans morsitans*) as it grows older. This can be used to estimate the age of insects caught in the field (Lehane and Mail, 1985).

ovary. This in turn can be influenced by the nutritional history of the insect in its immature stages (Colless and Chellapah, 1960). In insects showing gonotrophic concordance it is sometimes possible, by the very careful dissection and observation of the ovaries, to estimate how many egg-laying (gonotrophic) cycles the insect has undergone. This in turn indicates how many blood meals have been taken by the insect and also allows us to estimate the insect's physiological age and possibly its calendar age (Detinova, 1962; Hoc and Schaub, 1996). Although this technique requires great skill to perform and considerable judgement in its use, the information gained can be of great value to the epidemiologist.

Not all blood-sucking insects show gonotrophic concordance. Many, including biting flies, tsetse flies, hippoboscids and lice, require several meals for the production of each batch of eggs. Even some mosquitoes show a degree of gonotrophic dissociation, and some may need more than one meal to produce a single batch of eggs. For these insects ovarian dissection techniques are generally less successful in determining insect age and another technique, dependent on the regular accumulation of fluorescent pteridines in the insect, can give good results (Lehane, 1985; Lehane and Mail, 1985; Msangi *et al.*, 1998) (Fig. 6.6).

6.4   Nutrition

Blood-sucking insects can be divided into three groups based on their feed-ing strategies. The first group is a minor one containing forms such as the human-feeding Congo floor maggot, *Auchmeromyia luteola*, in which only the larval stage is blood-feeding. The second group contains insects that feed exclusively on blood throughout their entire life cycle and includes tsetse flies, streblids, hippoboscids and nycteribiids, triatomine and cimi-cid bugs, and lice. The third group comprises insects in which the adults are blood feeders but the larval stages are not, and includes mosquitoes, blackflies, ceratopogonids, sandflies, biting muscoids, horseflies and fleas (although the young stages of fleas may feed upon semi-digested faeces derived from blood-feeding adult fleas). The third group can be further subdivided into three categories: (i) species with adults who are obligatory haematophages, such as fleas; (ii) forms such as stableflies in which adults are optional blood feeders with both sexes also taking non-blood meals; and (iii) forms in which only the adult female is an optional blood feeder but the adult male never feeds on blood, such as mosquitoes.

Insects relying solely on blood as a food source throughout their life har-bour specialized symbiotic micro-organisms. These are usually not found in insects that use food sources other than blood at some stage in the life cycle. This strongly suggests that blood is not a complete food source and that the symbionts supplement the insect's nutrition in some way. Var-ious methods have been developed for the removal of symbionts from living insects so that their role in nutrition can be determined. Some of these methods depend on the physical location of the symbionts or their means of transfer from one insect to the next, both of which may differ among insect groups (Table 6.3). Symbiont-free reduviid bugs have been produced by physically preventing passage of the symbiont to the next generation (Brecher and Wigglesworth, 1944) or by antibiotic treatment (Beard *et al.*, 1992). In the louse *Pediculus humanus*, the organ containing the symbionts (the mycetome) has been physically removed (Aschner, 1932; Aschner, 1934; Baudisch, 1958). In tsetse flies, antibiotics have been used against the symbionts (Hill *et al.*, 1973; Pell and Southern, 1976; Schlein, 1977; Southwood *et al.*, 1975) or they have been attacked with lysozyme or with specific antibodies (Nogge, 1978).

Symbionts are often absent in adult males. For example, in the lice and the nycteribiid *Eucampsipoda aegyptica*, despite being present in the juvenile stages, symbionts are not present in the adult males (Aschner, 1946). Sup-pression of the symbionts in the juvenile stages impairs subsequent insect development. Loss of symbionts from the adult does not affect the gen-eral health of the individual measured, for example in terms of longevity. However, loss of symbionts from adults does affect the reproductive

Table 6.3 *Symbions are common in insects relying on blood as the sole food source throughout their lives. An outline is given of their anatomical locations and the means of transmission from one generation to the next in different insect groups.*

Insect group	Location of symbionts	Means of transmission between generations
Anoplura	Intracellular in midgut epithelium, between the epithelial cells of the midgut or in a mycetome beneath the midgut epithelium in mycetomes in the haemocoel	Infection of the egg during its intra-uterine development in the female
Cimicidae	Mycetomes in the haemocoel	The oocyte is infected and so the new individual contains symbionts at conception
Reduviidae	Partly intracellular in the midgut epithelial cells and partly in the gut lumen	The first-stage nymph acquires the infection from egg shells or the faeces of infected bugs
Hippoboscidae	Intracellular in a specialized zone of cells in the gut epithelium or in a mycetome adjacent to the gut epithelium	Transmitted to the intra-uterine larva in the secretion of the mother's milk gland
Streblidae	Small numbers of symbionts are found in a variety of different cells of the gut, Malpighian tubules, fat body and milk glands	Transmitted in the secretion of the mother's milk gland
Nycteribiidae	In mycetomes which may lie at various sites in the haemocoel symbiont-containing cells may also occur in the fat body and in the female, intertwined with the tubules of the milk glands, or only cells of the milk gland may be infected	Transmission is via the secretion of the mother's milk gland
Glossinidae	Intracellular in a specialized zone of cells in the gut epithelium	Transmitted to the intrauterine larva in the secretion of the mother's milk gland

Information from Buchner, 1965.

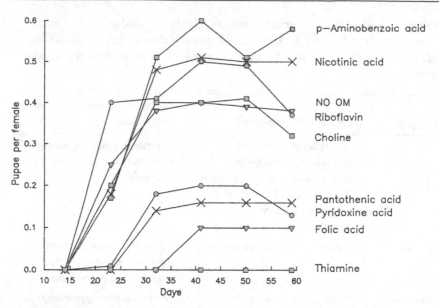

Figure 6.7 The diet of symbiont-free female tsetse flies was supplemented with a range of vitamins (no omission). In a series of experiments, the named vitamin was then omitted from the full diet. The effect on fecundity (expressed as puparial output per surviving female per nine-day period) is shown (Nogge, 1981).

performance of the adult female. The general situation is well illustrated in tsetse flies, in which the number of symbionts halves in the emerging adult males while it doubles in emerging adult females to give a complement of four times that seen in the mature male (Nogge and Ritz, 1982). This increase in symbiont numbers is necessary for female tsetse flies to reproduce normally as symbiont-suppressed females are sterile (Nogge, 1978). Sterility can be partially reversed by feeding the symbiont-free flies on blood supplemented with various vitamins (Nogge, 1981). The essential supplements are the B group vitamins thiamine and pyridoxine, along with biotin (vitamin H), folic acid and pantothenic acid (Fig. 6.7).

The anti-bacterial enzyme lysozyme is abundant in insect body tissues and it may be used to regulate the body areas that can be colonized by symbionts because, in tsetse flies, only the mycetome region of the midgut lacks this enzyme (Nogge and Ritz, 1982). Because there are such clear-cut differences between the numbers of symbionts in male and female tsetse flies, there obviously must be other factors at play within the mycetome cells themselves that regulate symbiont numbers (Nogge and Ritz, 1982).

The genome of one of the three symbionts of tsetse flies – *Glossinidia wigglesworthia* (Aksoy, 2000) – has been fully sequenced (Akman *et al.*, 2002) and full sequencing programs are under way for the other two symbionts. This will facilitate detailed molecular experimentation which will lead to rapid progress in understanding the interactions of symbiont, host insect and parasites transmitted. For example, molecular approaches have shown that the tsetse endosymbiont *Sodalis glossinidius* utilizes a type III secretion system for cell invasion (Dale *et al.*, 2001) suggesting that *Sodalis* may have evolved from an ancestor such as *Salmonella* with a parasitic intracellular lifestyle. That lends credence to the idea that vertically transmitted mutualistic endosymbionts may have evolved from horizontally transmitted parasites through a parasitism–mutualism continuum (Dale *et al.*, 2001).

These symbionts may prove a useful means of expressing transgenes in insects, such as the tsetse fly, which cannot themselves be transformed using technologies involving injection of embryos. It is possible that such transgenes might kill the parasites the insect transmits, thus changing the vectorial capacity of whole populations of insects (Aksoy, 2000; Beard *et al.*, 1992; Durvasula *et al.*, 1997).

The health of the host can influence the nutritional quality of its blood. Theoretically, insects achieve maximum protein intake in a blood meal with a high haematocrit (Fig. 6.8), although slight anaemia has been experimentally shown to increase erythrocyte intake (Taylor and Hurd, 2001). The normal human blood haematocrit is about 40 per cent, but many blood parasites cause a reduction in this value, either by direct destruction of red blood cells or indirectly by interfering with the physiology of the host. So, in theory at least, blood-sucking insects might improve their nutritional status by selectively seeking out and feeding on uninfected hosts, but other factors also need to be considered (see Chapter 3).

Particular species of blood-sucking insect find some hosts are 'better' food sources than others. For example, some insects when they feed on a different host species may show reduced fecundity generated by a reduced rate of development, a skewed sex ratio, reduced food intake or rate of digestion, or reduced longevity (Nelson *et al.*, 1975; Rothschild, 1975). This can be seen in the mosquito *Culex pipiens*, which produced 82 eggs mg^{-1} of ingested canary blood compared to 40 eggs mg^{-1} of human blood (Woke, 1937). Similarly, the tsetse fly, *Glossina austeni*, despite showing similar survivorship when fed on rabbit or goat blood, showed a consistently higher fecundity when fed on rabbit blood (Fig. 6.9) (Jordan and Curtis, 1968). Both the quantity and the quality of the offspring can be affected. *Glossina austeni* and *G. m. morsitans* produce the heaviest puparia when fed on pig blood, followed closely by puparia produced by flies fed on goat blood – with the puparia produced from flies fed on cow blood being considerably lighter. The differences in the suitability of hosts as food sources are not just interspecific. For example, *Xenopsylla cheopis* fed on baby mice produce far fewer eggs than fleas fed on adult mice (Fig. 6.10). From the studies

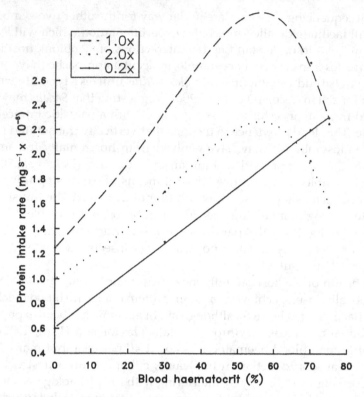

Figure 6.8 The rate of protein intake from a blood meal depends on the proportion of cells in the blood. The normal haematocrit for human blood is about 40 per cent, but this is lowered by many parasitic infections. If no other factors are considered, lowering of the haematocrit will lower the rate of protein intake, a factor that will favour blood-sucking insects feeding from uninfected hosts. Natural selection has ensured the evolution of mouthparts showing efficient mouthpart design. We can see this if we look at pharyngeal height in *Aedes aegypti*. The broken line shows protein intake calculated for this mosquito. If pharyngeal height were either increased or decreased then protein intake rate would fall. This is illustrated here: protein intake rates are calculated for theoretical pharyngeal heights of 0.2 times and 2.0 times that in the mosquito. (Information kindly supplied by T. L. Daniel and J. G. Kingsolver.)

performed to date, it is often not clear what the particular differences between various blood sources are. The different effects observed may be a function of reduced food intake, due to immune or other defence responses of the different hosts, or to the ratio of specific elements in the meal such as the amino acids. It has also been suggested that the levels of host hormones in the blood meal can have a profound impact on the biology of some blood-sucking insects, and we will look at that next.

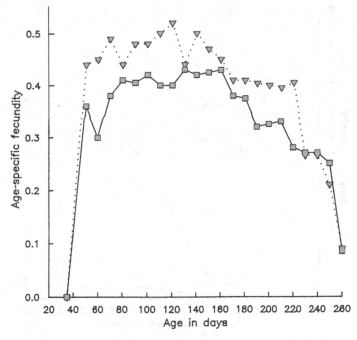

Figure 6.9 *Glossina austeni* shows increased fecundity when fed on rabbit blood (▼) rather than goat blood (■) (Jordan and Curtis, 1968).

6.5   Host hormones in the blood meal

Work done on the rabbit flea, *Spilopsyllus cuniculi*, feeding on the European rabbit *Oryctolagus cuniculus* provides evidence for a strong relationship between the reproductive cycles of the insect and host hormonal levels (Mead-Briggs, 1964; Rothschild and Ford, 1973). Throughout the rabbit's non-breeding season the distribution of fleas on its surface is related to skin temperature. The fleas are usually grouped on the rabbit's ears, where they are attached by their mouthparts. In very cold weather the fleas move onto the body of the animal and in hot weather the fleas leave the rabbit completely if they cannot hold their temperature below about 27 °C. This description of flea distribution on the body of the host is true only in a snapshot sense. If the association is continuously observed, fleas are seen constantly moving between rabbit and nest or directly from one rabbit to another.

The distribution described above, modulated by a range of environmental and physiological factors, persists throughout the rabbit's non-breeding season. With the onset of mating, in March and April, changes begin to occur. During mating, the rabbits' ear temperature rises by up to 7 °C and this stimulates the movement of fleas from one host to another. It

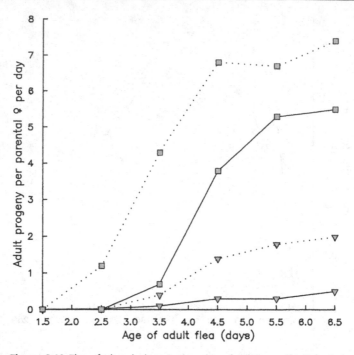

Figure 6.10 Fleas fed on baby mice produce far fewer offspring than those fed on adult mice. ▫, Adult mouse, fleas held at 30 °C; – ▫ -, adult mouse, fleas held at 22 °C; ▽, young mouse, 30 °C; -▽-, young mouse 22 °C (Buxton, 1948).

is suggested that the changing blood levels of various hormones such as progestin and oestrogen that occur in the female rabbit (doe) when she copulates induce the fleas to attach firmly to the doe. The effect of the increased interchange of fleas between hosts, combined with the increased attachment to female rabbits, is that fleas tend to accumulate on the does during the mating season. About 10 days before the doe gives birth there is a large rise in the corticosteroid hormone levels in her blood. It is suggested that the raised levels of these hormones in the flea's blood meal, combined with other hormonal influences from the rabbit, cause a series of dramatic changes in the fleas. Ovarian and testicular maturation take place, accompanied by hypertrophy of the midgut, salivary glands and fat body. The fleas also increase their feeding rates from about once every 15–30 minutes to once every 4 minutes in males and to once every minute or so in females. More changes occur in the hormonal levels in the doe's blood when she gives birth, and it has been suggested that the fleas move from the ears to the face of the doe in response to these. From there about 80 per cent of the doe's fleas will eventually move onto the new litter of rabbits.

As the fleas feed on the young rabbits they encounter a new range of hormones in their food. They are also exposed to kairomones, which are probably associated with the urine of the young rabbits. It is suggested that the combined effect of these factors is to stimulate the fleas into an orgy of mating and egg-laying over the eight-day period following the birth of the litter. By this time the hormonal levels in the litter have changed, and it is suggested that this induces the fleas to move back to the doe over the next 14 days. It is then suggested that under the influence of the new levels of luteinizing hormone and progestins in the doe's blood, the somatic changes the flea underwent on the pregnant doe are reversed, with the flea reverting to its non-reproductive phase until it encounters another pregnant rabbit.

While the events outlined above indicate the typical response of most fleas in a given population, not every flea responds to the changing physiological status of the rabbit in exactly the same way. This variation is to be expected as it will have survival value for the flea population, allowing it to withstand potential disasters such as the death of a litter or the abandonment of a burrow. Another such adaptation is seen in the development of the flea larvae in the burrow. One cohort develops rapidly and leaves the burrow with the does and the young rabbits. The second cohort contains individuals that are larger, better stocked with reserves, than those in the first, and these larger larvae develop more slowly, remaining in the burrow to await the reoccupation of the nest. In this way a disaster befalling one group of rabbits will not necessarily mean a catastrophe for that particular population of fleas.

Linking reproductive effort to that of the host is clearly advantageous to the insect as its offspring have the maximum chance of successfully finding a host that is not already saturated with parasites. This is a common strategy in many blood-sucking insects and also in other parasites. What is unusual here (and in a few other ectoparasitic forms) is the very complex physiological link between host and parasite ensuring the precise timing of reproduction in the invertebrate partner. Complex reproductive control mechanisms may have evolved in fleas because of the special feeding requirements of the larvae, which need the supply of dried blood found in adult flea faeces. Clearly this food resource is likely to be generally available only when rabbits habitually rest in a single place over a considerable period of time, the most reliable time being when a new litter is produced in the nest.

Attractive as the above story may be, it should be pointed out that studies in other flea–host associations have not shown these close links between host hormone levels and flea reproduction (Lindsay and Galloway, 1998; Reichardt and Galloway, 1994). This is a fascinating experimental system deserving more study.

6.6   Partitioning of resources from the blood meal

How are the resources gained from the blood meal partitioned into the various activities that the blood-sucking insect has to perform? Tsetse flies provide a well-studied example. The newly emerged adults use much of the resources of the first blood meals to build up their thoracic muscula-ture (Bursell, 1961). Subsequently resources are partitioned between basal metabolism, flight and reproductive effort. An energy budget is shown in Figure 6.11 (Adlington *et al.*, 1996; Bursell and Taylor, 1980; Gaston and Randolph, 1993; Loder *et al.*, 1998; Loke and Randolph, 1995). Clearly, under each of the different conditions shown in Figure 6.11, the reproduc-tive effort in the female consumes a considerable proportion of the avail-able energy, while it appears to be of negligible importance in the male. Again, regardless of the different conditions, nearly half the resources of the blood meal are devoted to excretion and the digestion of the meal. As basal metabolism is directly related to temperature, at lower temperatures more energy is left over each day for flight. The length of the interlarval period is inversely related to temperature. In consequence at lower temperatures fewer resources are put into the female reproductive effort each day and so again more energy is available for flight. In reality, it is likely that these theoretical effects are offset by the increasing intermeal period as the tem-perature falls, which will reduce the total energy available each day.

The females have considerably less energy available for flight than the males because of the major investment of their resources in reproduction (Fig. 6.11). Both sexes probably expend similar amounts of energy in host location and seeking out refuges in the intermeal period. As additional flights in the males are sexually appetitive ones (active seeking of receptive virgin females), they also invest any spare energy in reproductive effort (Adlington *et al.*, 1996; Bursell and Taylor, 1980; Loke and Randolph, 1995).

Diptera and Hymenoptera typically use carbohydrate, and Lepidoptera and Orthoptera carbohydrate and lipid as fuel for flight (Beenakkers *et al.*, 1984). Tsetse flies are unusual in using the amino acid proline as a substrate for flight (Bursell, 1975; Bursell, 1981). Female *Aedes aegypti* can also uti-lize proline as an addition to their usual carbohydrate sources of energy (Scaraffia and Wells, 2003). Use of proline is possibly a biochemical adap-tation to the high-protein diet associated with the blood-sucking habit, a suggestion supported by the fact that non-blood-feeding male *Aedes aegypti* do not use proline as a fuel for flight (Scaraffia and Wells, 2003). However, blood feeding cannot be the only cause because phytophagous beetles also utilize proline in flight (Gade and Auerswald, 2002). A range of Diptera has been classified on the ability of the insects' flight muscle mitochondria to oxidize either the amino acid proline or pyruvate and glycerophosphate (both of which are markers of carbohydrate as a fuel source) (Bursell, 1975).

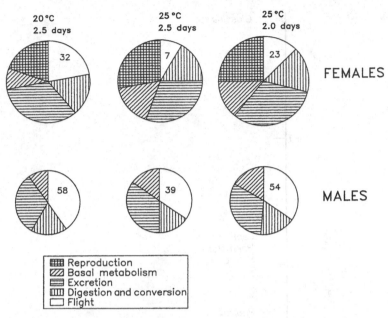

20 °C
2.5 days

25 °C
2.5 days

25 °C
2.0 days

FEMALES

MALES

Reproduction
Basal metabolism
Excretion
Digestion and conversion
Flight

Figure 6.11 Tsetse flies partition the energy derived from the blood meal into the various activities the insects must perform. The daily input of energy is represented by the area of the circle. Partitioning is affected by the sex of the fly and the conditions under which it is held (the temperature and feeding intervals are given). The numbers in the flight sectors of the pie charts are the daily duration of flight in minutes that these reserves would permit (Bursell and Taylor, 1980).

The insects were found to fall into three categories (Fig. 6.12). In the first category are flies that utilize carbohydrate as an energy source for flight. Flies in this category either do not feed on blood, or only the female feeds on blood. In the second category are facultative haematophages, both sexes of which feed on blood as well as other foods; these insects can efficiently use both proline and carbohydrate for flight. In the third category are the tsetse flies, both sexes of which are obligate haematophages and use only proline as the energy source for flight. This analysis shows that as blood increases in prominence in the diet, there is a greater tendency to use proline as an energy source, but proline is also used as an energy source in non-blood-sucking insects such as the colorado beetle. Clearly the biochemical capability for the use of proline is widespread in insects and the blood-sucking habit is only one lifestyle promoting its use over other energy sources. Why at least some blood-sucking insects choose to utilize amino acids as an energy source instead of conventional carbohydrates or lipids has been considered by Bursell (1981), whose arguments are outlined below.

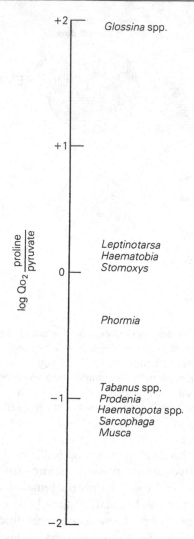

Figure 6.12 When compared on the ability of their flight muscle mitochondria to utilize either the amino acid proline or pyruvate (a marker for the use of carbohydrates as a fuel source), blood-sucking insects fall into three groups, as shown here. (Redrawn from information in Bursell 1975, 1981.)

About 90 per cent of the dry weight of the blood meal is protein and so the most abundantly available resources are amino acids (Table 6.2). These could be converted to carbohydrate through the gluconeogenic pathway to provide a good energy source, one that is far more easily stored than amino acids and, because it is highly soluble, could be easily translocated to the flight muscles when required. Gluconeogenesis would be best performed on the amino acids glycine, alanine and serine, to yield three carbon fragments, but these amino acids are drained from the available pool by the need to dispose of nitrogen in uric acid (McCabe and Bursell, 1975a; McCabe and Bursell, 1975b). The other amino acids produce two or four carbon fragments, which it would be energetically uneconomical to convert to carbohydrates.

Amino acids could also be converted to lipids to provide an excellent energy source that could be readily stored in the body. Mobilization of lipids in the haemolymph is difficult, but this difficulty has been overcome in all insects. In some insects mobilization is so efficient that lipids form a fuel for flight, but this is not the case in Diptera.

So the direct use of an amino acid as a fuel source may be an energetically useful alternative in insects with a high proportion of protein in their diets. This would be particularly true if an amino acid is available that is a good energy source, is low in nitrogen atoms to minimize the energy used for its excretion, and that is highly soluble to permit easy translocation to the flight muscles during times of high demand. Figure 6.13 shows that proline is not only highly soluble, but also has a high net energy yield, carrying only a single nitrogen atom per molecule. It has been firmly established that tsetse flies do indeed use proline as the prime energy source for flight (see Bursell, 1981). The energy is supplied by the partial oxidation of proline to alanine in the flight muscle, with the subsequent reconversion of alanine to proline in the fat body at the expense of stored triglyceride (Fig. 6.14). The synthetic pathway in the fat body cannot keep pace with the rate of proline oxidation occurring during flight, and so tsetse flies are restricted to fairly short flight periods. The 'proline battery' is then recharged during the enforced rests between flights.

Partitioning of amino acids from the blood meal has also been well studied in mosquitoes. Partitioning between energy storage and reproduction is influenced by the size of the female mosquito, which in turn depends on the success of larval feeding (Briegel et al., 2001; Takken et al., 1998), on adult nutrition including sugar feeding (Naksathit and Scott, 1998; Naksathit et al., 1999), on the age of the mosquito (Naksathit et al., 1999), on the species of host from which the blood is taken (Harrington et al., 2001), on whether or not the mosquito is mated (Klowden, 1993) and on the stage of the gonotrophic cycle the female is at (Briegel et al., 2002).

6.7   Autogeny

Autogeny, the capacity to produce at least one egg batch without the need for a blood meal, is found in several blood-feeding insects, including some mosquitoes, ceratopogonids, sandflies, blackflies and tabanids. Autogenous insects carry over the nutritional requirements to produce the

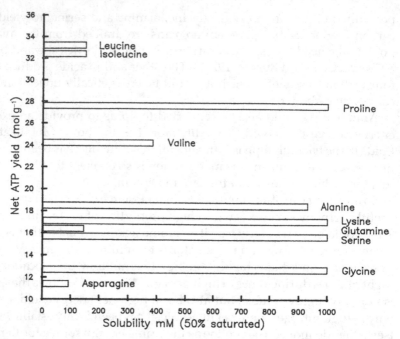

Figure 6.13 The energy yield (ATP yield from total oxidation of the amino acid minus the energy required to detoxify its nitrogen) is expressed against water solubility for a range of amino acids that might potentially be flight energy substrates. (Drawn from data in Bursell, 1981.)

egg batch either largely or wholly from the immature feeding stages. The adult female may supplement these reserves from several non-blood food sources (see below). Both facultative and obligate autogeny are known. Facultatively autogenous insects have the choice of producing an egg batch autogenously or of taking a blood meal and producing a larger egg batch. This flexibility allows the insect to determine whether autogeny is the best strategy considering both the quality of its larval feeding and the availability of hosts for the adult. If no hosts are available to the adult, some eggs can still be produced. If hosts are available, the adult female can choose to blood feed and many more eggs can be produced.

Facultative autogeny is clearly a useful strategy. Insects for which autogeny is obligatory do not have this choice. Even if feeding is poor for the immature stages of obligate forms, they are nevertheless obliged to produce an autogenous egg batch before blood feeding. This usually means fewer eggs in the first batch. Some insects appear to have gone further still and are unable to take blood at all, with egg production being entirely reliant on larval feeding. So autogeny, in one of its various forms, may be

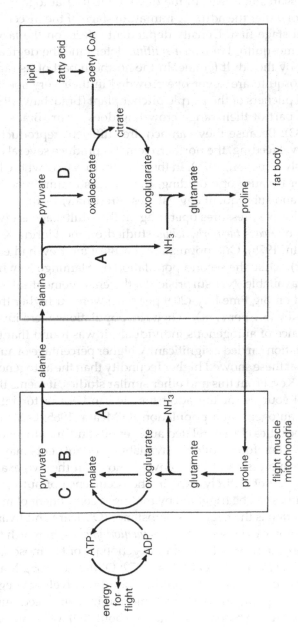

Figure 6.14 Proline is used as an energy source for flight in the tsetse fly and by adult female *Aedes aegypti*. The metabolic pathways involved in both the derivation of energy from the amino acid and the subsequent recharging of the 'proline battery' from triglycerides in the fat body are shown. Key enzymes are: A, Alanine aminotransferase; B, malate dehydrogenase; C, NAD-linked 'malic' enzyme; D, pyruvate carboxylase. (Modified from Bursell, 1977; Scaraffia and Wells, 2003.)

a useful adaptation by insects to suit the quality of larval and/or adult feeding encountered by them.

The selection pressures that lead to the development of autogeny can operate at the level of either the adult or immature stage of the insect. Let us look at the larval stage first. Density-dependent effects on the larvae of the pitcher plant mosquito, *Wyeomyia smithii*, determine the degree of autogeny displayed by the adult female. In the northern part of its range the larvae of this mosquito are never overcrowded in their very specialized habitat, the leaf pitchers of the purple pitcher plant (Bradshaw, 1980), but in the southern part of their range crowding does occur (Bradshaw and Holtzapfel, 1983). Because they can acquire sufficient reproductive resources during larval feeding, the northern females produce several egg clutches autogenously (Hudson, 1970). In their southern range, where larval feeding is poorer because of crowding, the females require blood to produce the second and subsequent egg batches (Bradshaw, 1980).

An example of selective pressures operating on the adult stage is given by two populations of *Aedes taeniorhynchus* studied on the Florida Keys (O'Meara and Edman, 1975). One population, on Big Pine Key, had easy access to hosts (deer), while the second population on Flamingo Key had few potential hosts available. Not surprisingly, the engorgement rates of mosquitoes sampled on Big Pine Key (20.9 per cent) were far higher than those on Flamingo Key (8.1 per cent). When the populations were investigated for the presence of autogenous individuals, it was found that the Flamingo Key population carried a significantly higher percentage of autogenous forms and that these showed higher fecundity than the autogenous forms from Big Pine Key. From this and other similar studies, it seems that availability of blood sources for the adult female can be a factor influencing the levels of autogeny in a population (O'Meara, 1985; O'Meara, 1987), but in many localities closely related autogenous and anautogenous forms are sympatric. Therefore, while the availability of nutrients may be a major factor in the expression of autogeny, as shown in the two examples given above, it cannot entirely explain the occurrence of autogeny. Autogeny in mosquitoes can be triggered by mating; a component of male accessory gland secretion is the triggering substance (O'Meara and Evans, 1976). This phenomenon can be seen in *Aedes taeniorhynchus*, in which the situation is a complex interaction between the genotype of the mosquito, mating and larval diet (Table 6.4) (O'Meara, 1979; O'Meara, 1985). Male-induced autogeny usually leads to the production of relatively few eggs (about 25) compared to populations displaying non-male-induced autogeny (about 65) or containing blood-fed females (about 150) (O'Meara and Evans, 1973). Females that can be stimulated into autogeny by mating actively seek a blood meal, which substantially increases their reproductive success.

Table 6.4 *Three types of female* Aedes taeniorhynchus *have been identified in terms of egg development: autogenous females (1), females that are autogenous if mated (2), and anautogenous forms (3). This pattern is influenced by the feeding success of the larval stage, as illustrated in this table.*

Larval diet	Type 1 female	Type 2 female	Type 3 female
High	autogenous	autogenous if mated	anautogenous
Intermediate	autogenous if mated	autogenous if mated	anautogenous
Low	autogenous	anautogenous	anautogenous

O'Meara, 1985.

Autogeny is, then, a back-up strategy that permits a few eggs to be produced even when the search for a host is unsuccessful. In Hemiptera (but not in the mosquitoes described above) the mated male may provide other factors (possibly nutrients) that assist in the development and increase the size of the egg batch. This is shown in both *Cimex* and *Rhodnius* females, which will produce up to 25 per cent fewer eggs if mated with a previously unfed male compared to a fed one (Buxton, 1930).

Interestingly, the male insect is rarely considered in discussions of autogeny, possibly because much of the work on autogeny has used species in which only the female feeds on blood. Like females, the male insect's capacity for successful reproductive activity can depend on its feeding history. Unfed male lice (Gooding, 1968) or unfed male *Glossina morsitans*, both of which will attempt to mate but are incapable of inseminating females (Foster, 1976), can be compared with both *Cimex* and *Rhodnius* males, which can successfully mate before they are fed (Buxton, 1930). The factors influencing these different evolutionary choices remain unexplored.

Some autogenous insects can increase the number of eggs they lay by feeding on sugar. Mosquitoes, for example, can efficiently convert sugar to fat (Briegel *et al.*, 2002) and feeding on sugar meals alone leads to the accumulation of fat in the body (Van Handel, 1984), especially in facultatively autogenous mosquitoes (O'Meara, 1985) in which sugar feeding may be used to boost the number of autogenously produced eggs (Table 6.5). The sugar is obtained from a variety of sources, including flower nectaries, overripe fruits and aphid honeydew. Not all insects are capable of efficient conversion of sugars to fats. *Stomoxys calcitrans* does not show a net increase in the fat reserves of the body after sugar meals (Venkatesh and Morrison, 1982; Venkatesh *et al.*, 1981), and so sugar meals will make little direct contribution to female fecundity. However, sugar may still make an indirect contribution to fecundity by increasing the energy available to the insect. For example, the energy available from the sugar meal increases the

Table 6.5 *Some mosquitoes can use sugar meals (10 per cent sucrose in this experiment) to increase the number of autogenously produced eggs.*

			Female	
Species (strain)	Treatment	n	% Autogenous	Eggs per female (mean ± S.E.)
Aedes taeniorhynchus (flamingo)	unfed	20	90.0	40.8 ± 4.6
	sugar	20	85.0	63.7 ± 7.4
Aedes bahamensis (Grand Bahama)	unfed	20	100.0	53.5 ± 2.1
	sugar	20	100.0	61.0 ± 1.7

Adapted from O'Meara, 1985.

insect's flight capacity and its chances of finding a host. It will also lower the demand on the female to use part of the blood meal as an energy source, thereby freeing more resources for egg development.

Feeding on more 'watery' foods (for example nectar) in addition to blood is common in several blood-sucking insect groups, including mosquitoes, blackflies, tabanids and blood-feeding muscoids. However, a blood meal is essential for the successful development of large egg batches in most of these insects. It is therefore important to these insects that they are not prevented from taking a blood meal because they have just fed on sugar solution. To avoid this, the insects display a dual sense of hunger. In other words, despite feeding to repletion on sugar-water, these optionally blood-sucking insects will still take a blood meal if it becomes available. The reverse is also true. Recently blood-fed *Aedes aegypti*, which had previously been water-starved, will still probe hot-water-soaked pads (Khan and Maibach, 1971). This dual sense of hunger is common to those Diptera, blood feeders and non-blood feeders alike, that will feed on sugar solutions but that require a protein meal before they can produce eggs. Regulation of this dual sense of hunger has been studied in the non-blood-feeding blowfly *Phormia regina* (Belzer, 1978a; Belzer, 1978b; Belzer, 1978c; Belzer, 1979). It has been shown that sugar and protein intake are regulated in two separate ways. Negative feedback from the recurrent nerve on the foregut regulates sugar intake, while feedback from stretch receptors in the abdomen controls the intake of protein. This bicameral control system enables the dual sense of hunger because the insect's hunger for sugar is satiated well before complete distension of the abdomen. This leaves room for a protein meal, which is stopped following full stretching of the abdomen.

The efficient use of this dual sense of hunger is also aided by the structure of the gut in Diptera that are optional blood feeders (Fig. 6.1). Under most circumstances sugar meals are sent to the crop and only regurgitated into the midgut as required. This leaves the midgut free to accept an immediate blood meal should one become available. So the insect can bypass the sugar meal and immediately digest the more important protein meal. Diversion of sugar to the crop may possibly have other functions. Nectar can contain inhibitors of insect proteinases (Bailey, 1952), so the sudden arrival of a full nectar meal in the midgut might impair enzyme activity. Diverting sugar to the crop and gradually regurgitating it in small packets may thus protect the enzymes needed for digestion of the blood meal. Sugar storage in the crop may also be important for water conservation in the insect because the crop is cuticle-lined and no absorption takes place from it.

Using this dual sense of hunger is clearly an advantage to the insect when the availability of blood meals is unpredictable. The sugar meals tide the insect over between blood meals, but do not seriously impair the insect's capacity to take a blood meal should one become available.

7

Host–insect interactions

Every animal needs to maintain a steady internal environment in order to carry out the various physiological processes that together allow life to continue. This maintenance is known as homeostasis. One of the key organs in homeostasis is the animal's surface covering, the skin, cuticle or other tegumentary substance. In mammals and birds the surface covering is extended to incorporate an outer insulating layer of hair or feathers. In many such animals this layer has proved to be an excellent home for permanent and periodic ectoparasitic insects, providing many different species with a relatively constant environment in which to live.

Vertebrate skin is formed of an inner dermis and an outer epidermis (Fig. 7.1). The thickness of these two layers, and the ratio between them, varies considerably between different parts of the body. The dermis is a connective tissue layer containing blood and lymphatic vessels, and nerves. Embedded in the dermis are the acini of the gland systems that open onto the skin's surface. There are two basic types of skin-associated gland in mammals. The sebaceous glands produce an oily secretion called sebum, which helps prevent the skin and hair from drying out and possesses anti-microbial components. The sweat glands (which are not present in carnivores or rodents) produce a watery secretion used in temperature regulation and in maintaining water and salt balance in the body. Birds have only the uropygial (preen or oil) gland, which is located in front of the tail and is used to provide oil for preening. The principal function of preening is to maintain the condition of the feathers, particularly their water-repellent properties.

The epidermis is formed of epithelial cells and does not contain nerves, blood or lymph vessels. The epidermal cells lying at the dermis–epidermis junction divide and give rise to a stream of new cells that gradually pass through the various levels of the epidermis, eventually arriving at the skin's surface. During this migration they synthesize and store increasing amounts of keratin, before eventually dying and forming the outer keratinized skin layer. They are ultimately lost from the surface of the skin as dry, scaly flakes. Keratinized cells also make up hair, feathers and nails. Hair and feathers serve a variety of useful functions such as extra mechanical

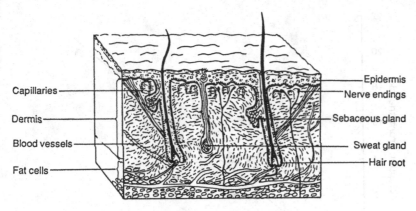

Figure 7.1 A section through mammalian skin.

protection, insulation, colouration (for camouflage or communication purposes), locomotion or an aid to buoyancy.

7.1 Insect distribution on the surface of the host

Permanent and periodic ectoparasites are distributed among the hair and feathers in a non-random fashion. One of the major influences on their distribution is the variation in the microclimate in different parts of the body covering. The most important microclimatic factor is the temperature gradient found in the covering; this has both a vertical and lateral component. The degree of shade and the humidity level in different parts of the hair or feather covering also appear to be important in some instances.

The temperature in the hair or feather covering is determined by ambient and core body temperature, the thickness of the covering and the degree of exposure to the Sun's rays. Insulation and ambient temperature are not the only variables here; core body temperature varies with the health of the animal, its activity patterns, age and species (birds commonly have body temperatures over 40 °C, while marsupials fall in the range 30–35 °C). By altering local blood flow, the animal can also alter the temperature of selected parts of its body surface to suit particular circumstances. This means that the temperature of the body's extremities (tail, feet, nose, etc.) is often considerably cooler than core body temperature. Over the trunk of the body, skin temperature is usually maintained nearer core body temperature. Here the density of hairs or feathers, their length and orientation to the skin (which the animal can alter) are normally the factors determining the temperature, rather than alterations in the local blood supply. Let us look at some examples of the effect of temperature gradients on insect distribution in the body covering.

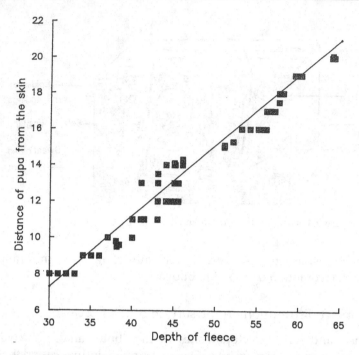

Figure 7.2 Fleece thickness in sheep affects microclimate in the fleece, which in turn controls the distance from the skin at which pupae of *Melophagus ovinus* are attached. (Redrawn from Evans, 1950.)

The sheep ked, *Melophagus ovinus*, migrates vertically within the sheep's fleece according to fluctuations in temperature (Evans, 1950). This has been shown by looking at the larviposition behaviour of the ked. As the thickness of the fleece varies, the fly moves nearer to or further from the skin until it finds the right temperature for puparial deposition (Fig. 7.2). The distribution of lice on sheep has also been partly attributed to temperature variation. *Linognathus ovillus* and *L. pedalis* both prefer hairs to fleece and both require similar, comparatively low temperatures for successful reproduction. *Linognathus ovillus* is most commonly found on the face of sheep, while the foot louse, *L. pedalis*, is restricted mainly to the legs (Murray, 1963). An extreme example of temperature-dependent site selection is seen in the anopluran louse, *Lepidophthirus macrorhini*, which is found on the southern elephant seal, *Mirounga leonina*. These lice are inactive below 5 °C and are most active at 20–30 °C. This essentially limits their distribution to the hind flippers because on shore, with an air temperature of 1.8 °C, the temperature of the back of the seal is 6 °C and that of the flippers is about 30 °C. The seal comes ashore only for two brief periods (three to five weeks) each year; at sea the flippers are still the chosen site for the lice because they

Table 7.1 *The choice of feeding site of* Aedes triseriatus *on eastern chipmunks and grey squirrels is influenced by body hair length and density. The different feeding patterns on the two hosts reflects the differences in hair cover between them.* N.S.= *non-significant*

Host	Back	Ear	Eyelid	Foot	Nose	Chi^2 test
Eastern chipmunk	0	27	13	2	8	P < 0.01
Mean hair density per 9 mm^2	(1583)	(153)	(254)	(49)	(58)	
Mean hair length (mm)	(10.5)	(1.6)	(2.9)	(2.1)	(0.9)	
Grey squirrel	0	19	11	16	4	P < 0.01
Mean hair density per 9 mm^2	(454)	(120)	(59)	(81)	(119)	
Mean hair length (mm)	(13.6)	(2.2)	(2.0)	(3.8)	(1.4)	
Chi^2 test		N.S.	N.S.	P < 0.01	N.S.	

(Edman *et al.*, 1985).

remain, periodically at least, the warmest skin areas as they are used to dissipate heat during bouts of strenuous activity and the seal may warm its flippers at other times by holding them out of the water. It is presumably during these warmer periods that the louse feeds while at sea (Murray and Nicholls, 1965).

Anyone who keeps a bird or a dog, or who has looked carefully at themselves naked in the mirror, knows that the texture, size, density and colours of feathers or hairs show considerable variation from one body area to another. Such variations in the physical nature of the habitat are a second factor leading to the concentration of permanent and periodic ectoparasites in specific areas. For example, the louse *Haematopinus asini* deposits its eggs on the coarse hairs of the tail, legs and mane of the horse but not on the finer body hair. The non-blood-feeding louse *Damalinia equi* is smaller than *H. asini* so it cannot use these large-diameter hairs and is confined to the horse's finer body hair (Murray, 1957). A similar differential distribution is seen in the cattle lice *Haematopinus eurysternus* and *Damalinia bovis* (Matthysse, 1946).

The feeding sites of temporary ectoparasites are also affected by the varying degree of cover and the different skin thicknesses on different parts of the host's body (Table 7.1) (a quick rule of thumb suggests that the thicker the overlying coat, the thinner the underlying skin). This is illustrated by different species of tabanid, which select markedly different landing sites on cattle (Fig. 7.3). A positive correlation has been shown between the hair depth at the landing site and the length of the mouthparts of the tabanid choosing to land there (Mullens and Gerhardt, 1979). Another example is furnished by the distribution of *Lipoptena cervi* on the host animal. The

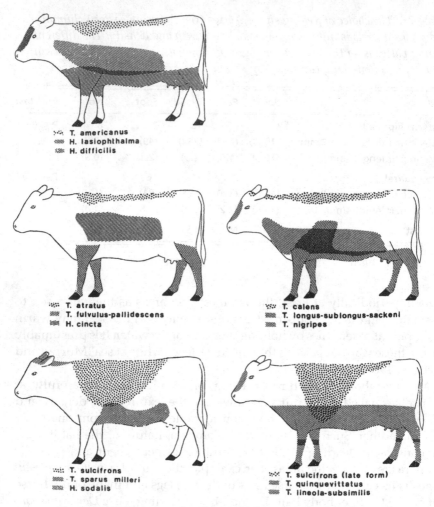

Figure 7.3 Different species of tabanid show a marked preference for different landing sites on the host. Some of these sites are illustrated here. There is a correlation between the length of the mouthparts of the various species and the thickness of the various landing sites ($r = 0.5513$). (Redrawn from Mullens and Gerhardt, 1979.)

total span of the tarsi of this hippoboscid is about 0.22 mm, limiting the fly to host regions with hair of this or a smaller diameter, such as the deer's groin, which is one of the fly's favoured sites (Haarlov, 1964). A second factor determining site selection by *L. cervi* appears to be skin thickness. The mouthparts are only 0.9 mm long and the fly needs to feed where the vascularization in the dermis is nearest to the skin's surface. Clearly the skin areas that have the greatest degree of vascularization near to the surface are probably also the warmest; thus the groin, which is the favoured site for *L. cervi*, is the warmest skin area on the deer. Heat may be an important factor guiding such insects to their preferred site.

The degree of protection offered by a particular site is another important physical characteristic affecting ectoparasite distribution. Thus, no lice are found on the heads of British grebes or divers as the head feathers offer little or no protection for the insects when the birds are underwater (Rothschild and Clay, 1952). Some aquatic mammals, such as the coypu, trap air in certain parts of the pelage when they are diving. The ectoparasites of such animals are often restricted to those parts of the body that remain dry throughout the dive (Newson and Holmes, 1968). But some ectoparasites can adapt to a more or less complete loss of skin covering. For example, elephants (Mukerji and Sen-Sarma, 1955) and pigs (Henry and Conley, 1970) have only sparse body hair; the lice found on these animals have adapted to living in folds in the skin.

7.2   Morphological specializations for life on the host

As with all living organisms, blood-sucking insects are continuously evolving to better fit their ever-changing niche. Many of the characters adopted are subtle and difficult to ascribe to the blood-sucking way of life, while others are more clear-cut, particularly those of the permanent ectoparasites with their highly specialized lifestyle. The most obvious specializations of blood-sucking insects, their mouthparts, are dealt with separately in Section 5.2.

Blood-sucking insects are far smaller than their hosts. There is a rough correlation between the size of permanent ectoparasites and that of their hosts which suggests there may be an optimum size ratio, but the adaptive significance of the relationship has not been explained (Kim, 1985). In the case of the permanent and periodic ectoparasites it is obvious that small size is a great advantage to the insect in helping it escape the host's grooming activities. In the temporary ectoparasites small size helps them to approach and escape with their blood meal unnoticed by the host. However, not all blood-sucking insects are 'small'. When viewed from a human perspective some of the larger tabanids and triatomine bugs are, in anyone's eyes, substantial insects. But the potential problems caused by their unusual bulk are acceptable only because they are offset by other factors in their relationship with the host. For example, an adult *Triatoma infestans* is about an inch in length, but because of its exceptionally stealthy approach to the host, commonly a sleeping human, and its use of a salivary anaesthetic (Dan *et al.*, 1999), it normally escapes with a full blood meal despite the fact that it often feeds around the host's face. Large tabanids prosper either because the host, commonly a large herbivore, is relatively insensitive to the attack or because it is unable to defend itself effectively.

Many blood-sucking insects have shapes clearly modified to suit a particular lifestyle. The lateral flattening of fleas and some streblids and the dorsoventral flattening of lice, polyctenid bugs and hippoboscid flies is a morphological adaptation to life in the covering layer of their hosts. Flattening of these insects allows them considerable freedom of movement among hair or feathers, and also allows adpression against the host or its covering. Both of these adaptations help the insect to evade the host's grooming activities. The flattening may even allow the insect to avoid the tines of comb-like devices used in grooming. The flattening of the triatomine and cimicid bugs allows them to retreat during the intermeal period to the safety of cracks and crevices in the home of the host.

Wings may also be a barrier to rapid movement within the covering of the host. Ectoparasites have dealt with this problem in several ways. Some, such as the hippoboscid *Melophagus ovinus*, have lost their wings entirely and rely on other means of transfer between hosts. Other forms lose their wings once the host is found. This may happen by the progressive abrasion of the wings, as occurs in some moths, or by the deliberate shedding of part or all of the wing, as occurs in females of the streblid *Ascodipteron* and in hippoboscids such as *Lipoptena*. As well as permitting easier passage through the covering of the host, loss of the wings also prevents the insect undertaking flights that may cause it to lose contact with the host. The female streblids of the genus *Ascodipteron* go even further. On contacting a host they shed not only their wings but also their legs. They then burrow deeply into the host's skin (during which they may lose the halteres as well) so that only the rearmost abdominal segments are exposed (Marshall, 1981).

Flattening is a general adaptation to life in the host's covering, but more subtle adaptations are often seen that suit the permanent ectoparasite to particular parts of its host's body. Consequently, different species inhabiting similar sites on an animal often adopt a similar shape and size. Conversely, large morphological differences may be seen between related species living on different parts of the same animal. Such subtle adaptations can often be seen in the surface covering of ectoparasites. Commonly the cuticle of permanent ectoparasites is covered with spines and bristles, which may be aggregated together in the form of combs. These cuticular extensions are seen in the streblids, but are particularly well developed in the polyctenids, nycteribiids and fleas. Other ectoparasites from the Anoplura, Staphylinidae, Pyralidae and Hippoboscidae possess analogous cuticular structures (Marshall, 1981). These cuticular extensions are usually associated with delicate appendages such as antennae, or weaker areas in the insect's cuticle such as the intersegmental membranes. Clearly the type and degree of development of the cuticular extension adopted by the ectoparasite are partly dependent on the type of host it is associated with. This can be seen in the remarkable degree of convergent evolution that has occurred among distantly related ectoparasites on different host types. At a macro level this is shown in bird-infesting forms, which tend to have much longer, more slender bristles than mammal-infesting close relatives. The nature of the association is also important. Nest-dwelling fleas that visit the host only briefly to feed tend to have a reduced cuticular embellishment compared to fleas that live for long periods in the host's fur. But the association between host type and the form of the cuticular covering of the insects ectoparasitic upon it can be traced down to a much more detailed level. Traub (1985) states:

> The chaetotactic modifications may be so diagnostic that infestation of a shrew can be recognised merely by examining the spines of the flea, irrespective of the taxonomic standing or geographic location of the flea.

The function of the tremendous array of combs, bristles and spines present on ectoparasites has been the subject of some debate. It has been proposed that the combs are used for attachment or to help prevent dislodgement (Amin and Wagner, 1983; Traub, 1985); this could be achieved by interlocking with host hair, particularly when there is backward movement of the insect. It has also been argued that the combs protect delicate body regions (Marshall, 1981; Smit, 1972). It seems improbable that the remarkable degree of convergent evolution in the shape, size and spacings of combs that is shown by poorly related fleas on the same host, or the correlation of the spacing of the cuticular extensions with the size of the host hair (Fig. 7.4) (Humphries, 1967) would be seen if the cuticular embellishments had only a protective function. In addition, there is a good correlation between the degree of development of these cuticular embellishments and the comparative risk to the flea if it lost contact with the host. Thus, the bristles, spines and combs are developed to the highest degree in ectoparasites of nocturnally active, flying or tree-dwelling hosts, while they are least developed in ectoparasites of diurnally active, surface-dwelling communal forms (Traub, 1985). Clearly the danger to the flea in losing host contact in the former case is far greater than in the latter, again indicating an anchorage function for the combs. An attachment function for the combs therefore seems hard to dispute. On the other hand, these embellishments are commonly associated with delicate parts of the insect's body, such as antennae, mouthparts or intersegmental membranes, pointing to a protective function (Marshall, 1981). In my opinion there is no biological reason why the embellishments could not perform both functions. The association with the weaker points of the cuticle suggests a possible protective origin with a later adaptive evolution of the size, shape and spacing of

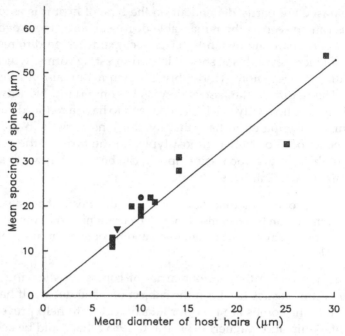

Figure 7.4 There is a significant correlation between the spacings of the spines on ectoparasites and the width of the host's hair. This suggests that the spines can interlock with the hair, helping prevent the dislodgement of the parasite ■, 15 spp. of flea; ●, *Platypsyllus castoris*; ▼, *Nycteribia biarticulata*. (Redrawn from Humphries, 1967.)

the cuticular extensions to meet the anchorage requirements demanded by the host's covering and lifestyle. The evidence suggests that the combs, bristles and spines seen today serve the dual purpose of attachment and protection.

An urgent problem that must be faced by all insects is water loss. The lipids of the cuticle play a vital role as a waterproof covering. Abrasion of the surface layers of the cuticle may cause a dramatic and fatal increase in water losses. Ectoparasites, particularly those that move rapidly through closely packed hair, are at risk because of the abrasive character of their surroundings. Waterproofing of the cuticle depends not only on the outer wax layer but also on lipid in the other constituent layers of the cuticle. For this reason, and because of the thinner separating layer between the haemolymph and the air, water loss from an abraded thin area of cuticle such as an intersegmental membrane is probably considerably greater than losses from abraded areas of thicker cuticle. Combs, spines and bristles that guide hair and feathers away from these membranes may well have an important role in protecting the insect from desiccation.

Blood-sucking insects, particularly the permanent and periodic ectoparasites, have developed specific means of attachment to the host. Tarsal claws are almost universal and are used for gripping the hair or feathers of the host animal. Hippoboscids possess paired tarsal claws, each of which operates against a basal thumb. Ornithophilic species tend to have lighter, more deeply cleft claws than mammophilic forms (Kim and Adler, 1985). There are similar arrangements in nycteribiids (Theodor, 1967) and the legs of anopluran lice also possess very well-developed claws. Some permanent ectoparasites effect an even stronger attachment to the host. The sticktight fleas, such as *Echidnophaga* spp., have anchoring mouthparts that allow them to attach firmly to the host for long periods. Burrowing fleas such as the chigoe flea of man, *Tunga penetrans*, actually tunnel into the skin (it is possible that the host immune response to the attached flea is of more importance in the embedding of the flea than the burrowing activities of the flea itself (Traub, 1985)). These burrowing forms are commonly neosomic (Audy *et al.*, 1972), showing a tremendous enlargement of the abdomen compared to the rest of the body after attachment of the adult to the host. Streblids of the genus *Ascodipteron* also attach firmly to the host and are neosomic. The adult females shed their wings and their legs once the bat host has been contacted, then burrow deeply into its skin.

To ensure that their offspring remain in contact with the host, lice possess special cement-producing glands and the cement is proteinaceous (Burkhart *et al.*, 1999). This cement is used by the female to glue her eggs to the host's hairs or feathers or, in the case of the human body louse, *Pediculus humanus*, to the host's clothes. One flea, *Uropsylla tasmanica*, is also known to glue its eggs to the fur of the host (Audy *et al.*, 1972; Dunnet, 1970), and females of the wingless hippoboscid, *Melophagus ovinus*, glue their viviparously produced offspring to the fleece of sheep.

As mentioned above, hair and feathers form an abrasive and intrusive environment. Ectoparasites, particularly the permanent ectoparasites, have an adapted body shape to minimize the damage that the covering causes. For this reason, a reduction in the size of antennae and/or their protection in refuges is common in permanent ectoparasites. For example, the more highly specialized forms of anopluran mammal-infesting lice, such as the Linognathidae, have even fewer than the five antennal segments seen in other Anoplura. Robert Hooke (1664) first described the antennal groove into which the flea's antennae can be lowered and raised again as required. Nycteribiids and streblids also hide their antennae in grooves. In hippoboscids the delicate arista of the flattened antennae can be hidden in a groove of the larger second antennal segment. Like the antennae, the potentially vulnerable mouthparts are also usually protected in some fashion. This may be achieved by folding them back under the head, as occurs in Polyctenidae; by folding them under and partly retracting them

into the head, as occurs in Siphonaptera; by retracting the delicate components fully into the protection of the head and its snout-like extension, as
occurs in Anoplura; or by heavy cuticularization of the complete head and
thorax, as occurs in hippoboscids. In nycteribiids the head is bent back and
lies protected in a groove on the dorsal surface of the thorax.

Adaptation of the sensory organs occurs in response to the blood-
sucking habit. Reduction in the size of the antennae and their protection
has been discussed above. The eyes are commonly absent in permanent
ectoparasites, particularly the smaller forms: eyes are minimal or absent in
polyctenids, apterous hippoboscids, nycteribiids, streblids and anopluran
lice. This may be another adaptive response to the abrasive surroundings
of these insects, permitting the thickening of the cuticle of the head capsule
and so minimizing abrasion damage. The absence or poor quality of receptors is compensated for by the continued close association of the insect
with the host.

7.3   Host immune responses and insect salivary secretions

After a few feedings by a particular blood-sucking insect species on a host
animal, a pruritic, red weal will start to appear at biting sites. This is the
basis of most people's earliest awareness of blood-sucking insects – their
bites itch . . . but what causes the itch?

Three possibilities immediately present themselves: (a) the response is a
localized traumatic reaction to the injury caused by the insect's mouthparts;
(b) it is a response to a toxin introduced into the wound in the insect's saliva;
or, (c) it is an immune response to an antigen in the saliva. It was shown not
to be a response to mechanical injury in a series of experiments in which the
salivary glands of mosquitoes were surgically severed. These mosquitoes
were then unable to introduce saliva into the wound during feeding and
no host reaction occurred when they subsequently fed on a sensitized host
(Hudson *et al.*, 1960). So the saliva causes the problem. Accumulated weight
of evidence, largely based around the passive transfer of reactivity from
sensitized to naive hosts, has subsequently shown that the response is
immunologically based and is due to an antigen rather than a toxin in the
saliva. Identification of these allergens has proved difficult, and it is likely
that each insect possesses several molecules in its saliva that are potential
allergens, and that individual hosts respond differently to these molecules
depending on host species, genetic makeup and physiological history
(Arlian, 2002; Belkaid *et al.*, 2000; McDermott *et al.*, 2000). For example,
in cat fleas a major allergen for dogs is an 18-kDa protein named Cte f 1
(McDermott *et al.*, 2000), but for humans the major cat flea allergen has
potentially been identified as a different protein of 34-kDa (Trudeau *et al.*,
1993).

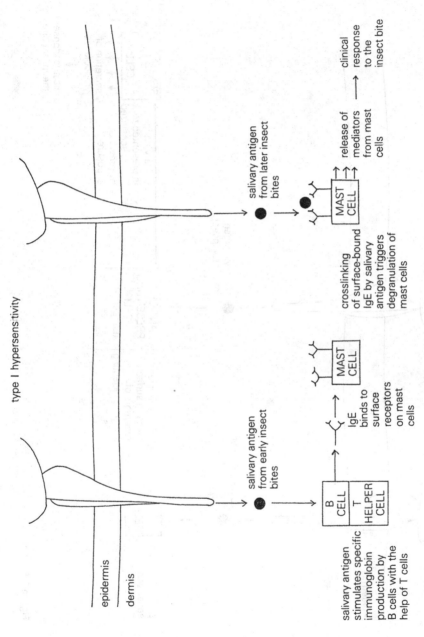

Figure 7.5 A summary of type I and type IV immune responses to insect bites.

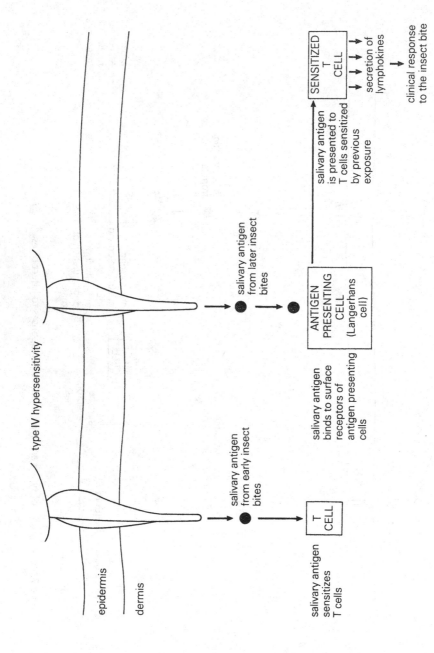

Figure 7.5 (cont.)

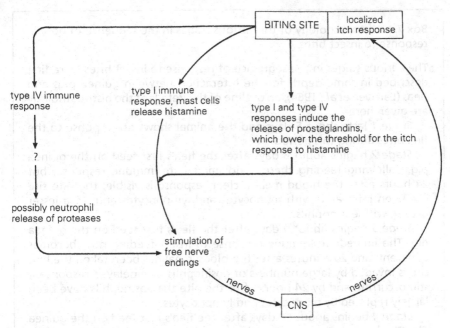

Figure 7.6 Type I and IV immune responses may follow an insect bite and both are associated with an intense itchiness (pruritus). The mechanisms leading to the itch are summarized.

Some of the details of the vertebrate response to insect saliva are known, and it seems that the vertebrate reaction often follows a five-stage career (Nelson, 1987; Reunala *et al.*, 1990) consisting of:

(1)   No response.

(2)   A delayed (type IV) immune response with associated pruritus (itchiness).

(3)   An immediate (type I) immune response followed, 24–48 hours later, by a delayed (type IV) immune response, with both stages having an associated pruritus.

(4)   Only the immediate (type I) immune response with an associated pruritus.

(5)   No response.

A detailed discussion of immunological mechanisms is outside the scope of this book, but the basic details of type I and type IV reactions are given in Figure 7.5 (Jones, 1996; Sandeman, 1996) and the main factors leading to pruritus are outlined in Figure 7.6.

The various stages in the sequence of response to insect bites given above were first described in some depth for the interaction of guinea pigs

---

Box 7.1 Histopathology of the various stages in the sequence of host response to insect bites.

The various stages in the sequence of response to insect bites were first described in some depth for the interaction between guinea pigs and fleas (Larrivee *et al.*, 1964); an outline of the details of the histopathology are given here.

Stage 1 lasts about 5 days and the animal shows no response to the flea bites during this period.

Stage 2 begins about 6 days after the flea's first feed on the guinea pig. Following feeding there is no immediate immune response, but 24 hours after the blood meal a clear response is visible; the bite site has been infiltrated with monocytes and lymphocytes and, to a lesser extent, with eosinophils.

Stage 3 begins about 9 days after the flea's first feed on the guinea pig. The immediate immune response to insect feeding now becomes apparent, and 20 minutes after the blood meal has been taken the bite site is invaded by large numbers of eosinophils. The delayed response is still occurring and by 24 hours after the bite the eosinophils have been largely replaced by monocytes and lymphocytes.

Stage 4 begins about 50 days after the flea's first feed on the guinea pig. The immediate response is in operation and the bite site is rapidly infiltrated by eosinophils. The delayed response has now largely disappeared and very few monocytes or lymphocytes appear in the wound.

Stage 5 begins about 80 days after the flea's first feed on the guinea pig. Desensitization to the bites occurs and almost no cellular response to the bite can be seen.

---

and fleas (Larrivee *et al.*, 1964). The details of the histopathology seen in this interaction, which are outlined in Box 7.1, are not standard for every insect bite, but are a good general guide to the type of response that has been reported in the few thorough investigations carried out to date (Jones, 1996; Reunala *et al.*, 1990; Sandeman, 1996).

The number of bites required before the onset of each of the five stages of the response is known to vary considerably among different individuals, with different insect–vertebrate combinations and with the conditions under which the study is made (Nelson, 1987). The timing seen in the highly controlled guinea pig and flea study cannot be used as a definitive guide to reaction times in the real world where widely varying responses occur. As a rough guide, regular exposure to bites commonly means stage 5 (desensitization) is reached within two years. With irregular exposure each stage can be very protracted and stage 5 may never be reached. Even when it does occur, the lack of response seen in stage 5 may be restricted to particular zones of the body that are receiving regular bites. If the biting insect moves just a few inches from this zone it may still induce a full reaction at the new biting site.

There are other factors defining the nature of the immune response and its timing in addition to the host's degree of prior exposure to antigen and the regularity of that exposure. These factors include the age and nutritional status of the host, the nature of the antigens introduced by the insect, and the route by which they are introduced (i.e. pool versus vessel feeders). There is also a genetic component in host responsiveness. In any given host population there is usually a spectrum of reactivity to bites. At one extreme, a few of the individuals being bitten may die from anaphylactic shock, others may develop massive oedema or an intense pruritis, while at the other extreme some individuals may fail to respond to bites at all; most hosts will fall somewhere in between (Arlian, 2002). Work on mice infested with the louse *Polyplax serrata* has illustrated the variation in resistance that can occur (Clifford *et al.*, 1967); this work is outlined in Figure 7.7.

In research in which mosquitoes were allowed to feed on the investigator, the time of onset of the pruritis associated with the immediate response (stages 3 and 4) was established to be about three minutes (Gillett, 1967). Because of the alerting effect of the pruritis on the host, the mosquito would be well advised to have completed the meal inside these three minutes and to have left the scene of the crime. Those still feeding after three minutes have a greater chance of being swatted (a most efficient agent of natural selection) or of being disturbed before a full meal has been taken (which may affect reproductive success). In this way the immediate response, and the irritability resulting from it, are likely to be strong factors selecting haematophagous insects for rapid feeding, especially if the host animal is an efficient groomer. This view is supported by comparative work on the rapidity of feeding of wild and colonized mosquitoes whose regular hosts include primates. The colonized mosquitoes had been fed on restrained hosts over a period of three years prior to the experiments, and this had effectively removed the selective pressure for rapid feeding. It was found during the experiments that these mosquitoes would often fail to feed within the 'safe period' (i.e. in the three minutes before the onset of the immediate response). In contrast, wild mosquitoes rarely failed to complete feeding within the 'safe period' (Gillett, 1967). These conclusions are supported by separate, more recent work on wild and colonized mosquitoes (Chadee and Beier, 1997).

Not all blood-sucking insects complete a blood meal in under three minutes. For example, adult triatomine bugs may take 20 minutes to complete a meal. Such insects are likely to require special measures to deal with the immediate hypersensitivity response and host irritability, so perhaps it is

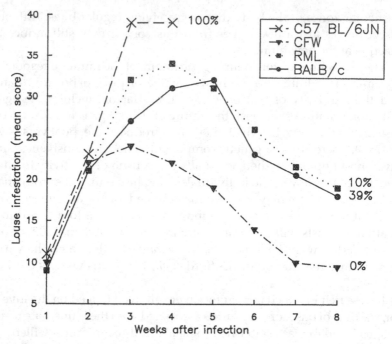

Figure 7.7 Mice prevented from grooming can control infestations of lice (*Polyplax serrata*) only by immunological means. By infesting different strains of mice (see key) the genetically determined variability in the immunological responses of the host becomes clear, as some strains develop greater infestations than others. Mortality (per cent) is greatest in those strains developing the highest infestations. Louse infestations are quantified as an average score. This is an arbitrary scale from 0 to 40 where 0 = no lice; 10 = rare; 20 = few to moderate; 30 = many; 40 = very many. (Redrawn from Clifford *et al.* (1967.)

not surprising that these longest feeding of blood-sucking insects are the only ones for which a local anaesthetic appears to be produced (Dan *et al.*, 1999). The consequences of host irritability are considered further below.

It is interesting to consider the concept of a safe period when, as happens under field conditions, a series of temporary ectoparasites attack a host over a period of time. In these circumstances only the first insect to feed has a safe period. Insects attacking subsequently are likely to be met by an irritable host. As argued above, an irritable host is likely to be dangerous, possibly killing the insect or impairing its reproductive success by interrupting its feeding. This series of events is to the reproductive advantage of the first insect that managed to feed within the safe period, because its offspring will have fewer competitors. In this sense stimulating an immune response in the host could be seen, not as an unfortunate consequence of the injection of anti-haemostatic factors in saliva, but as a positive advantage to particularly stealthy insects or to an insect that can arrive early enough and feed quickly enough.

It is clear that the safe period is not the only thing to consider when looking at host immune mechanisms and insect saliva, because insect saliva has a wide variety of molecules capable of modulating the host immune response on a wide scale (Gillespie *et al.*, 2000; Schoeler and Wikel, 2001). Some insects live for long periods with the same host. For example, 58 per cent of female *Ctenocephalides felis felis* live on the same host for up to 113 days (Dryden, 1989). In these cases the insect may benefit greatly from modulating the immune response of its long-term host. Thus *Phlebotomus papatasi*, which lives in the nests of rodents, causes a strong delayed-type hypersensitivity response. If insects subsequently feed at a delayed-type hypersensitivity site, they can ingest a meal much faster than at normal skin sites because of the increased blood flow at these hypersensitivity sites. So an argument can be made for *P. papatasi* and other arthropods that feed regularly on the same host (e.g. fleas, bed bugs) that the strong, saliva-induced immune response may reflect an adaptation of the insect to manipulate host immunity for the insect's own advantage (Belkaid *et al.*, 2000). Interestingly, and seemingly in contrast, another sandfly, *Lutzomya longipalpis*, may inhibit delayed-type hypersensitivity responses. Maxadilan produced in saliva of this insect inhibits nitric oxide production by macrophages (Gillespie *et al.*, 2000), secretion of TNFα and augments production of prostaglandin E_2, IL-6 and IL-10 (Bozza *et al.*, 1998; Soares *et al.*, 1998). This may lead, for example, to the inhibition of activation of delayed-type hypersensitivity responses. Again, for this insect feeding regularly upon the same host, a benefit may be gained from minimizing the host's irritability by modulating the allergic response to previous bites. This is a complex field of investigation where much remains to be learned, and it is possible that each insect–host salivary / immune system relationship may function in subtly different ways.

While arguments such as those given above can be made for insects regularly in contact with the same host, the selective advantage of long-term modulation of the host immune system to insects not feeding regularly on the same host is not obvious. For example, mosquitoes and blackflies produce salivary molecules that downregulate the production of pro-inflammatory cytokines (Bissonnette *et al.*, 1993; Cross *et al.*, 1994). The downregulated immune response will appear in the host long after the insect has paid probably its only visit to that host. How this could be selectively advantageous to the insect is unclear. It seems possible that these phenomena may merely represent side effects of biologically very active molecules whose role is in the short-term modulation of haemostasis (e.g. anticoagulation, vasodilation) and that these side effects may in many cases have no selective advantage to the insect. Whether they are side effects or not, these phenomena are of major importance in disease transmission (see Section 8.7).

7.4   Behavioural defences of the host

In New York State, at the height of the fly season, a horse may receive up to 4000 bites per day from tabanids, with blood loss of up to half a litre (Tashiro and Schwardt, 1953). Triatomine bugs feeding on humans in houses in Latin America may cause blood loss of 2–3 ml per day, contributing to chronic blood loss and iron-deficiency anaemia (Schofield, 1981). While these might be exceptional cases, they still clearly explain why vertebrates make efforts to defend themselves from blood-sucking insects. Apart from direct loss of blood, there are also costs to the host from immunological responses to the insect bite (Lochmiller and Deerenberg, 2000) and, obviously, from any parasites transmitted to the host. On the other hand, behavioural defences against insect attack will also be a cost and, in an evolutionary sense, the host must balance one cost against the other. Our understanding of the interplay between the two is still rather poor.

Given the correct microclimatic and physical conditions, insects usually choose to feed (and reside in the case of permanent and periodic ectoparasites) at host body sites where they will be least disturbed. Because the host grooms or preens and shows a variety of other defensive behaviours towards ectoparasites, this means, in practice, that ectoparasites are usually found in only a restricted part of their potential range. This is most obviously seen in permanent and periodic ectoparasites. The distribution of the louse *Polyplax serrata* on the mouse is a well-described example. Normally, an infested mouse bears about 100 lice, which are congregated on its head and neck. The mouse grooms its head and neck with its toes, whereas it grooms the rest of its body by using its two lower incisors in a comb-like manner. If the mouse is prevented from grooming with its teeth, lice move onto the body and increase in numbers to over 2000 in three weeks. Clearly, grooming with the teeth is a more efficient method of removing lice than grooming with the toes. Grooming with the toes does nevertheless kill lice, as demonstrated by the increased louse populations on the head and neck when the mouse is prevented from using its feet for grooming (Murray, 1987).

Mutual grooming is another important factor limiting ectoparasite numbers. This can be seen in stable hierarchies of mice, in which, if they were prevented from grooming with their feet, louse burdens on the head and neck areas were still kept low because mice groomed each other using their teeth (Lodmell *et al.*, 1970). Supporting data for wild populations of rodents is available (Stanko *et al.*, 2002). Interestingly, in the laboratory study if the hierarchy of the group was not stable, grooming did not control louse numbers and mice died from the excessive burdens of lice that developed (Lodmell *et al.*, 1970). Mutual grooming is also seen in social birds such as penguins (Brook, 1985) and, of course, in the primates, including man. Mutual grooming to regulate ectoparasite numbers may be an important factor determining social interactions (Bize *et al.*, 2003).

Grooming, as a means of permanent ectoparasite control, decreases in efficiency with increasing host size. A healthy mouse can restrict lice to the head and neck. On the rat *Rattus norveigicus* lice are to be found widely distributed on the trunk. On the vole *Microtus arvalis*, which is intermediate in size between the mouse and the rat, lice are again restricted to the neck and head, with another colony at the base of the tail (Murray, 1987). But, although larger animals may be unsuccessful in restricting parasites to particular parts of the body, grooming by the host is still important in restricting the total number of ectoparasites on the body as a whole. For example, the ox can use the coarse surface of its tongue as a comb in grooming and can significantly reduce the numbers of lice on its body as a consequence (Lewis *et al.*, 1967).

Conventional wisdom suggests that when the host moults the number of ectoparasites falls significantly. Moulting takes a variety of forms in different animals, but in most birds and mammals it is usually a gradual process. However, it probably causes most damage to ectoparasite populations when it is sudden, is widespread on the body surface and involves the loss of a considerable proportion of the body's protective covering. Observation suggests that there is some direct parasite loss with the covering itself, while changes in the microclimatic and physical characteristics of the insect's environment also probably reduce parasite numbers. Other insects die due to the increased efficiency of grooming in the thinner outer covering of hair or feathers. Moulting may also occur artificially, as it does each season in the shearing of sheep, when it causes a dramatic fall in the population of the sheep ked *Melophagus ovinus* (Evans, 1950). Although humans do not moult, it has been argued that hairlessness in humans is virtually a state of permanent moult that has evolved to reduce parasite loads (Pagel and Bodmer, 2003).

The conventional wisdom concerning moulting and ectoparasite numbers outlined above is based on observational evidence, but recent experimental evidence on the effect of feather moult on ectoparasites calls this conventional wisdom into question (Moyer *et al.*, 2002). In this work feral pigeons were induced to moult by altering day length. The visual data indicated a significant effect of the moult on lice numbers. However, using a more robust body-washing method for counting ectoparasites, the authors showed that the moult in fact had no effect on louse abundance. The lice were actively seeking refuge inside the sheath that encases developing feathers, where the lice cannot be seen. These data suggest that more rigorous experimental work is needed in this area.

Some lifestyles mean that the negative effects of moulting may be avoided by some insects. The lice *Haematopinus asini* and *H. eurysternus* both deposit their eggs on the coarse hairs of the tail, legs and mane but not the finer body hair of their respective hosts, the horse and cow. The non-blood-feeding lice *Damalinia equi* and *D. bovis* are smaller and egg attachment is confined to the finer hair covering of the horse and cow, respectively (Matthysse, 1946; Murray, 1957). Observational evidence suggests that both *Damalinia* spp. are seriously affected by moulting, as large numbers of eggs will be shed along with the body hair of the animals, but the two *Haematopinus* spp. are hardly affected because, unlike the body hairs, the long hairs of the mane and tail are not shed wholesale in the moult.

Even though permanent ectoparasites are normally best adapted for life on one particular part of a host animal, and are restricted further by the grooming activities of that animal, given the right circumstances they can often spread widely over the animal's surface. This is often seen in sick or injured hosts. In such animals ectoparasites are dispersed over an unusually large host body area, often in abnormally high numbers. The principal reason for this is probably the inability of the host to groom itself efficiently, but delay in moulting and the possibility that an unhealthy host is more attractive or available to the insect (see Section 8.4) may also be contributory factors.

The number of temporary ectoparasites that successfully feed on a host is also affected by host defensive activity. One detailed study has been conducted on the feeding success of mosquitoes on a selection of ciconiiform birds (herons and egrets). The birds were held overnight in test cages containing mosquitoes and the feeding success of the insects was determined the following morning. It was clear from the results that the black-crowned night heron and green heron were bitten far more frequently (by three to eight times) than the other five species used. This variability of biting frequency was not related to size, colour, weight or smell of the birds. Experiments in which restrained birds were exposed to mosquitoes made it clear that different levels of anti-mosquito behaviour among the different birds determined which species were bitten most often. The types of anti-mosquito behaviour seen in these birds is summarized in Table 7.2. Foot-pecking and foot-slapping appeared to be most effective in species where mosquitoes attack the exposed leg (Webber and Edman, 1972). The five species that achieved a degree of success in protecting themselves from mosquito attack showed an average of about 3000 movements per hour, so they were virtually in perpetual motion! The most frequently bitten species, the green heron and black-crowned night heron, also displayed

Table 7.2 *The anti-mosquito behaviour of a range of ciconiiform birds, showing that different host species display various types and degrees of defensive behaviour against blood-sucking insects.*

Anti-mosquito behaviour	Night heron	Green heron	Little blue heron	White ibis	Louisiana heron	Cattle egret	Snowy egret
Using head and bill							
Head shake	+	+	+	+	+	+	+
Head rub (body)			+	+	+	+	+
Bill snap or jab	+	+	+	+	+	+	+
Bill rub (body)			+	+	+	+	+
Bill rub (legs)				+	+	+	+
Bill rub (perch)			+				
Bill peck (body)			+	+	+	+	+
Bill peck (legs)			+	+	+	+	+
Bill peck (perch)			+			+	
Using legs and feet							
Foot shake				+			
Foot stamp (perch)			+		+	+	+
Foot slap (other foot)			+		+	+	+
Head scratch				+	+	+	
Using body							
Wing flip or flap			+		+	+	+
Body fluff				+		+	+

From Edman and Kale (1971).

anti-mosquito activity, but at the lower level of about 650 movements per hour (Webber and Edman, 1972).

These studies also showed that host anti-mosquito behaviour could have a significant impact on the amount of blood a feeding mosquito obtained. Less than 2 per cent of the mosquitoes feeding on the black-crowned night heron or the green heron failed to get a full blood meal, but between 15 per cent and 31 per cent of those feeding on bird species that showed more efficient anti-mosquito behaviour obtained less than half of a complete blood meal.

The degree of host defensive behaviour seen is directly related to the number of insects attacking the host (Edman *et al.*, 1972; Waage and Nondo, 1982), is variable among different individual hosts (Anderson and Brust, 1996). It may be related to host size, smaller animals grooming more than large animals to make up for the increased costs of parasitism due to the larger surface to volume ratios in small animals (Mooring *et al.*, 2000).

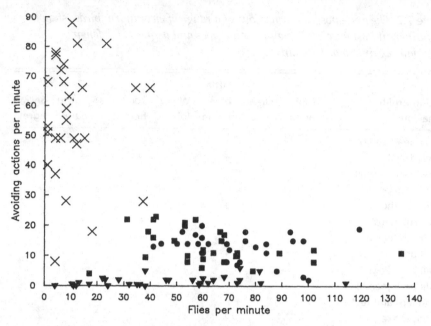

Figure 7.8 Calves displaying the highest levels of behavioural defences have the fewest numbers of stableflies settling on them. Different calves are represented by different symbols. (Redrawn from Warnes and Finlayson, 1987.)

In consequence, the number of mosquitoes successfully obtaining a blood meal, and the amount of blood obtained by a population of mosquitoes, is regulated by the number of mosquitoes attacking the host and the efficiency of the host's anti-mosquito mechanisms. Interactions such as this may constitute an efficient density-dependent means of limiting the size of blood-sucking insect populations, particularly those closely associated with a single host species (Schofield, 1982) (see Section 7.5).

Mammals also engage in defensive behaviour against temporary ectoparasites. Most of us have seen cows flicking their ears or swishing their tails in response to the attentions of the large numbers of insects that are attracted to them. The numbers of ear flicks and tail swishes are directly related to the number of flies that are present on the cattle (Harris *et al.*, 1987), and these mechanisms, together with head swings, leg kicks and shuddering of the skin, are an effective means of reducing annoyance from these insects (Fig. 7.8) (Warnes and Finlayson, 1987). The efficiency of these mechanisms in reducing fly attack can be considerable, as can be seen from field work in Zimbabwe, in which 15 times more tsetse flies fed on sedated goats that were unable to show defensive behaviours than on unsedated animals. Responses of game animals to tsetse and other biting flies are also directly correlated with the number of flies attacking the animal. The responses seen include neck shuddering, tail lashing and scraping of the body with the hooves (in giraffe), and distressed lions rolling on their backs, hiding in hyena holes or climbing trees to avoid attacks from *Stomoxys* spp. (Kangwangye, 1977). Mice will bury themselves (Kavaliers *et al.*, 2001), and elephants even use tools to protect themselves (Hart and Hart, 1994).

As well as these physical attempts to ward off the insects, large mammals show aggregation behaviour as a defence against temporary ectoparasites. This operates through selfish herd and encounter-dilution effects much as these behaviours do in the protection of animals from predators (Mooring and Hart, 1992). Grouping behaviour is clearly seen in caribou and reindeer populations, which are gregarious throughout the year, but in the post-calving season the formation of particularly large herds occurs. These large aggregations of animals coincide with the seasonal peaks of blood-sucking insects on these northern ranges. The attack rate from blood-sucking insects on these herded rangifers is about ten times higher for animals on the periphery of the herd compared to those at the centre, demonstrating the selfish herd effect (Breev, 1950). When carbon dioxide-baited silhouette traps were substituted for real animals and placed in a herded pattern, the same was true. Even though the 'herd' size was only 24, the traps on the periphery received more attention from blood-sucking insects than did traps at the centre of the 'herd' (Helle and Aspi, 1983).

Encounter-dilution effects occur when blood-sucking insect encounters with a host group are fewer than the sum of encounters of the insects with distributed hosts. These effects will occur if groups are detected proportionately less often than distributed hosts and provided that blood-sucking insects do not increase their rate of attack on the group compared to distributed hosts. Such an effect is suggested by several studies on horses that have shown that the number of flies on a horse depends upon how many horses are grouped together (Duncan and Vigne, 1979; Rubenstein and Hohmann, 1989; Rutberg, 1987). In one of these studies, when horses were in groups of 8–32 individuals they had fewer than a third of the flies *per capita* than horses in groups of 3. Horses moving from small groups to large groups showed the expected fall in fly numbers, discounting the possibility that they were inherently more attractive for flies (Duncan and Vigne, 1979). A particularly convincing demonstration of this effect was achieved using carbon dioxide-baited silhouette traps: when these were placed either in a herded pattern or to represent individual animals, insects were found to be more attracted to individual traps (Helle and Aspi, 1983).

Temporary ectoparasites appear to be attracted to single animals with a good area of clear space around them, a situation in which they are less likely to be crushed or swatted and which provides easier access to the lower halves of the animals, the preferred feeding sites of many temporary ectoparasites. Indeed, if insect attack is particularly high, cattle crowd closely together and eventually lie down, exposing only their backs to the attackers. As a final response to intense insect activity, cattle will stampede. The efficiency of herding as a mechanism for reducing the attentions received from blood-sucking insects may be increased by the movement of the herd onto selected sites (Downs *et al.*, 1986; Keiper and Berger, 1982). At times of peak fly activity Camargue horses aggregate at sites known locally as chomadous. These are sites exposed to maximum wind velocity, which acts to minimize insect activity. Similar sites are chosen by cattle subjected to severe attack by blackflies. Reindeer also congregate at specific areas known as tanders, and it is suspected that local climatic or biotic factors at these sites minimize insect activity.

The accounts of selfish herd and encounter-dilution effects given above refer to the degree of exposure of hosts to temporary ectoparasites in cases when herding can significantly decrease the number of bites an individual host receives. But epidemiological models suggest that herding may lead to increased infestation with periodic and permanent ectoparasites such as fleas and lice that rely on contact for transmission (Anderson and May, 1978; Morand and Poulin, 1998). Field data support this conclusion for some ectoparasite–host interactions (Krasnov *et al.*, 2003) but not for all (Sorci *et al.*, 1997; Stanko *et al.*, 2002). The models mentioned above do not account for variations in host behaviour, and so a possible explanation for the latter findings might be increased mutual grooming in larger groups. Clearly more experimental data are needed before we have a clear understanding of host group size and ectoparasite exposure rates.

Laboratory studies have also shown that infection of the host can have an impact on the number of successful attacks by temporary ectoparasites. Many rodent species show highly efficient anti-mosquito behaviour and for this reason are only rarely fed on by mosquitoes. It is also known that rodent populations may be enzootic for various mosquito-borne diseases such as the arboviral disease Venezuelan equine encephalitis. How can these seemingly conflicting pieces of information be reconciled? Under experimental conditions the mouse *Mus musculus*, when fit and healthy, displays a series of very efficient behavioural mechanisms that prevent feeding by mosquitoes. Malaria-infected mice show the periodic peaks of parasitaemia that are typical of this disease. During, and particularly just after, these peaks the sick mouse is less able to defend itself against mosquitoes, which feed readily from it (Fig. 7.9). These feeding opportunities for the mosquito, which occur a day or two after peak parasitaemia in the mouse, coincide with a peak in the number of gametocytes (the stage of the malaria parasite that is infectious for the mosquito) in the mouse's blood (Day and Edman, 1983). In other words, the malaria parasite appears to have modulated the behavioural activity of the host in such a way

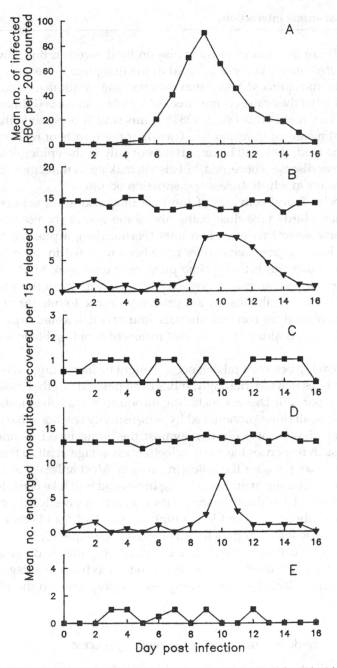

Figure 7.9 Feeding of mosquitoes on restrained mice infected with malaria is continuous and shows no peaks (B, upper plot; D, upper plot). Mosquito feeding on unrestrained mice infected with malaria shows clear peaks (B, lower plot, *Aedes aegypti*; D, lower plot, *Culex quinquefasciatus*) which occur just after the occurrence of maximum parasitaemia (A). Virtually no feeding occurs from uninfected, unrestrained mice (C, *Aedes aegypti*; E, *Culex quinquefasciatus*) which show efficient anti-mosquito behaviour. (Redrawn from Day and Edman, 1983.)

as to maximize its chances of transmission by the vector. But how does the uninfected mouse become infected in the first place if it is only rarely fed upon by mosquitoes? Mice often eat attacking mosquitoes, and it has been shown that they can become infected with both malaria (Edman *et al.*, 1985) and La Crosse virus (Yuill, 1983) in this way. It is possible that this is a normal means of transmission. Clearly if these sorts of relationships occur in the field, they will be important not only in the epidemiology of the particular diseases concerned, but also in making us think more deeply about the ways in which disease transmission occurs.

Other behavioural activities of the host may also affect ectoparasitic insects. Many birds take dust baths and some anoint themselves with the defensive secretions of certain ants (Formicidae), a practice known as anting. Both of these behaviours have been reported to kill ectoparasites, but whether this is their prime purpose is unknown. Mud-bathing, which many large mammals enjoy, partially protects them from temporary ectoparasites as the insects are probably unable to bite through the dried crusts of mud seen on such animals. Immersion in water, as practised by hippopotami, is also a very efficient means of avoiding blood-sucking insects.

Before leaving behavioural defences, it might be interesting to consider why insect bites itch so irritatingly. It has been pointed out that most temporarily ectoparasitic Diptera, including mosquitoes, blackflies, tsetse flies and *Culicoides*, all cause pronounced hypersensitivity reactions that do not provide any significant protection against the biting insect (Sandeman, 1996). Do such responses have any selective advantage at all to the host? I think they may have for the following reason. Most animals, including humans, have a strong aversion to being bitten and will take considerable measures to avoid it – this can be seen in the energetically costly behavioural defences described above and in the high sales of DEET and other repellents in our societies. Taking such measures and consequently being bitten less means a reduction in the chance of acquiring parasitic disease from vector insects and this will have a very strong selective advantage – one that may have selected for our strong allergic responses to the bites of insects.

7.5   Density-dependent effects on feeding success

Population limitation may be achieved by single, but more normally by multiple, factors and the limiting mechanisms that operate may change in time and space. These factors are commonly subdivided into density-dependent and density-independent categories. Separating the effects of one from the other, under natural conditions, is often a complicated process, not least because density-dependent factors will always influence the fitness of individuals to withstand density-independent pressures. But to generalize, in rapidly fluctuating environments or environments that the insect finds harsh, density-independent factors such as temperature and humidity are often of greatest importance in limiting population numbers. In more congenial circumstances, where populations will tend to expand more smoothly and continuously, insect populations are more likely to be limited by density-dependent factors such as competition for food or space, or by increased exploitation of the population by parasites or predators. Blood-sucking insects show a range of lifestyle strategies. At one end of this spectrum are insects such as the mosquitoes that are r-selected, having high reproductive rates adapted to maximize the instantaneous rate of population increase in unstable habitats and showing strong dispersal capacity. At the other end of the spectrum are those insects, like reduviid bugs and tsetse flies, that are K-selected, have low reproductive rates, and are adapted to succeed under highly competitive conditions in stable habitats. To generalize once again, density-dependent effects are likely to be more important in the limitation of population size in K rather than r strategists. In line with the approach taken in this book, I intend to discuss only density-dependent effects on feeding success as these present some circumstances peculiar to the blood-sucking habit.

Food availability can be a density-dependent factor limiting population size in blood-sucking insects, as it can in all other animals. In many, if not most, circumstances blood-sucking insects will differ from most other animals in the nature of their density-dependent control by food availability. This is because it is not normally a shortfall in the sheer physical quantity of food (i.e. blood) that limits their population growth, but the increasing difficulty of obtaining this food as the density of the insect population increases. There are two clear ways in which this may happen, both of which depend on the increased stimulation of the host's immune system as the number of attacking insects rises.

The first way the host's immune system can influence the feeding success of insects is through acquired resistance (Nelson, 1987; Ratzlaff and Wikel, 1990), the most convincing examples of which occur in permanent ectoparasites. When acquired resistance is seen in a host immune response appears to bring about a change in the physical nature of the feeding site such that the insect finds it more difficult to feed. The circulating antibodies produced against the salivary antigens of the insect appear to have little impact (Fig. 7.10). Acquired resistance rarely, if ever, reaches the status of full immunity. It normally results in just a lowering of the numbers of insects that can maintain themselves on the host rather than their complete elimination.

A characteristic pattern of population growth and decline is seen when permanent ectoparasites move onto a naive host which then develops an

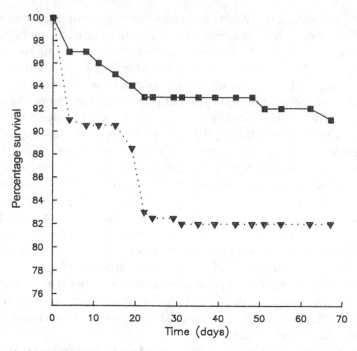

Figure 7.10 When tsetse flies continually feed at high densities on the previously exposed ear of a rabbit (▼), their percentage survival decreases compared to flies that feed on the naive ear of the same rabbit (■). This suggests that localized immune responses and not circulating antibodies are implicated in acquired resistance. (Redrawn from Parker and Gooding, 1979.)

acquired resistance (Fig. 7.11). For the first four weeks following infestation of a hindfoot-amputated mouse with *Polyplax serrata*, the numbers of lice on the mouse increase. During this time the skin becomes infiltrated with large numbers of lymphocytes, eosinophils and neutrophils which peak in the second week and decline thereafter. These changes are accompanied by hyperplasia of the skin and vasoconstriction. This immune response to the presence of the lice then leads to the development of acquired resistance, which appears over the next eight weeks. Acquired resistance is characterized by an increase in the numbers of lymphocytes and monocytes in the skin as well as increasing numbers of fibroblasts and mast cells. As these changes occur so the numbers of lice on the hindfoot-amputated mouse decrease until eventually very few, or even none, remain. This second phase of response, which coincides with the decline in the numbers of lice on the host, is not considered part of the immune response because polymorphonuclear granulocytes are low throughout. It is a chronic response resembling that to topically applied chemical irritants. Further evidence that this is an acquired host response, induced by the feeding activity of the lice, is shown by the fact that the degree of resistance

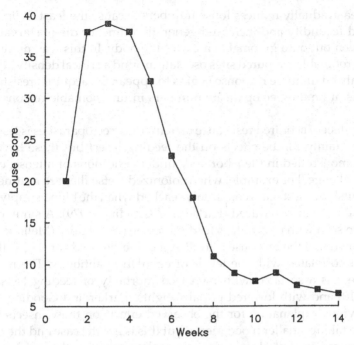

Figure 7.11 A hindfoot-amputated mouse infected with the louse *Polyplax serrata* will show localized acquired resistance and the numbers of lice will fall with time. This type of cycle can only be produced on an infested area of a mouse once in its lifetime. The decrease in the number of lice is caused by starvation. (Redrawn from Nelson *et al.*, 1977.)

expressed is directly correlated to the intensity and duration of the infestation on the hindfoot-amputated mouse. The response of the mouse to the lice also appears to include some progressive, physiological adaptation to the increasing burden, because the sudden transfer of large numbers of lice from an infested mouse to a healthy, naive, hindfoot-amputated mouse can lead to the death of the new host within 24 hours, possibly from toxic or anaphylactic shock.

Further work on the louse/hindfoot-amputated mouse combination showed that when resistant skin from sensitized mice was grafted onto naive, athymic mice it continued to show resistance to louse feeding. Lice fed happily on adjacent skin, or skin grafted from non-resistant mice, but not on the grafted resistant skin (Bell *et al.*, 1982; Nelson and Kozub, 1980). So, acquired resistance is a local phenomenon restricted to the area of skin exposed to feeding lice. The operating factor seems to be the impairment of louse feeding caused by the restriction in blood flow to that skin region.

What is the importance of this localized response to the host? Permanent ectoparasites tend to accumulate in the places that the host finds the most difficult to groom. The increasing development of acquired resistance in that area gradually reduces louse numbers because the least fit lice show reduced fecundity and increased generation time, or die of starvation or are driven out onto groomable areas of the body. In this way numbers in the ungroomable, favoured sites oscillate around a critical density at which the localized immune response begins to appear. So, acquired resistance is a means of limiting ectoparasite numbers in ungroomable regions of the body.

The effect of acquired resistance on temporary ectoparasites is less clear. There certainly can be effects on the feeding insect but these have only been demonstrated in the laboratory under conditions of intense, continuous challenge. For example, when colonized tsetse flies are routinely fed on rabbits' ears, a stage is commonly reached when the blood supply to the ears is significantly reduced (Parker and Gooding, 1979). A similar effect has been seen with mosquitoes feeding on the ears of mice (Mellink, 1981). It is known that the response is localized in the exposed ear (Fig. 7.10) and it is not correlated with the levels of circulating antibody. The response in the ear is associated with increased mortality of feeding tsetse flies (Fig. 7.10) and with lowered pupal weights (Parker and Gooding, 1979). The obvious explanation for the observed effects on these insects is that they are taking smaller blood meals, but this is not the case and the underlying causes are not clear. It is also unclear whether this evidence can be used to argue for naturally occurring acquired resistance to temporary ectoparasites, because the artificially high feeding levels seen in these experiments are unlikely to occur in field situations. No record of a naturally occurring acquired resistance to temporary ectoparasites has yet been recorded.

The second way that the immune response can diminish the feeding success of blood-sucking insects is through its stimulation of increased defensive behaviour by the host (see above). The level of defensive behaviour in the irritable host is dependent on the density of attacking insects and is provoked by the pruritis induced by insect bites. In this way the immune response limits the feeding success of the insect population as a whole, even though early assailants may obtain a full blood meal before pruritis and defensive behaviour are stimulated. Reduced feeding success in the insect population probably leads to increased generation times, reduced fecundity and increased mortality. An effect of this kind has been recorded under experimental conditions in the reduviid bug *Triatoma infestans* (Schofield, 1982). As the number of bugs in a population increases, 'scramble competition' comes into operation – the number of bugs successfully feeding and the amount of blood they are ingesting declines (Fig. 7.12). There is also evidence for density-dependent effects on feeding success in tsetse flies (Schofield and Torr, 2002; Torr and Mangwiro, 2000; Vale, 1977), mosquitoes (Waage and Nondo, 1982; Webber and Edman, 1972), horseflies (Waage and

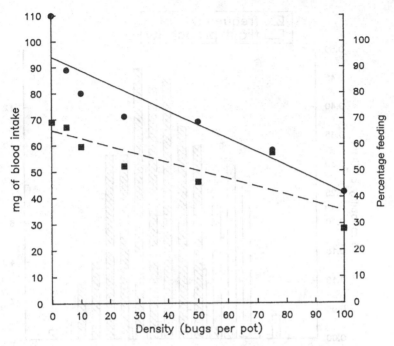

Figure 7.12 Increasing the density of fifth-instar nymphs of *Triatoma infestans* leads to a decrease in both the intake of blood (●) and the number of bugs successfully feeding (■). This is a factor in the density-dependent limitation of bug populations. (Redrawn from Schofield, 1982.)

Davies, 1986) and sandflies (Kelly *et al.*, 1996). In *Triatoma infestans* reduced feeding success leads to an increase in the development times of the immature stages, a decrease in fecundity and an increase in the likelihood of adult flight (Fig. 7.13) (Lehane and Schofield, 1981; Lehane and Schofield, 1982). Each of these effects serves to adjust the population to a lower stable density, which occurs, in this instance, without an increase in mortality. There is evidence from South America that the size of reduviid bug populations in dwellings shows a strong positive correlation with the number of occupants (see Schofield, 1985), so these limiting mechanisms may be operating in the field. The potential effects of density-dependent feeding success on bug populations have been modelled (Castanera *et al.*, 2003).

Under field conditions the relationship between feeding success and the number of attacking insects could be very complex. We know that there are interspecific differences in host tolerance to insect biting. There are also intraspecific differences which can depend on genetic makeup, host health, age and other factors. Tolerances also probably change in a complex manner with changing insect density. There are likely to be interspecific effects on feeding insects when mixed insect species are feeding on the same animal

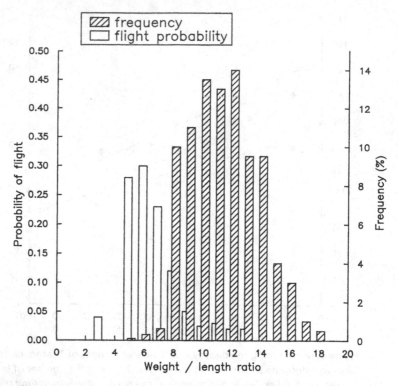

Figure 7.13 The probability of flight for female *Triatoma infestans* increases markedly as the bug gets hungrier and its weight/length ratio decreases. The frequency of weight/length ratios of a population of bugs taken from rural houses in Brazil shows that flights would be expected under natural conditions. Such flights may serve to reduce population density in the house, increasing the feeding success of the remaining individuals and regulating population density. (Redrawn from Lehane and Schofield, 1982.)

(Schofield and Torr, 2002). For example, large numbers of insects present on a host may disturb each other when feeding. Extensive field work is required before we will begin to clarify how important density-dependent effects are on feeding success as a means of limiting blood-sucking insect populations. Density-dependent feeding success is most likely to act as a factor limiting population size when the insects have access to only a limited number of host animals and/or a limited number of host species. Such circumstances regularly arise for arctic insects, and for nest-dwelling forms (including the domestic bug populations discussed above). Studies on these forms are most likely to improve our knowledge in this area.

The life histories of many blood-sucking insects (for example mosquitoes and blackflies) are such that the intraspecific competition for blood is minimized and as a consequence the reproductive potential of the available females is maximized. This is achieved by having larval forms that feed on non-blood food sources, and this is the case for most holometabolous blood-sucking insects. Many blood-sucking forms take this further and reserve blood feeding for the adult females, males normally feeding from a variety of sugar sources. Both strategies limit the attention paid to the host, the dividend for the insect being the reduced response it promotes from the host and the improved chance of successful feeding by the mature females. But not all blood-sucking insects adopt this dual feeding strategy. In hemimetabolous insects, such as lice and the reduviid bugs, both adult and juvenile stages compete for the same blood food sources. Although this is clearly a successful life history strategy for these species, it must limit the theoretical numbers of successful adult females compared to a situation in which juvenile stages and males are nourished by an alternative food.

As mentioned above, in most circumstances the total volume of available blood is not a limiting factor on the population size of blood-sucking insects, but blood is not a resource that is evenly spread throughout an ecosystem; it is collected together in 'packets' (hosts). In some situations, where hosts are scarce, the difficulty of finding such a 'packet' may be a factor limiting population size. In circumstances where the availability of a blood meal is unpredictable, many blood-sucking insects cease to rely entirely on the adult obtaining a blood meal in order to lay eggs. Instead they concentrate on maximizing larval nutrition and carry over reserves to the adult stage, permitting them to produce a batch of eggs autogenously (see Section 6.7).

Humanity's manipulation of the planet has had a profound impact on the availability of food for blood-sucking insects. This in turn has had a tremendous influence on the distribution and abundance of blood-sucking insects and on the epidemiology of the diseases they transmit. A good example is furnished by the changing feeding patterns of the anopheline mosquitoes of Europe discussed in Section 3.1. Another example is the growth of unplanned tropical urbanization, which has greatly increased the available food resource for some blood-sucking insects. Partly as a result of this, populations of insects such as the yellow fever mosquito, *Aedes aegypti*, have boomed. Originally thought to be a rot hole-breeding species limited to the forests of east Africa, this mosquito has followed humans into unplanned urban developments throughout the tropics where, along with other exploiters of these conditions such as *Culex quinquefasciatus*, it is responsible for a considerable degree of disease transmission.

8

Transmission of parasites by blood-sucking insects

Like all other organisms, blood-sucking insects have their own array of parasites. Many of these parasites are common to a range of different insects, blood-sucking and non-blood-sucking alike. Others are transmitted between the vertebrate host and the insect and so are peculiar to blood-sucking insects, and the parasites normally depend on the blood-sucking habit for their existence. In line with the approach taken in this book, it is the relationships between blood-sucking insects and this latter group of parasites on which I will concentrate in this chapter.

8.1   Transmission routes

Table 8.1 shows that blood-sucking insects are responsible for the transmission of many important disease-causing organisms. At its simplest, transmission may involve the insect as a mechanical bridge between two vertebrate host species. At its most complex, transmission involves an obligatory period of replication and/or development by the parasite in the vector insect. A division is often drawn between 'mechanical' and cyclical or 'biological' transmission.

Mechanical transmission is said to occur when the blood-sucking insect is no more than a flying pin, transferring pathogens from one vertebrate host to another on contaminated mouthparts. Relatively little work has been carried out on the relationships grouped together as mechanical transmission and the possibility of more complex interactions should not be ignored. For example, it has been suggested, on the basis of epidemiological evidence, that some mechanism for the concentration of parasites in the vector's mouthparts may occur in the mechanical transmission of trypanosomes such as *Trypanosoma vivax viennei* and *T. brucei evansi* (Wells, 1982). As our understanding of the subtleties and complexities of the interactions between parasite and vector increases, the collection of parasites placed in the 'mechanical' group will probably decrease.

Clearly any blood-sucking insect is a potential mechanical transmitter, but, because the pathogens cannot survive for long outside the host's body, insects that habitually take a succession of partial meals from several vertebrate hosts are probably the most efficient mechanical transmitters. Larger

Table 8.1 *Some of the most important associations of disease-causing organisms carried to humans and other animals by blood-sucking insects: (a) viruses, (b) rickettsia and bacteria, (c) protozoa and (d) nematodes.*

	Major vectors	Major hosts	Geographical distribution
(a) VIRUSES			
Semliki Forest virus	mosquitoes	humans	N. Africa, S. Africa, Amazon, Philippines, Madagascar
chikungunya	mosquitoes	humans, monkeys	India, Africa, E. Asia
Venezuelan equine encephalitis	mosquitoes	humans, equines, rodents, birds	S. America, southern USA, Europe
Western equine encephalitis	Culex tarsalis	humans, equines, birds, reptiles, amphibians	America, Czech Republic, Italy
Yellow fever	Aedes aegypti, several other mosquito species	humans, monkeys	Africa, tropical America
Dengue	Aedes aegypti Aedes spp.	humans	S.E. Asia, Caribbean
St Louis encephalitis	Culex tarsalis Culex spp.	humans, birds, rodents, bats	America
Japanese encephalitis	Culex tritaeniorrhynchus Culex spp.	humans, equines, pigs, birds	S.E. Asia
Murray Valley encephalitis	Culex annulirostris	humans, birds	E. Australia, Philippines, New Guinea
wesselsbron	mosquitoes	humans, sheep	Central and S. Africa, Thailand
Ilheus	forest mosquitoes	humans, monkeys, birds	Central and S. America
West Nile virus	Culex spp.	humans, birds	Central and N. Africa, India, Mediterranean, former Soviet Union
Sandfly fever	Phlebotomus papatasii	humans	Mediterranean, Near East, India, Sri Lanka, southern China, Central Asia
African horse sickness	Culicoides spp.	equines	S., Central and E. Africa, India
Bluetongue	Culicoides spp.	sheep, rodents	worldwide

(cont.)

Table 8.1 (*cont.*)

	Major vectors	Major hosts	Geographical distribution
Rift Valley fever	*Aedes* spp. *Eretmopodites* spp. *Ochlerotatus* spp.	humans, domestic and feral animals	Central, E. Africa and southern Africa
California encephalitis	*Aedes* spp. *Culex* spp.	humans	N. America
Myxomatosis	mosquitoes *Spilopsyllus cuniculi* *Culicoides* spp.	rabbits	America, Europe, Australia
(b) RICKETTSIA AND BACTERIA			
Rickettsia prowazekii	*Pediculus humanus humanus* *Pediculus humanus capitis*	humans, possibly cycles in domesticated animals	worldwide
Rickettsia typhi	*Leptopsylla segnis* *Xenopsylla cheopis* *Nosopsyllus fasciatus*	humans, rats, mice	worldwide
Rochalimaea quintana	*Pediculus humanus humanus*	humans	Europe, Mexico, China, Ethiopia, Algeria
Bartonella bacilliformis	*Phlebotomus verrucarum*	humans	S. America
Anaplasma marginale	biting flies	cattle, zebra, water buffalo, bison, antelope, deer, elk and camels	tropics and subtropics
Eperythrozoon suis	*Haematopinus suis*	pigs	worldwide
Borrelia recurrentis	*Pediculus humanus humanus*	humans	Ethiopia, Eritrea
Francisella (Pasteurella) tularensis	*Chrysops* spp. *Tabanus* spp.	humans, rodents, rabbits, birds	northern hemisphere
Yersinia (Pasteurella) pestis	*Xenopsylla cheopis* *Synosternus pallidus* *Nosopsyllus fasciatus*	humans, rodents	Manchuria, S.E. China, Thailand, Java, Burma, E. Africa, India, Iran, Madagascar
Treponema pertenue	*Aedes aegypti*	humans	northern South America

(c) PROTOZOA

Plasmodiidae

Plasmodium falciparum	*Anopheles* spp.	humans	tropics, subtropics and some temperate countries
Plasmodium ovale	*Anopheles* spp.	humans	tropical Africa
Plasmodium vivax	*Anopheles* spp.	humans	some temperate and tropical areas worldwide
Plasmodium malariae	*Anopheles* spp.	humans, chimpanzees	tropical Africa, India and S.E. Asia
Plasmodium gallinaceum	*Aedes* spp.	domestic fowl	India
	Armigeres spp.		
Plasmodium relictum	*Culex* spp.	pigeons, anatidae, passerines	worldwide
	Anopheles spp.		
	Aedes spp.		
Plasmodium cynomolgi	*Anopheles* spp.	monkeys, macaques, occasionally humans	Asia
Plasmodium knowlesi	*Anopheles hackeri*	monkeys, macaques, occasionally humans	S.E. Asia
Plasmodium berghei	*Anopheles* spp.	tree rats	Central Africa
Haemoproteus columbae	hippoboscids	various birds, particularly pigeons and doves	worldwide
Leucocytozoon simondi	*Simulium* spp.	ducks and geese	N. America, Europe, Asia
Leucocytozoon smithi	*Simulium* spp.	turkeys	N. America, Europe

Trypanosomatidae

Leishmania tropica	*Phlebotomus* spp.	humans, dogs, rodents	Asia, Middle East, Mediterranean
Leishmania donovani	*Phlebotomus* spp.	humans, rodents, serval, genet cats, dogs, foxes, jackals	S. America, tropical and N. Africa, Asia Mediterranean
Leishmania brasiliensis	*Lutzomyia* spp.	humans, rodents, monkeys	S. America, Iran

(cont.)

Table 8.1 (*cont.*)

	Major vectors	Major hosts	Geographical distribution
Trypanosoma lewisi	*Glossina* spp.	equines, cattle, sheep, dogs, cats, many wild game animals are reservoirs	Africa from latitudes of 15° N to 25° S
Trypanosoma gambiense	*Glossina* spp.	humans, cattle, sheep, goats, horses, dogs, cats, pigs	W. and Central Africa from latitudes of 15° N to 18° S
Trypanosoma rhodesiense	*Glossina* spp.	humans, wild game animals are reservoirs	tropical E. Africa
Trypanosoma congolense	*Glossina* spp.	a very wide range of domestic and wild game animals	tropical Africa
Trypanosoma vivax	*Glossina* spp. biting flies	cattle, water buffalo, sheep, camels, goats, horses, antelope, deer	tropical Africa, Caribbean, Central and S. America
Trypanosoma uniforme	*Glossina* spp.	sheep, goats, cattle, antelope	tropical Central Africa
Trypanosoma simiae	*Glossina* spp. biting flies	wart-hog, pigs, camels	tropical E. and Central Africa
Trypanosoma suis	*Glossina* spp.	pigs	Zaire
Trypanosoma evansi	biting flies	camels, horses and dogs, a wide range of domestic and feral animals are reservoirs	India, Far East, N. Africa, Near East, Central and S. America
Trypanosoma equinum	biting flies	equines, dogs, cattle, sheep, goats	Central and S. America
Trypanosoma theileri	*Tabanus* spp. *Haematopota* spp.	cattle	worldwide
Trypanosoma melophagium	*Melophagus ovinus*	sheep	worldwide
Trypanosoma lewisi	*Ceratophyllus fasciatus*	rats	worldwide
Trypanosoma rangeli	reduviid bugs	humans, dogs, cats, opossums, monkeys	S. America

Trypanosoma cruzi	reduviid bugs	humans, opossums, armadillos, wide range of feral and domestic reservoir hosts	S. America
(d) NEMATODES			
Onchocercidae			
Onchocerca volvulus	*Simulium* spp.	humans	tropical Africa, Central America
Onchocerca gutturosa	*Simulium* spp.	cattle, buffalo	worldwide
	Odagmia spp.		
	Friesia spp.		
Onchocerca gibsoni	*Culicoides pungens*	cattle, zebu	Asia, Australasia, southern Africa
Onchocerca lienalis	*Simulium* spp.	cattle	Australia, N. America
Onchocerca cervicalis	*Anopheles* spp.	equines	worldwide
	Culicoides spp.		
Filariidae			
Wuchereria bancrofti	*Culex* spp.	humans	tropics, subtropics and some temperate countries
	Aedes spp.		
	Mansonia spp.		
	Anopheles spp.		
Mansonella ozzardi	*Culicoides furens*	humans	S. America, Caribbean
	Simulium amazonicum		
Brugia malayi	*Aedes* spp.	humans, leaf monkeys, cats, dogs, civet cats, pangolins	India, S.E. Asia
	Mansoni spp.		
	Anopheles spp.		
Brugia pahangi	*Mansonia* spp.	cats, dogs, civet cats, leaf monkeys, tigers, slow loris	Malaysia
	Armigera spp.		
Brugia patei	*Mansonia* spp.	cats, dogs, genet cats, bush babies	Africa
	Aedes spp.		

(cont.)

Table 8.1 (*cont.*)

Brugia timori	*Anopheles* spp.	humans	Indonesia
Loa loa	*Chrysops silacea*	humans, baboons, monkeys	W. and Central Africa
	Chrysops dimidiata		
Dirofilaria immitis	mosquito spp.	dogs, foxes, wolves, cats and occasionally humans	tropics, subtropics, some temperate countries
Dirofilaria repens	mosquito spp.	dogs and occasionally humans,	Europe, Asia, S. America E. Europe
Parafilaria multipapillosa	*Haematobia atripalpis*	equines, anatid birds, sheep,	N. America N. America
Ornithofilaria fallisensis	*Simulium* spp.	deer, elk	
Elaeophora schnedieri	*Hybomitra* spp.		
	Tabanus spp.		
Setariidae			
Setaria equina	*Aedes* spp.	equines	worldwide
	Culex spp.		
Setaria labiatopapillosa	*Anopheles* spp.	cattle, deer, giraffe, antelope	worldwide
Dipetalonema reconditum	*Ctenocephalides* spp.	dogs	N. America, Africa, S. Europe
	Pulex spp.		
Dipetalonema evansi	*Aedes detritus*	camels	Egypt, Far East, E. former Soviet Union
Dipetalonema streptocerca	*Culicoides* spp.	humans	Central and W. Africa
Dipetalonema perstans	*Culicoides* spp.	human, anthropoid apes	Africa, S. America
Stephanofilaria stilesi	*Stomoxys calcitrans*	cattle	former Soviet Union, N. America
Spiruridae			
Habronema majus	*Stomoxys calcitrans*	equines	worldwide

insects that give painful bites, such as biting muscids and tabanids, are often disturbed by the vertebrate host before completing their meal. As they commonly feed on large sociable herbivores and are highly mobile insects, the unfinished meal is often quickly completed on a second vertebrate host. Another important factor in mechanical transmission is the amount of blood that can be transferred between animals by the insect. This is because, for any given level of parasitaemia, the more blood transmitted, the greater the chance of disease transmission. Large tabanids, such as *Tabanus fuscicostatus*, may transmit as much as a nanolitre of blood on their sponging mouthparts. Probably for these reasons, tabanids are the mechanical transmitters of trypanosomes such as *Trypanosoma evansi* (= *T. equinum*, *T. hippicum* and *T. venezuelense*) and on occasion *T. equiperdum* (Soulsby, 1982) which cause serious and often fatal diseases in equines in many parts of the world. There is strong experimental evidence for the mechanical transmission of *Trypanosoma vivax* by the tabanid *Atylotus agrestis* (Desquesnes and Dia, 2003). Smaller insects that have less painful bites and that are capable of moving smaller quantities of blood between vertebrates are still important mechanical vectors. Thus myxoma virus, which causes myxomatosis in rabbits, is mechanically transmitted by the rabbit flea, *Spilopsyllus cuniculi*, in Britain and by other fleas and mosquitoes in Australia.

Cyclical or biological transmission is said to occur when there is a biological dependency between the vector and parasite, with the pathogen undergoing a period of growth and development and, in some instances, multiplication in the insect. Biological transmission is the most common and important means of pathogen transmission by insects.

Each parasite has its own peculiarities of lifestyle in the insect. Some of these characteristics can be used to subdivide biological transmission into the following categories (Huff, 1931):

(a) *Propagative transmission*
Propagative transmission occurs when the pathogen multiplies in the insect host but undergoes no development. This occurs with bacterial pathogens such as *Yersinia pestis*; the large numbers of pathogens leaving the flea are essentially the same as those ingested. Arboviruses are usually included in this category, although this is not strictly correct as these pathogens acquire a portion of the plasma membrane from the cell in which they were formed. This coat is species-specific and in this respect the virus that leaves the insect is different from the one that entered it. In essential details, however, such as the nature of the envelope and coat proteins and the genetic information contained in each virus particle, they are members of this category.

(b)   *Cyclo-propagative transmission*
       Cyclo-propagative transmission occurs when the pathogen not
       only multiplies in the insect but also changes its form in some
       manner. A large number of parasites are included in this group,
       such as many of the trypanosomes, leishmaniases and the malaria
       parasites.
(c)   *Cyclo-developmental transmission*
       Cyclo-developmental transmission occurs when the pathogen
       undergoes a developmental transformation in the vector but does
       not multiply, which is true of helminths such as the filarial worms.

The details of the sojourn in the insect tend to vary not only from one
group of pathogens to another but also from species to species. As space
does not allow a detailed discussion of the passage of each important parasite, some generalizations based on major parasite groups are given below.
In addition, an outline of the main routes taken by parasites transmitted
by insect vectors is given in Figure 8.1.

Leishmania can be subdivided into three groups (suprapylarian, peripy-
larian and hypopylarian) according to their distribution in their sandfly
hosts (Lainson and Shaw, 1987). The suprapylaria are limited to the portion
of the gut anterior to the pylorus of their sandfly vectors. The peripylaria
occur in the posterior area of the gut, in the abdominal midgut and in the
pylorus. The hypopylaria are restricted to the hindgut. In the subgenus
Leishmania, amastigotes are released in the midgut of the sandfly from
the vertebrate macrophages ingested in the blood meal. Division of these
amastigotes occurs in the blood meal in the first two to three days following
feeding. The free-swimming promastigotes produced remain constrained
within the peritrophic matrix. The surface molecules of insect forms of the
parasite include lipophosphoglycan (LPG) which appears to be involved
both in defence against insect proteases and in attachment to the gut surfaces of the sandfly (Handman, 2000). As the peritrophic matrix breaks
down at the completion of blood-meal digestion, parasites escape and
shortened haptomonad-like promastigotes attach to the stomadaeal valve.
Colonization of the oesophagus and pharynx by small, rounded, sessile,
flagellated paramastigotes follows. A similar picture is seen in the subgenus
Viannia, except that the ileum and pylorus regions of the hindgut become
colonized with small, rounded, haptomonad-like promastigotes and paramastigotes (Molyneux and Killick-Kendrick, 1987). Infective metacyclic
stage parasites finally form and are transferred to the vertebrate from the
sandfly's mouthparts when it feeds (Sacks, 1989; Saraiva *et al.*, 1995). Intracellular forms have been reported in insect guts, but their importance is
unknown (Molyneux *et al.*, 1975). *Leishmania* can undergo sexual recombi-
nation at a very low frequency (Dujardin *et al.*, 1995; Panton *et al.*, 1991),

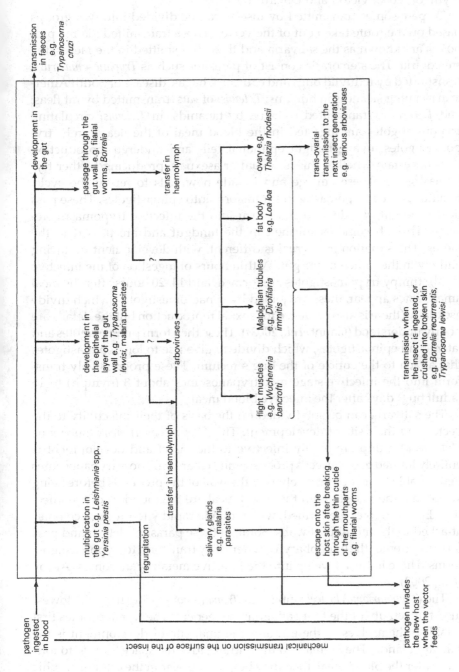

Figure 8.1 An outline of various routes parasites take in their vector hosts. The box represents the insect.

but the importance of sex for this and some other parasites is debatable (Lythgoe, 2000; Victoir and Dujardin, 2002).

Trypanosomes transmitted by insects can be divided into two groups based on the route taken out of the vector. Those transmitted via the proboscis are known as the salivaria and those transmitted in the faeces, the stercoraria. The stercoraria consist of parasites such as *Trypanosoma cruzi*, transmitted by reduviid bugs and causing Chagas' disease or South American sleeping sickness in humans; *T. lewisi* of rats, transmitted by rat fleas; and *T. theileri*, transmitted to cattle by tabanids. In *T. lewisi* circulating trypomastigotes are ingested in the blood meal of the flea. Rarely, trypomastigotes invade midgut epithelial cells and undergo reproduction (it is uncertain whether this is sexual or asexual), producing further trypomastigotes. These emerge and invade new cells to repeat the cycle. Eventually free trypomastigotes transform into epimastigotes. These pass down the gut, divide, and transform into the infective trypomastigote-stage. These become established in the hindgut and are passed in the faeces. The situation in *T. cruzi* is different, with development occurring entirely in the lumen of the gut. Within hours of ingestion of the infective meal, stumpy trypomastigotes are formed, and 14–20 hours after the meal amastigotes appear, these are joined by sphaeromastigotes, which divide asexually (there is also evidence for sexual reproduction but the details are poorly understood (Gaunt *et al.*, 2003)). These then form promastigotes and later small epimastigotes, which divide to give rise to long epimastigotes that attach to the cuticle of the bug's rectum. These progressively transform into the infective-stage metatrypanosomes about 8 (nymph) to 15 (adult bug) days after the infective blood meal.

The salivaria can be subdivided on the basis of their infectivity to the vector and their site of development. The (*Trypanosoma*) *vivax* (subgenus *Duttonella*) group are highly infective to the insect and development is entirely limited to the insect's proboscis and cibarium. Those trypanosomes that are able to anchor themselves to the wall of the proboscis before being swept into the midgut with the blood meal are the founders of the infection. Epimastigotes are formed, which divide rapidly to produce colonies attached to the mouthpart walls. Some of these parasites detach and pass forward to enter the hypopharynx, where they transform to trypomastigote forms. These in turn develop into the infective metatrypanosomes (Aksoy *et al.*, 2003).

The (*Trypanosoma*) *brucei* (subgenus *Trypanozoon*) group have the lowest insect infectivity of the four salivarian subgenera. The trypanosomes first establish themselves in the gut and the final site of development is the salivary glands. They remain in the endoperitrophic space for two to four days after the blood meal, then they begin to appear in the ectoperitrophic space. It is uncertain if the trypanosomes achieve this by penetrating the peritrophic matrix (Ellis and Evans, 1977) or by passing down the gut to the rectum, where the peritrophic matrix is torn up by the rectal spines, and passing around the open ends of the matrix back up the gut in the ectoperitrophic space. Trypanosomes are believed to subsequently leave the ectoperitrophic space by penetrating the peritrophic matrix where it emerges from the proventriculus, an area of the membrane thought to be 'soft'. Once back in the endoperitrophic space, the trypanosomes travel up the foregut to the end of the food canal. They then emerge from the food canal and enter the hypopharynx and travel to the salivary glands via the salivary ducts, where they develop into vertebrate-infective metacyclic forms (Aksoy et al., 2003).

The (Trypanosoma) congolense (subgenus Nannomonas) group occupies an intermediate position between the vivax and brucei groups. It displays an intermediate degree of infectivity for Glossina. The life cycle is similar to the brucei group, but final development is in the mouthparts not the salivary glands. In the mouthparts the trypanosomes attach to the walls of the labrum and some enter the hypopharynx, which is where the infective metacyclics are found (Aksoy et al., 2003).

The fourth group of the salivaria (subgenus Pycnomonas) has only a single representative, Trypanosoma suis, which has a very limited distribution; accounts of its life cycle are incomplete (Aksoy et al., 2003).

In the Haemosporidia, male (micro-) and female (macro-) gametocytes are taken up in the blood meal by the vector, commonly a mosquito. Almost immediately, the male gametocyte produces numbers of flagellate microgametes by a process termed exflagellation. These fertilize the female gametocyte and the resulting ookinete penetrates the midgut wall. Most Haemosporidia, including the human malaria parasites, then produce a thin-walled oocyst that becomes established on the outer wall of the gut epithelium beneath the basement membrane and muscle layers. Exceptionally, oocysts of Hepatocystis break free of the gut, and travel to the head capsule of their Culicoides hosts before encysting. Rapid division occurs in the oocyst, with the production of fusiform sporozoites some time between the 4th and 15th day following the infected blood meal. The number of sporozoites produced varies with species: Plasmodium falciparum produces about 10 000 and some haemoproteids as few as 30 or 40. The oocyst ruptures, releasing the sporozoites, which become distributed throughout the insect's body. Some penetrate to the salivary gland lumen and, during insect feeding, are passed to the next vertebrate host in the saliva. Hepatocystis are believed to take a slightly different route, with the sporozoites directly entering the mouthparts.

Adult filarial worms produce live young called microfilariae. These first-stage larvae are ingested along with the blood meal of the vector insect, which may be a mosquito, muscid fly, flea, ceratopogonid midge, tabanid or blackfly, depending on the species of worm. Microfilariae of some filarial worm species possess a sheath consisting of the elongated remnant of the eggshell. This sheath is normally shed once the worm is in the vector's intestine. Successful microfilariae usually pass through the intestinal wall within an hour of blood meal ingestion. Once in the haemolymph, they pass to one of the insect's internal organs, inside which they develop through the second larval stage to the infective third stage. The choice of internal organ for development varies among species of filarial worm. The filarial worms of humans, *Onchocerca volvulus*, *Brugia malayi* and *Wuchereria bancrofti*, all take about 8–12 days to develop in the thoracic flight muscles of the vector. Another filarial worm of humans, *Loa loa*, develops in the fat body of *Chrysops* spp., while the filarial worms of dogs, *Dirofilaria immitis* and *D. repens*, both develop in the Malpighian tubules of their mosquito vectors. The infective, third-stage larva moves via the haemolymph to the insect's mouthparts. It escapes from the insect's haemolymph by breaking through the cuticle at the time of the next blood meal to become deposited on the skin of the vertebrate host, which the worm enters through the puncture hole left by the insect's mouthparts.

Arboviruses are ingested by the vector in the blood meal. Infection of the insect occurs only if the virus is ingested in sufficient concentration to overcome the insect's gut barrier. The infection threshold level varies among different arbovirus insect interactions. The nature of the gut barrier is not understood but may centre around the number of available attachment sites for the virus on the surface of the midgut cells. The gut barrier may be intimately associated with virus–vector specificity as it is often possible to infect otherwise refractory insect species by the direct injection of the virus into the insect's haemocoel. In relation to the gut barrier, it is notable that mosquitoes are more easily infected with arboviruses when they are concurrently infected with microfilariae, which probably assist the passage of the virus through the intestinal barriers (Mellor and Boorman, 1980; Turell *et al.*, 1984a). Even after successful infection of the gut epithelium, the virus may be prevented from infecting the haemocoel by the 'mesenteronal escape barrier', and the salivary glands by the 'salivary gland infection barrier' (Kramer *et al.*, 1981). The virus can escape the salivary gland barrier by concurrent infections with *Plasmodium* parasites (Vaughan and Turell, 1996). These barriers are also dose-dependent, but the mechanisms involved are not understood. Even if the arbovirus achieves the infection of the salivary glands, transmission will only occur if the virus overcomes the 'salivary gland escape barrier' (Fu *et al.*, 1999; Grimstad *et al.*, 1985).

Once in the haemolymph, arboviruses infect other tissues in addition to the salivary glands. Infection of the ovaries is particularly important in some virus–vector associations because transovarial transmission can occur to the next generation of vectors (DeFoliart *et al.*, 1987; Watts *et al.*,

1973). Not only can the female progeny infect new vertebrate hosts, but the infected (non-blood-feeding) male progeny may increase the population of infected vectors by venereally transmitting the arbovirus to previously uninfected females (Thompson, 1977).

8.2   Specificity in vector–parasite relationships

Each parasite transmitted by blood-sucking insects is normally associated with a restricted number of vector species. Even mechanically transmitted organisms such as the myxoma virus show a degree of specificity in their vector associations. When myxomatosis escaped from an experimental area of the Murray Valley in Australia in 1950, it first moved along major watercourses and was transmitted by the locally abundant, river-haunting mosquito, *Culex annulirostris*. In subsequent seasons it managed to move out across vast areas of semi-arid country by using another mosquito, *Anopheles annulipes*, which survived the harsh local conditions by using rabbit burrows for shelter. In Europe the myxoma virus uses yet another carrier, the European rabbit flea, *Spilopsyllus cuniculi*. Specificity in this instance is largely determined by details of local ecology, the myxoma virus making use of any appropriate 'flying pin' for its mechanical transmission from one vertebrate host to another. Relationships in which the parasite has a biological dependence on the host insect are more specific. Thus, the four malaria parasites of humans are only transmitted by anopheline mosquitoes, *Culicoides* spp. are the vectors of bluetongue virus and Chagas' disease is transmitted by reduviid bugs.

There are several ways in which a specific relationship between a pathogen and its vector insect can be mediated. Of prime importance is vector–pathogen coincidence. Except for transovarial transmission, if the insect does not feed on a host containing the parasite it cannot become a vector. So a vital factor determining vector–parasite relationships is the choice of host the potential vector feeds on. Host choice is discussed in detail in Chapter 3.

The importance of physiological factors in determining vector–parasite specificity can be shown under experimental conditions when abnormal combinations of pathogen and vector are contrived. Under these circumstances we find that, like the human malaria parasites, which die if ingested by any insect other than an anopheline mosquito host, most parasites will only develop successfully in a very narrow range of insect species. Physiologically based susceptibility to particular parasites, as well as varying among different insect species (interspecifically), also differs among individuals of the same species (intraspecifically). This has been clearly shown in the relationship of mosquitoes with both malaria parasites (Collins *et al.*, 1986; Huff, 1929; Huff, 1931; Kilama and Craig, 1969; Ward, 1963) and filarial worms (Kartman, 1953; Macdonald, 1962a; Macdonald, 1962b;

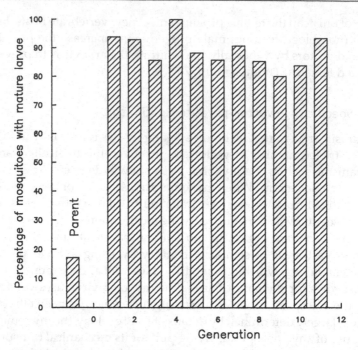

Figure 8.2 The susceptibility of blood-sucking insects for the organisms they transmit can be genetically determined. Here a west African strain of *Aedes aegypti* was selected for susceptibility to the filarial worm *Brugia malayi* over five generations. The selected mosquitoes were then maintained as a colony for several generations without an appreciable loss in susceptibility. (Drawn from data in Macdonald, 1962b.)

Macdonald, 1963). In both associations there is considerable variation in the ability of individual mosquitoes of a given species to permit the development of the parasites. This can range from complete refractoriness to complete susceptibility. Selection experiments have shown that vector susceptibility is usually genetically based (Beerntsen *et al.*, 2000) (Fig. 8.2), but extrachromosomal inheritance has been reported (Maudlin and Dukes, 1985; Trpis *et al.*, 1981). It should also be borne in mind that parasite populations display a spectrum of infectivities for available vectors (Laurence and Pester, 1967), which further complicates the situation found in the field. Our understanding of the mechanisms involved in determining the success or failure of vector parasite interactions is still far from complete, particularly at the molecular level. However, some information, particularly at the genetic level, is available and is described below.

Anopheles gambiae is the major vector of human malaria in Africa. Experimentally it can also transmit a range of other malaria parasites of simians, rodents and birds. A strain of the mosquito selected for refractoriness to the simian parasite *Plasmodium cynomolgi* B is also refractory to many, but not all, other malaria species (Collins *et al.*, 1986; Severson *et al.*, 2001) and also reacts to inanimate objects (Gorman *et al.*, 1996), suggesting a general refractory mechanism. Refractory mosquitoes of this strain defend themselves against the parasites by encapsulating and melanizing ookinetes (see Section 8.6). Investigations have revealed one major (*Pen1*) and two minor (*Pen2* and *Pen3*) quantitative trait loci (QTL) for the refractory encapsulation reaction (Severson *et al.*, 2001; Zheng *et al.*, 1997), and it now appears that different QTL may be involved in responses to different strains of *Plasmodium* (Zheng *et al.*, 2003).

Other refractory mechanisms also operate in the *Anopheles–Plasmodium* relationship. *Plasmodium gallinaceum* may be killed by a non-melanizing response, causing the lysis of ookinetes in the midgut wall of *Anopheles gambiae* (Vernick *et al.*, 1995). A similar mechanism may operate in *Aedes aegypti*, controlled largely by a single gene, *pls*, on chromosome 2 (Kilama and Craig, 1969; Severson *et al.*, 1995; Thathy *et al.*, 1994). The mechanism in *Anopheles* seems to be controlled through a single dominant locus on chromosome 3L (Severson *et al.*, 2001). Field studies have identified two more resistance markers *Pfin1* and *Pfin2* and more are probably waiting to be discovered (Niare *et al.*, 2002).

Work on the refractoriness of mosquitoes to filarial worms also shows a wide range of genes controlling susceptibility. Susceptibility of *Aedes aegypti* to *Brugia malayi* (and some other filarial worms developing in thoracic musculature) is controlled by two QTL, a major one at *fsb*[1, *LF178*] (probably the same as the f^m locus originally identified in the 1960s) and a minor one, *fsb*[2, *LF98*] (Macdonald, 1962a; Macdonald, 1963; Macdonald and Ramachandran, 1965; Severson *et al.*, 1994). The modulating impact of the minor genes is suggested by laboratory studies. For example, looking again at *Aedes aegypti* that are fully susceptible to *Brugia pahangi* (i.e. f^m/f^m), we find that about a quarter of the larvae that enter the thoracic flight muscles still die in the first few days (Beckett and Macdonald, 1971). Susceptibility of the same mosquito to another filarial worm, *Dirofilaria immitis* (developing in Malpighian tubules) is controlled by a different sex-linked recessive gene designated f^t (McGreevy *et al.*, 1974). Another QTL, *idb*[2, *LF181*] appears to influence the number of microfilaria ingested by the mosquito and the number penetrating the midgut epithelium (Beerntsen *et al.*, 1995). Susceptibility of *Aedes aegypti* to *Plasmodium gallinaceum* and yellow fever virus also appears to be associated with the region of chromosome 2 containing *idb*[2, *LF181*] and *fsb*[2, *LF98*]. From the data above it might be thought that each gene is acting on a particular organ system and conferring refractoriness to anything attempting to develop there. This is not the case in the following example in *Culex pipiens*. In this mosquito a sex-linked recessive gene *sb* controls susceptibility to *Brugia pahangi*, but does not control susceptibility for another filarial worm, *Wuchereria bancrofti*, even though both follow the same route into the mosquito and both develop in the insect's flight musculature (Obiamiwe and Macdonald, 1973).

Susceptibility of mosquitoes to arboviruses is also genetically determined. Refractoriness operates as a series of barriers at various levels in the passage of the virus through the mosquito. Thus, dengue-2 virus development in *Aedes aegypti* can be determined by a midgut barrier controlled by two QTL on chromosomes 2 and 3. Another QTL on chromosome 3 then regulates escape of the virus from the midgut and dissemination around the body of the mosquito (Bosio *et al.*, 1998; Bosio *et al.*, 2000).

The molecular nature of the genetic factors described above will be fascinating to discover. It has been speculated that in the *Plasmodium*–mosquito relations at least, many may represent pattern recognition molecules (Niare *et al.*, 2002; Zheng, 1999).

Non-Mendelian inheritance of susceptibility to *Brugia malayi* and *B. pahangi* has been reported in mosquitoes of the *Aedes scutellaris* complex (Trpis *et al.*, 1981). It has also been reported to govern the susceptibility of tsetse flies for trypanosomes (Maudlin and Dukes, 1985). In tsetse flies susceptibility appears to depend upon the numbers of rickettsia-like organisms (RLOs) that are present in the female parental fly (Maudlin and Ellis, 1985). (Subsequent work has shown that the term RLO was used to refer to both *Sodalis* and the *Wolbachia*-related symbionts of tsetse flies (Aksoy *et al.*, 2003).) Under natural conditions only about 10 per cent or fewer of tsetse fly populations are susceptible to trypanosome infection (Jordan 1974). It has been suggested that refractory tsetse defend themselves against invading trypanosomes by using gut-based lectins (see Section 8.7). Lectins bind to specific sugar residues on the surface of more than one trypanosome, thereby linking them together (agglutination) or directly inducing cell death (Pearson *et al.*, 2000; Welburn, 1987; Welburn and Murphy, 1998). The suggestion is that in susceptible flies RLOs interfere with these lectins by producing an enzyme called chitinase. This enzyme attacks chitin in the midgut, releasing glucosamine, which binds to specific lectins, neutralizing them and thus inhibiting their action on trypanosomes (Welburn *et al.*, 1993).

An interesting example of how a very intimate relationship can be built up between a particular parasite and a particular vector can be seen in the relationships of *Leishmania* and sandflies (Mbow *et al.*, 1998; Theodos *et al.*, 1991; Theodos and Titus, 1993; Titus and Ribeiro, 1990). When small numbers of *Leishmania major* promastigotes are artificially inoculated into a vertebrate host they fail to establish an infection. In contrast, the same number of parasites injected by needle along with salivary extracts from sandflies successfully establish an infection and undergo an increase in numbers of two to three orders of magnitude. In the vertebrate host *Leishmania* are obligate intracellular parasites of macrophages. The macrophage can kill *Leishmania* by the production of nitric oxide (NO) and other oxygen-derived metabolites. The macrophage is stimulated to do this by IFN-γ (gamma interferon) produced by activated parasite-specific T cells. The *Leishmania* enhancing factor (LEF) of the salivary glands of sandflies inhibits this IFN-γ-stimulated production of superoxide (Hall and Titus, 1995). In addition LEF inhibits the ability of the macrophages to present leishmanial antigens to parasite-specific T cells (Theodos and Titus, 1993). Immunomodulation also appears to be intimately linked with induction of the cytokine IL-4 (Mbow *et al.*, 1998). The immunomodulatory properties of sandfly saliva are not specific to *Leishmania* because they influence responses to antigens not related to the parasite (Titus, 1998).

What are the molecular components of LEF? Maxadilan, a potent vasodilatory substance (see Section 5.1), has immunomodulatory effects on human cells and is at least a component of LEF (Morris *et al.*, 2001; Rogers and Titus, 2003). Maxadilan is present only in New World sand-flies (*Lutzomyia*). The components of LEF in Old World sandflies (*Phlebotomus*) are less clear but may involve adenosine, adenosine deaminase and hyaluronidase (Kamhawi, 2000; Sacks and Kamhawi, 2001).

The associations described above are elegant but are based on experimental co-injection of saliva extracts and parasites by needle. Natural transmission by sandfly bite gives a different or at least modified picture, and further work will be necessary before the differences are explained (Kamhawi *et al.*, 2000). Also, the picture above refers to vertebrate hosts that did not have pre-exposure to sandfly saliva. Pre-exposure, by being bitten by uninfected sandflies, abrogates many of the phenomena described above and indeed gives some protection against infection with *Leishmania* (Kamhawi *et al.*, 2000). Nevertheless, LEF has not been found in blood-sucking insects other than sandflies and it may prove to be an important factor in the co-evolution of *Leishmania* and their sandfly vectors.

8.3 Origin of vector–parasite relationships

There has been considerable debate concerning the origin of vector–parasite associations. Because parasites leave few, if any, fossils, attempts to describe the evolutionary development of the associations has, in the past at least, been largely speculation. The arrival of phylogenetics based on nucleic acid or protein-coding gene sequence comparisons has provided more solid data that enable these arguments to be moved forward. Two conflicting evolutionary pathways have been proposed for many of the associations. In one route, the parasites are seen as being originally associated with the invertebrate, the vertebrate becoming involved when a lasting interaction between vertebrate and invertebrate has developed. The second route proposes an origin of the parasite in the vertebrate, with transmission to the invertebrate occurring when the vector developed the blood-sucking habit. Two such possible origins have been suggested for the family Trypanosomatidae, which contains the important genera *Leishmania* and *Trypanosoma* (Hughes and Piontkivska, 2003; Lainson and Shaw, 1987; Molyneux, 1984; Stevens *et al.*, 2001).

Monogenetic flagellates are common intestinal parasites of invertebrates. Majority opinion suggests that present-day parasitic forms had their origins in similar intestinal forms. Regular transmission can only have been established some time after the evolution of a recurring, close association between the insect and vertebrate. Initially transmission was probably by contamination in the insect faeces. Stercorarian (posterior station) trypanosomes, such as the causative agent of Chagas' disease, *Trypanosoma cruzi*, are still found today. Transmission via the mouthparts is seen as a later development as the salivarian trypanosomes adapted to life in the anterior parts of the insect's intestine. Molecular evidence for evolution of the digenetic (two-host) lifestyle from an initial invertebrate-only infection is seen in an extensive phylogenetic analysis showing a clade formed by the genera *Leishmania* and *Endotrypanum*, which are digenetic, along with insect-only parasites in the genera *Leptomonas*, *Crithidia*, *Blastocrithidia* and *Wallaceina* (Hughes and Piontkivska, 2003).

There is also a minority view that vector-transmitted trypanosomes of vertebrates did not develop along this route, but instead developed from vertebrate gut-dwelling forms that invaded the bloodstream. Certainly if this occurred, and the levels of parasitaemia achieved in the blood were sufficiently great, then the chances of immediate transmission by blood-sucking insects would be high (present-day forms can be mechanically transmitted by insects). In support of a vertebrate origin, it is pointed out that when flagellates parasitic in invertebrates are experimentally transferred to vertebrates they do not survive well. Conversely, vertebrate parasites successfully survive in a wide variety of insects, including species that are not vectors.

Phylogenies based on rRNA sequence data suggest that the Trypanosomatidae is polyphyletic (Hughes and Piontkivska, 2003) and that parasitism may have arisen independently on several occasions within the group. This presents the possibility that both hypotheses outlined above may be true, with some digenetic Trypanosomatids being first parasitic in insects while others were first parasitic in vertebrates.

It has often been said that there is an inverse relationship between the pathology caused to the host, and the length of association of that parasite and host, and this has been invoked to claim an invertebrate origin for most insect-transmitted parasites, including the Trypanosomatidae. Although this principle is regularly repeated, it is not supported by the available data (Ball, 1943; Schall, 2002).

The occurrence of sexual reproduction in the insect host has occasionally been suggested as powerful evidence for an invertebrate origin (Lainson and Shaw, 1987), but it is certainly not universally true. As we shall now see, the majority opinion is that malaria parasites originated in vertebrates even though sexual reproduction occurs in the insect. It is generally accepted that malaria parasites evolved from gut-dwelling parasitic apicomplexans (Baker, 1965). There are no gut-dwelling monogenetic, monoxenous parasitic apicomplexans in present-day Diptera (although they are represented, if rather poorly, in other insect orders). There are many such forms in vertebrates, some with a tendency to leave the gut and to become tissue-dwelling, pointing to a vertebrate origin for malaria parasites. The strongest argument in favour of a vertebrate beginning is the observation that if malaria parasites were derived from invertebrate species, then it would be expected that the life cycle would be the inverse of that actually found. So Bray (1963) argues, by reference to coccidian life cycles, that exocoelomic schizogony would be expected to occur in the insect gut, and schizogony and gametogony in the haemocoel, with sporogony occurring in the vertebrate host. Arguments have also been put in favour of an invertebrate origin: greater apparent pathogenicity of malaria parasites in their vertebrate hosts and sexual reproduction occurring in the insect have both been cited in this cause. As suggested above, neither of these is a strong basis for argument. More convincingly, it has been pointed out that malaria parasites are found only in one order of insects, the Diptera, while they are found in three classes of vertebrates. This might suggest an origin in invertebrates, but equally may reflect a later radiation into various vertebrate groups once the highly efficient vectorial link had been forged. The molecular phylogenetic data does not help to distinguish between an insect or vertebrate origin. The data suggests apicomplexa are extremely ancient and that apicomplexans may have parasitized ancestors of the chordates and co-evolved along with the vertebrates into the *Plasmodium* species we see in modern-day birds and mammals (Escalante and Ayala, 1995).

It is suggested that the filarial worms may have evolved from gut-dwelling vertebrate parasites (Anderson, 1957). These parasites could have reached the orbit of the eye by migration via the oesophagus, nasal cavities and nasolacrimal duct. Once in the eye, the larval stages could be picked up by eye-feeding insects. Present-day, orbit-dwelling spirurids, such as the ruminant parasite *Thelazii gulosa*, are transmitted in this way by non-blood-sucking flies such as *Fannia* and *Musca*. The next stage could have been the invasion of other tissues, either because the infective-stage larva were returned to a wound away from the eye or, more likely, because of migration by the adult nematode. These early tissue-dwelling forms are likely to have caused wounds in the skin that attracted insects, which then became the vectors of their larvae; modern examples of this sort of life cycle exist. The bovine-infesting filarial worm *Parafilaria bovicola* produces subcutaneous nodules that burst and bleed, attracting vectors such as *Musca lusoria* and *M. xanthomelas*, insects that are not normally blood-feeding. Both *Parafilaria multipapillosa* and *Stephanofilaria stilesi* have a similar life cycle, but this time the vector may be a blood-sucking insect such as *Haematobia atripalpis* or *Stomoxys calcitrans*. The next step on the evolutionary path is the abandonment of lesion formation as the larvae come to depend on insects that can readily break the skin to obtain their blood meal. Initially the microfilariae are likely to have stayed in the dermal tissues and to have relied on pool feeders for their transmission. This is still seen in current-day *Onchocerca* spp. with their reliance on simuliid and ceratopogonid flies as vectors. Finally, it is suggested, microfilariae moved to the peripheral bloodstream and became dependent on capillary-feeding insects such as the mosquitoes, a situation seen in the important parasites of humans *Wuchereria bancrofti* and *Brugia malayi*. Molecular phylogenetic studies have yet to cast light on the ancient origins of this group (Xie *et al.*, 1994).

8.4   Parasite strategies for contacting a vector

Vector-borne parasites adopt strategies that enhance their chances of encountering a vector. There are several examples. Blood-sucking insects take small blood meals compared to the total volume of blood in the host animal (Table 8.2). This means that the chances of any single infective stage encountering a vector insect are slim – as high as two million to one against for one example given in Table 8.2. To overcome this problem, vector-borne parasites produce very large numbers of offspring; malaria gametocytes, microfilariae, trypanosomes and arboviruses are all good examples. Failure of some parasites to produce sufficiently high parasitaemias may partly help explain why more of them are not transmitted by insects. For example, human immunodeficiency virus (HIV), the causative agent of acquired immune deficiency syndrome (AIDS), produces a level of viraemia in patients that is estimated to be about six orders of magnitude too low for it to be a candidate for regular mechanical transmission by arthropods (Piot and Schofield, 1986; Webb *et al.*, 1989). In addition to being present in large numbers, the offspring of many vector-borne parasites concentrate at sites that give them the optimum chance of encountering a suitable insect. For example, the gametocytes of both *Plasmodium inui* and *P. yoelii* are not all equally infective to the vector. The younger, slightly larger gametocytes are more infective than the older, smaller forms. It is the younger gametocytes that are preferentially ingested by blood-feeding mosquitoes because they

Table 8.2 *Blood-sucking insects commonly take meals that are only a small proportion of the total blood present in the host animal (the ratio between total blood in the host and size of the insect's blood meal is given). This minimizes the chances of the insect ingesting any individual parasite during feeding. One strategy adopted by insect-borne parasites to overcome this problem is to produce large numbers of infective stages that circulate in the blood of the host.*

Host	Approximate total blood volume in host adult (ml)	Ratio	Vector	Approximate blood meal volume (ml)
Human	4000 (47.9 ml kg^{-1})	2×10^6	*Anopheles*	0.002
Cow	34 200 (57 ml kg^{-1})	1.14×10^6	*Glossina*	0.03
Chicken	140 (95.5 ml kg^{-1})	4.6×10^4	*Culex*	0.003

are concentrated in the peripheral capillary beds of the skin from which the mosquitoes feed (Dei Cas *et al.*, 1980). There is also evidence for concentration of *Plasmodium falciparum* gametocytes in human blood (Pichon *et al.*, 2000).

Many microfilariae are also found concentrated in precise locations in the vertebrate host in order to increase their chances of encountering a suitable vector. *Onchocerca volvulus* microfilariae are limited to the dermis, and consequently they are not ingested by vessel-feeding insects such as mosquitoes, in which they cannot develop; instead, they are ingested by pool-feeding insects such as their blackfly vectors. In another filarial worm, *Parafilaria multipapillosa*, the adults produce nodules beneath the skin in which microfilariae congregate. These nodules burst to produce a bloody wound that is attractive to blood-sucking flies such as the vector *Haematobia atripalpis*. Other microfilariae are found in the blood circulation, from where they are picked up by vessel-feeding vectors. In addition to this general localization at the feeding site of their vector, mosquitoes pick up more microfilariae than is to be expected from the microfilarial density measured in host blood and the size of the blood meal ingested. This suggests a concentration effect is occurring at some level in the system. For example, *Culex quinquefasciatus* feeding on hosts showing the very low microfilarial density of three microfilaria per millilitre of blood ingest 30 times the expected number of parasites. The probable explanation for the apparent anomaly is that microfilaria concentrate in those parts of the peripheral circulation most likely to be lanced by the vector.

Filarial worms also display microfilarial periodicity, which results in the periodic concentration of parasites in the peripheral circulation in order to

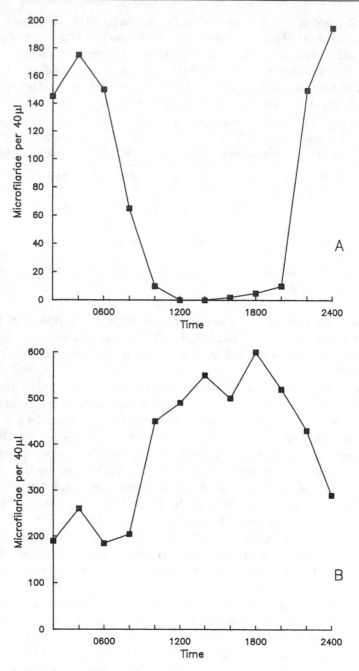

Figure 8.3 Microfilariae of *Wuchereria bancrofti* appear in their largest numbers at the peak biting times of their vectors. The periodic strain (A) is carried by night-biting vectors such as *Culex quinquefasciatus*, and the subperiodic strain (B) by day-biting mosquitoes such as *Aedes polynesiensis* (Hawking, 1962).

Table 8.3 *The microfilariae of many filarial worms display a pronounced periodicity, with microfilarial numbers in the peripheral blood coinciding with the peak biting time of locally abundant vector species.*

Periodicity	Parasite	Major host	Major vector
Nocturnal	*Wuchereria bancrofti* (periodic form)	humans	*Culex quinquefasciatus*
	Brugia malayi	humans	*Mansonia* spp.
	Brugia pahangi	cats	*Mansonia* spp., *Armigeres* spp
	Brugia patei	cats	*Mansonia* spp., *Aedes* spp.
	Dirofilaria corynodes	monkeys	*Aedes* spp.
	Dirofilaria repens	dogs	*Mansonia* spp., *Aedes* spp.
	Loa loa	monkeys	*Chrysops* spp. (night-biting)
Diurnal	*Loa loa*	humans	*Chrysops* spp. (day-biting)
	Wuchereria bancrofti (subperiodic form)	humans	*Aedes polynesiensis*
	Dipetalonema reconditum	dogs	*Ctenocephalides canis*
	Ornithofilaria fallisensis	ducks	*Simulium* spp.

optimize their chances of encountering a suitable vector insect. There are two distinct strains of the filarial parasite of humans *Wuchereria bancrofti*. In the periodic strain, the microfilariae are found in the lung during the day and only at night do they appear in numbers in the peripheral circulation (Fig. 8.3). The appearance of these microfilariae in the peripheral blood coincides with the peak biting time of the major vector (which is *Culex quinquefasciatus* over a large part of the parasite's range). Over a limited part of the parasite's range, particularly in the South Pacific, it is transmitted by day-biting mosquitoes such as *Aedes polynesiensis*. In these areas the parasite is diurnally subperiodic, with microfilariae present in the blood throughout the 24-hour period but showing a clear peak in the afternoon. Similar synchronization between the appearance of microfilariae in the peripheral blood and the peak biting time of the major vector is seen in many other filarial worm–vector associations (Table 8.3). Clearly the appearance of the microfilariae in the blood at particular times of the day is a means of maximizing the chances that the microfilariae will be ingested by a vector insect. But why not stay in the peripheral blood throughout the day? There must be a selective advantage in moving from the peripheral blood. The most likely explanation is that a microfilaria showing periodicity greatly reduces its chances of being ingested by a vector in which it

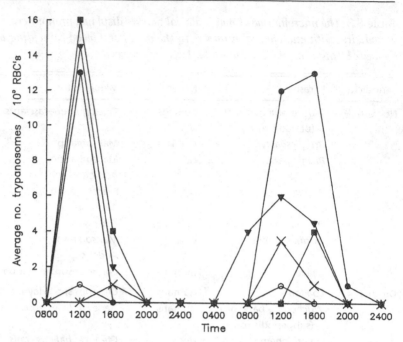

Figure 8.4 The number of trypanosomes found in the peripheral circulation of the frog *Rana clamitans* shows a circadian rhythm. The fluctuations for five separate frogs are shown (Southworth *et al.*, 1968).

cannot develop. It is also possible that outside the peak biting times of their vectors, microfilariae move into a physiologically more favourable part of the body, a move that will increase their longevity and, consequently, their chances of transmission.

Although the rhythmicity of trypanosomes has not been extensively investigated, 24-hour periodicities are also known from this group. Indeed, it has been suggested that periodicity may be widespread among these protozoans (Worms, 1972). In the frog *Rana clamitans*, individuals of the *Trypanosoma rotatorium* complex congregate in the kidney and appear as a flush in the peripheral blood around midday (Fig. 8.4) (Southworth *et al.*, 1968). *Trypanosoma minasense* shows a mid-afternoon peak in the peripheral blood of the Brazilian marmoset. *Trypanosoma lewisi* and *T. duttoni* also show circadian periodicity in rats and mice, respectively (Cornford *et al.*, 1976), with peak parasitaemia occurring soon after dusk. *Trypanosoma congolense* will also display periodicity under laboratory conditions (Bungener and Muller, 1976; Hawking, 1976), although other workers have been unable to show such clear rhythmicity in cattle under field conditions. It would be interesting to know if these peaks enhance the chances of trypanosome transmission to their vectors.

Circadian variation in the infectivity of gametocytes for mosquitoes has been reported for several malaria parasites of birds and mammals (Coatney et al., 1971; Hawking et al., 1972; Hawking et al., 1971; Hawking et al., 1968; Hawking et al., 1966). Periodicity also occurs in Plasmodium falciparum, but its functional significance in the field is still not clear (Bray et al., 1976; Gautret, 2001; Gautret and Motard, 1999; Magesa et al., 2000). The following hypothesis has been proposed to account for the periodicity observed in the laboratory. The period of ripeness of the gametocytes is short (much shorter than 24 hours). The gametocytes arise at schizogony from the same merozoites as the asexual forms, with the timing of peak gametocyte maturity regulated by the timing of schizogony. The timing of schizogony in turn is almost constant for particular species or strains of parasite. In most Plasmodiidae it occurs either at 24-hour intervals or at multiples of 24 hours. This would enable the parasite to produce maximum numbers of mature gametocytes at precise times of the day. Another possible explanation of the circadian variation in infectivity may be the differential availability of gametocytes at various times of the day. This has been reported in a related species, Leucocytozoon smithi, in the turkey (Gore and Pittman-Noblet, 1978); this parasite changes the peripheral distribution of its gametocytes in parallel with diurnal changes in host body temperature (under the influence of hormones and the day–night cycle).

Parasites are also capable of manipulating the behaviour of the vertebrate host to facilitate their transmission to vector insects. The mouse Mus musculus, in common with many other rodent species, shows very efficient anti-insect behavioural defences. In consequence, it is only rarely fed upon by temporary ectoparasites. However, if the mouse is infected with malaria there is a period during and just after peak parasitaemia when the mouse is disabled and unable to defend itself, during which time mosquitoes can feed readily from it (Day and Edman, 1983). Under experimental conditions most mosquitoes feed on these mice about two days after peak parasitaemia, and this coincides with a peak in gametocyte numbers (see Fig. 7.9). So the parasite has modulated the host's behaviour in such a way, and at a precise time, that will maximize the chances of parasite transfer to a vector insect.

Parasites can also affect haemostatic mechanisms in the vertebrate host in ways that enhance their chances of transmission. The longer an insect takes to obtain its meal, the more likely it is to be swatted or disturbed by the host. To minimize host contact time, insects produce various anti-haemostatic factors in the saliva (see Section 5.2). Parasites within the vertebrate host animal can also interfere with host haemostasis. Mosquitoes feeding on mice or hamsters infected with Plasmodium chabaudi or Rift Valley fever virus, respectively, had their probing times reduced by at least one minute (Fig 8.5) (Rossignol et al., 1985). It is also known that tsetse flies can

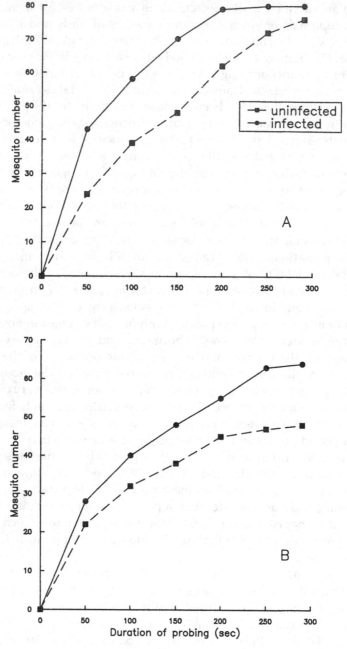

Figure 8.5 Mosquitoes feed more quickly on infected hosts. This can be seen in the trends in probing times observed in groups of mosquitoes variously fed on uninfected hosts (A ■ and B ■), mice infected with *Plasmodium chabaudi* (A ●), or hamsters infected with Rift Valley fever virus (B ●) (Rossignol *et al.*, 1985).

feed more rapidly on trypanosome-infected oxen than uninfected beasts (Moloo *et al.*, 2000). In both cases this would increase the chances of the insect successfully escaping with a blood meal (and the parasite). These and other changes in the vertebrate host blood system caused by vector-borne parasites may well be a parasite strategy to increase its chances of being successfully transmitted by a vector insect (Aikawa *et al.*, 1980; Halstead, 1990; Wilson *et al.*, 1982). The selective advantage to the insect, from the reduced dangers in feeding rapidly, might lead to vectors choosing to feed on infected hosts. Whether this short-term advantage would outweigh the longer-term disadvantage of becoming parasitized is not clear (see below).

Another intellectually attractive proposition is that the fevers so typical of many vector-borne illnesses are caused by the parasite as a means of attracting vectors to the infected host. This possibility has been proposed on many occasions (Gillett and Connor, 1976), but experimental work on malaria-infected mice has shown no significant increase in numbers of feeding mosquitoes when the host is hyperthermic (Day and Edman, 1984). Work on Sinbis virus-infected chickens has shown that there are attractive factors other than heat associated with infected hosts that have not yet been identified. In these experiments, infected chickens held inside cages to which mosquitoes have no access attract more mosquitoes than uninfected controls, despite there being no consistent differences in temperature or carbon dioxide output from the traps (Mahon and Gibbs, 1982). It is possible that these attractive factors are indirectly generated by the parasite to enhance its chances of transmission. Similarly, it is known that trypanosome-infected oxen are more attractive to tsetse flies than uninfected oxen, although the reasons are not known (Baylis and Mbwabi, 1995). A similar situation occurs in *Leishmania*-infected hosts, and it has been shown that there are odours particular to the infected host. If these odours attract more sandflies, clearly that is likely to enhance chances of the parasites' transmission (O'Shea *et al.*, 2002).

8.5   Parasite strategies for contacting a vertebrate host

Parasites may alter the behaviour, particularly the feeding behaviour, of the vector insect to enhance their chances of entering the vertebrate host (Hurd, 2003). In terms of feeding behaviour, development of a mature infection in the vector often leads to increased feeding activity by the vector, enhancing the chances of parasite transmission – examples are given below. In many instances, it will be equally important to the parasite that the vector is not manipulated to undergo increased feeding activity before the parasite has reached maturity and is ready for transmission because that will increase the risk of the vector's death without increasing the chances of parasite transmission.

Table 8.4 *Tsetse flies infected with trypanosomes feed more readily and probe more often than uninfected flies' thereby increasing the chances of parasite transmission (Jenni et al., 1980).*

	Starving for 24 hours			Starving for 48 hours		
	Number feeding	Number feeding on 1st probe	Mean no. probes ± S.E. before feeding	Number feeding	Number feeding on 1st probe	Mean no. probes ± S.E. before feeding
40 infected flies	40	3	5.08 ± 0.40	40	4	4.68 ± 0.37
40 uninfected flies	28	13	1.80 ± 0.21	38	24	1.53 ± 0.13

Infected sandflies have difficulty in feeding, and will often bite a host repeatedly before a blood meal is taken (Beach *et al.*, 1985; Rogers *et al.*, 2002). The phenomenon has been known for many years and is often referred to as the 'blocked fly hypothesis'. The problem occurs because of the blockage of the stomodeal valve by a promastigote-derived secretory gel (PSG) plug. This is a filamentous gel matrix that is partly formed of promastigote proteophosphoglycan (fPPG) (Stierhof *et al.*, 1999). Embedded in it are the majority of the metacyclic promastigote *Leishmania* population in the sandflies, along with leptomonad promastigotes. Parasites are regurgitated into the wound during repeated feeding attempts. This a clear example of a parasite-derived product (fPPG) altering the behaviour of the vector and thus enhancing the chances of transmission of the parasite from the vector insect to the vertebrate host.

In a similar story, the plague bacillus *Yersinia pestis* also interferes with the feeding of its host the plague flea, *Xenopsylla cheopis*. Blockage of the alimentary canal by the parasite leads to the regurgitation of the bacillus into the wound and its consequent transmission (Bacot and Martin, 1914). Early reports suggested that blockage was due to blood coagulation induced by a bacterial plasminogen activator (Cavanaugh, 1971), but this is now known not to be the case (Hinnebusch *et al.*, 1998). The success of midgut blockage decreases with increasing temperature and this explains the pattern of bubonic plague epidemics, which are always abruptly terminated by the arrival of the hot season (mean monthly ambient temperature $> 27\,°C$).

It has been reported that the tsetse flies *Glossina morsitans morsitans* and *G. austeni* will probe the host more frequently and feed more voraciously when they are infected with the salivarian trypanosome *Trypanosoma brucei* (Table 8.4) (Jenni *et al.*, 1980). Similar results have been reported in tsetse infected with *Trypanosoma congolense* (Roberts, 1981). The colonization of the foregut by the trypanosomes may possibly interfere with the feeding process, in particular with the function of those labral mechanoreceptors that are responsible for detecting blood flow rate in the food canal. Also, colonies of trypanosomes in the gut reduce the cross-sectional area of the food canal, which will significantly interfere with the feeding process (see Section 5.6). To ingest the same size of meal the fly probably has to feed longer, or to take more, smaller meals, both of which may help in parasite transmission. Finally, colonies of trypanosomes occurring in the hypopharynx mean that considerably higher pressures have to be applied by the insect to secrete its saliva. This again is likely to increase the chances of trypanosome infection of the vertebrate host. However, the situation is not clear-cut because other workers have been unable to repeat these results (Jefferies, 1984; Moloo, 1983; Moloo and Dar, 1985).

Malaria parasites have also been reported to interfere with the feeding process, increasing the number of feeding attempts made by some (Rossignol et al., 1984; Wekesa et al., 1992) but not all mosquitoes (Li et al., 1992). Infected mosquitoes are more persistent in their biting activities (Anderson et al., 1999) and bite more hosts than uninfected mosquitoes, both of which are to the advantage of the parasite in its attempts to reach a vertebrate host (Koella et al., 1998). Interestingly, increased persistence was seen only when sporozoites were in the salivary glands and ready to be transferred to the vertebrate host. At earlier stages of infection, when increased vector–host contact would raise the chances of death of the vector (and parasites), infected mosquitoes actually showed reduced feeding persistence (Anderson et al., 1999). Disruption of feeding activity occurs this time not because of blockage of the gut, but probably as a result of the sporozoite-induced pathology in the salivary gland and the attendant four- to five-fold decrease in the levels of apyrase produced in the saliva. Without sufficient apyrase the insect has trouble in feeding rapidly (see Section 5.5). This increase in feeding time and/or number of hosts bitten is likely to result in an increase in the number of parasites transmitted. It is quite possible that other parasites that invade the salivary glands, such as *Trypanosoma brucei* or the arboviruses, might also impair the function of the glands, causing a similar increase in the duration of vector–host contact. Reduced feeding success and increased probing times have also been seen in some, but not all, female *Aedes triseriatus* mosquitoes infected with La Crosse virus and in *Aedes aegypti* infected with dengue virus (Grimstad et al., 1980; Platt et al., 1997).

8.6 Vector pathology caused by parasites

Given the brief association of parasite and vector during mechanical transmission, it is not surprising that little, if any, pathological effect is seen. However, considerable damage can be caused during the more extensive associations seen during biological transmission. In the most extreme cases, the ingested parasite can lead to the death of the vector insect. For example, *Rickettsia prowazekii* and *R. typhi* can both cause the death of the louse *Pediculus humanus* (Jenkins, 1964; Weyer, 1960). Normally, increased mortality occurs when the insect has ingested either an unusually large number of parasites, or an uncommonly large number of parasites have survived entry and are developing successfully in the insect. This is well illustrated in the infection of *Aedes trivittatus* with the filarial worm *Dirofilaria immitis*. These mosquitoes can tolerate low numbers of microfilariae, but when they feed on dog blood showing microfilaraemia as high as 347 microfilariae per 20 ml, a significantly increased rate of mortality occurs in the vector population (Fig. 8.6) (Christensen, 1978). Reduced longevity of the vector can also be seen in some malaria infections of mosquitoes (Gad *et al.*, 1979; Klein *et al.*, 1982), although whether this is true for natural vector–parasite combinations is still far from clear (Ferguson and Read, 2002b). Trypanosome infections of salivary glands in tsetse fly, but not infections of the midgut, also reduce vector longevity (Maudlin *et al.*, 1998). Increased mortality in the vector population, particularly if it occurs early, before the infection is mature, is not only harmful for the insect but also to the parasite, reducing overall levels of transmission. Also, if this increased mortality significantly reduces the reproductive success of the vector population, then it may lead to the selection of more resistant strains of the vector species. For these reasons it seems probable that the parasite, in an evolutionary sense, will try to avoid reducing vector longevity.

Parasites also have sublethal effects on vector insects. In particular, they may impair the reproductive potential of the vector (Hurd *et al.*, 1995). This has been shown in laboratory experiments in which infected mosquitoes produce reduced numbers of eggs (Freier and Friedman, 1976; Hacker and Kilama, 1974; Hogg and Hurd, 1995; Maier and Omer, 1973). While this may happen by decreasing the longevity of the vector, as mentioned above, it is likely that more subtle mechanisms are also operating (Hurd, 2003). Invasion of the midgut by ookinetes coincides with resorption of eggs in the ovary caused by apoptosis of the follicle cell (Ahmed *et al.*, 2001; Carwardine and Hurd, 1997; Hopwood *et al.*, 2001). *Plasmodium* infection of the mosquito also leads to reduced fat body production of vitellogenin, the major yolk protein (Ahmed *et al.*, 2001), and subsequent egg batch size and egg hatch rates are significantly reduced (Ahmed *et al.*, 1999). These changes could be caused by the stress and damage associated with infection, but there are a number of other possible explanations. Firstly, it is possible that reproductive success may be impaired because of competition with the parasites for available nutrients. A second possibility is that the vector is deliberately conserving resources for use in the immune response to the parasite (Ahmed *et al.*, 2002; Moret and Schmid-Hempel, 2000). A third

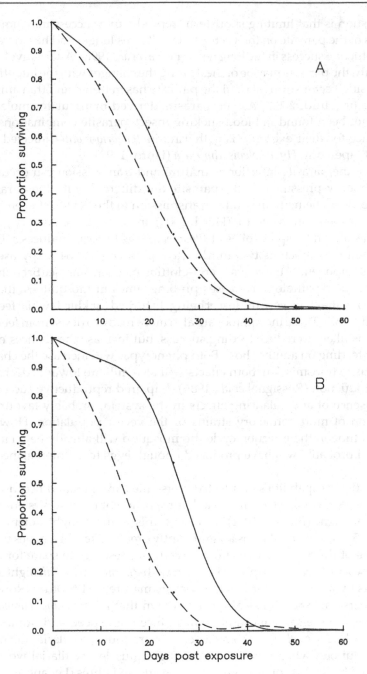

Figure 8.6 Mosquitoes ingesting low numbers of microfilariae (A, broken line) show little increased mortality over uninfected controls (A, solid line), but the ingestion of abnormally high numbers of microfilariae (B, broken line) leads to significant changes (Christensen, 1978).

suggestion is that limiting short-term reproductive success may limit the effects of the parasite on the vector so that it lives longer and has increased reproductive success in the long term (Perrin *et al.*, 1996). Alternatively, and fourthly, the parasite may be orchestrating the changes with the intention of increasing vector survival until the parasite has matured and transmission can occur (Hurd, 2003). As yet, parasite-derived manipulator molecules have not been found in blood-sucking insect–parasite combinations, but evidence for their existence is gathering in *Tenebrio molitor* infected with the rat tapeworm *Hymenolepis diminuta* (Hurd, 1998).

Once the parasite infection is mature and transmission can occur, the evolutionary pressures on the parasite are different, with the parasite's main concern being to maximize transmission to the next host even if this is at the expense of the vector (Hurd, 2003). Invasion of the salivary glands of *Aedes aegypti* by sporozoites of *Plasmodium gallinaceum* reduces salivary apyrase by as much as two thirds (Rossignol *et al.*, 1984). Apyrase is a major component of the salivary anti-clotting mechanisms (see Section 5.5). This salivary pathology increases probing times in mosquitoes, thereby increasing the chances of the insect being disturbed (or killed) while feeding (Gillett, 1967). These mosquitoes are also more ready to desist from feeding, which is likely to reduce feeding success, but increase the chances of the vector feeding on another host. Both phenotypes will increase the chances of parasite transmission; both effects will also militate towards decreased egg production (Rossignol *et al.*, 1986). Impaired reproductive success, in the absence of any balancing effects in the system, probably favours the selection of more refractory strains of the vector population. However, by this time in the infection cycle the mosquito will already be relatively old and probably will have produced enough eggs to counterbalance this effect.

The flight capabilities of infected tsetse flies are impaired because trypanosomes have used a considerable proportion of the reserves normally used for flight (Bursell, 1981). In males, this reduces flight potential by about 15 per cent and there is a significantly greater effect in females which, because of the reproductive effort, have fewer reserves to spare for other purposes. In *Anopheles stephensi* infected with *Plasmodium yoelii*, flight activity falls by about one third as the oocysts mature and rupture (Rowland and Boersma, 1988). It has also been shown that *Plasmodium cynomolgi* B infections in the same mosquito will reduce both speed and duration of flight (Schiefer *et al.* 1977). Similar effects are found in filarial infections of mosquitoes, when invasion of the flight muscles by filarial worms is followed by the loss of glycogen from the affected fibres (Lehane and Laurence, 1977). *Aedes aegypti* parasitized by *Brugia pahangi* fly for a significantly shorter time and will not show the characteristic increase in flight time with increasing hunger typical of uninfected mosquitoes (Fig. 8.7)

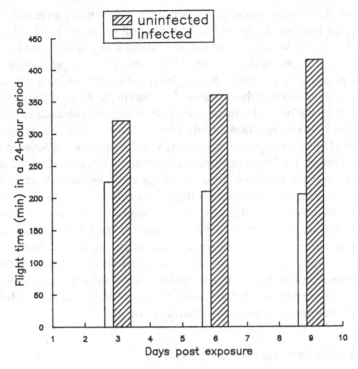

Figure 8.7 Infection of *Aedes aegypti* with the filarial worm *Brugia pahangi* reduces the insect's flight capacity. It also inhibits the gradual build-up in flight time after the blood meal that is characteristic of uninfected flies (Hockmeyer *et al.*, 1975).

(Hockmeyer *et al.*, 1975). Filarial worms interfere with flight at two stages of development (Rowland and Lindsay, 1986). Flight is temporarily diminished during the first two days after feeding but then recovers to normal levels. After this, insects harbouring small numbers of worms show normal flight activity. In mosquitoes harbouring 13 or more filarial worms, flight activity shows a second and dramatic fall 7 to 8 days after the infective blood meal, to about 10 per cent of that seen in control insects. This coincides with the emergence of the infective-stage larvae from the flight muscles and their migration to the mouthparts. Arbovirus infection of mosquitoes can also impair flight activity (Lee *et al.*, 2000). Results on the effects of parasites on vector flight are not all in one direction. In some studies parasitism has been seen to have no effect on flight activity (Wee and Anderson, 1995); in others an increase in the flight range of a parasitized vector has even been reported (Alekseyev *et al.*, 1984).

Common sense suggests that vectors with decreased flight capabilities will be less successful in reaching a host and transmitting the disease. If this is true, it is another good reason for the parasite to try to minimize the damage that it causes to its vector, but it is still too early to come to such hard and fast conclusions about the interactions of parasites and vectors. Under some circumstances a limited amount of vector pathology might be to the parasite's advantage. Taking lymphatic filariasis as a speculative example, it is known that the number of worms in the vertebrate host can be modulated by the host's immune response. Hosts with elephantiasis show the greatest immune reaction and display the lowest number of parasites. The greatest number of parasites is seen in vertebrate hosts displaying little or no pathology. Children of infected mothers are exposed in the womb to filarial antigens and may come to accept these as 'self'. Such children react little to filarial worms when they themselves become infected and so are excellent hosts for the worms. So lymphatic filariasis is a familial infection (Haque and Capron, 1982; Malhotra et al., 2003) in which filarial worms prosper by being transferred to the offspring of infected mothers. In such circumstances there might be a selective advantage in causing limited damage to the flight capacity of the vector because this may localize transmission with the result that transfer is more likely to occur to a tolerant host (P. A. Rossignol, personal communication, 2003).

8.7   Vector immune mechanisms

I have outlined above some of the ways in which parasites can damage the insect, and I have also pointed out selective pressures that may favour the emergence of refractory vectors. At first glance, as with other parasitic diseases, it is hard to see why completely refractory populations of vector do not appear. The fact is they do not, and parasites continue to exist; there must be reasons for this. The most obvious reason may be that insects cannot produce refractory strains against their parasites because as quickly as they adopt new defence mechanisms, the parasites evolve strains capable of avoiding them. Another suggestion is that, under some circumstances, susceptible strains are more fit than refractory ones. For example, any physiologically based defence mechanism will entail energy costs to the vector that will presumably affect other fitness-relevant traits such as growth and reproduction (Ferdig et al., 1993; Koella and Boete, 2002; Lochmiller and Deerenberg, 2000; Schmid-Hempel, 2003; Webster and Woolhouse, 1999). The costs may be associated with actually mounting an immune response to the parasite or merely retaining the capacity to produce the immune response. In most vector populations the rate of infection is normally less than 10 per cent. Hence the low probability of being infected in such populations may mean that adoption of all-out refractoriness is not the best selective strategy and a less radical solution may be more advantageous. For example, there is some evidence that insects show specific immune responses to particular pathogens (Ferrari *et al.*, 2001; Mallon *et al.*, 2003), and such a narrow response to a regularly encountered parasite may be highly efficient. In other instances the value of the immune response to the insect (in terms of reproductive success) may vary according to the insect's age. Here a sensible option may be to alter life history strategies, and this has been demonstrated in the infection of the snail intermediate hosts of *Schistosoma mansoni*. Infection of snails will eventually lead to a reduction in, or possibly complete inhibition of, egg-laying. Snails continuously exposed to infection partly overcome this by a significant increase in the level of egg-laying in the prepatent period (Minchella, 1985; Minchella and Loverde, 1983). This insurance against the possibility of being parasitized has the advantage that if the snail does not become infected it has minimized wastage of resources that could have been used in the reproductive effort. We might make a similar argument for the association between vector and malaria parasites discussed above. The parasites do not deleteriously affect the vector insect until the infective stages are produced. This takes between 8 and 14 days from the infective blood meal. During this time the mosquitoes are likely to have produced two or more egg batches. Given the heavy bias towards young insects in mosquito populations, the production of two to three egg batches may well represent a good reproductive effort; resources put towards refractory mechanisms might be selectively disadvantageous by reducing this early reproductive effort. Questions such as these are beginning to attract the attention of evolutionary ecologists (Schmid-Hempel, 2003), and a much clearer understanding of these relationships can be expected.

A second way in which susceptible strains may be more fit than refractory ones is if a degree of mutualism has evolved between vector and 'parasite'. For example, if being parasitized causes only small losses to the vector then they may be outweighed by any of the following advantages: an incapacitated host is unlikely to damage or kill the vector (see Chapter 7); a parasitized host's haemostatic mechanisms may be impaired by the infection, so decreasing the time necessary for feeding and increasing the chances of successful feeding (see Section 8.5); the viscosity of the host's blood may be lowered by the parasite, this again may reduce the time required to complete feeding (see Section 5.6).

All of the factors above would also have to be viewed in the light of different virulence levels seen in different parasite populations (Ferguson and Read, 2002a; Williams and Day, 2001). Undoubtedly, natural populations of vector also vary in their susceptibility and other physiological responses to parasites. How all these interacting factors combine under field conditions to influence vector and parasite success is a considerable challenge to evaluate. These factors have been discussed (Hurd, 2003) and some attempts have been made to produce mathematical models of these interrelated and counterbalancing effects that deal with the impact either on the vector population (Dobson 1988) or on disease transmission (Kingsolver, 1987; Rossignol and Rossignol, 1988). More basic data and more comprehensive models are required before we will have a clearer picture of these complex interactions.

An area where rapid progress has been made is in understanding the insect immune system itself. Not surprisingly, the insect immune system plays a key role in the relationship of the vector and the parasites it transmits. While some work has been done on the immune system in blood-sucking insects, most of the key work has been performed on model insects, particularly *Drosophila*. Because of the undoubted importance of the insect immune system in vector–parasite interactions, I will combine information from model insects and blood-sucking insects in this section in order to try to give a fuller picture than would be possible by concentrating on blood-sucking insects alone.

Insects possess an immune system that protects them from the potentially damaging effects of biological invaders. To give a broad view of the function of insect defence mechanisms, let us first compare them with the vertebrate immune system, which has been studied in immense detail. Perhaps the overriding feature of the vertebrate system is its ability to step up both the speed and intensity of the immune response on a second and subsequent exposure to a pathogen, a capacity termed acquired immunity. Vertebrate immune mechanisms displaying acquired immunity are characterized by their memory, the specificity of the response and its widespread dissemination and amplification as a result of challenge (Janeway *et al.*, 2001). Like vertebrates, insects clearly have the ability to distinguish self from non-self. They are also capable of amplification of both cellular and humoral responses to infection, and are able to disseminate these responses throughout their bodies. But insects appear to lack the memory and specificity so characteristic of the vertebrate system. (Some evidence for memory in invertebrate immunity is beginning to appear (Kurtz and Franz, 2003).) Although insect defence mechanisms may lack some of the more sophisticated components seen in vertebrates, the tremendous success of insects as a group is proof that their immune mechanisms meet their survival needs. The difference in the degree of sophistication required in the two immune systems is probably explained by the short generation time of most insects where a memory component to immunity would add little to the insect's reproductive success (Anderson, 1986). So what is the insect immune system like?

As well as acquired immunity, vertebrates also have an innate (non-adaptive) immune system that forms the first line of host defence, brought into action immediately following infection (Janeway *et al.*, 2001). Unlike the acquired system described above, which is modified and moulded by the immunological experience of the animal, the innate system relies on hardwired, germline-encoded systems of pathogen recognition and destruction. Insects possess an immune system that has remarkable parallels to the innate immune system seen in vertebrates and similar systems in plants, suggesting innate-type defence systems arose very early in evolutionary history (Hoffmann and Reichhart, 2002; Menezes and Jared, 2002).

Insect innate immunity is a complex, interacting system of several components. It comprises physical barriers, including the cuticle and peritrophic matrix, and cellular components involved in phagocytosis and cellular encapsulation responses. It also has humoral components such as anti-microbial peptides and cytotoxic free radicals; enzyme cascades leading to coagulation or melanization for wound healing and non-cellular encapsulation of invaders; and lectins, which may be important because of their agglutinating activity or may have more subtle recognition and regulatory roles in the immune response. It is certain that the insect immune system is tightly coordinated, and it is probable that it has considerable redundancy built into it. For example, *Drosophila* mutants such as *domino* that do not produce normal haemocytes, *Black cell* that do not perform melanization responses, or *imd* that do not produce normal anti-microbial peptide responses all survive as well as wild-type flies when bacterially challenged in the laboratory. But if another mutation is introduced to the insects so that, for example, they lack both haemocytes and anti-microbial peptides (*domino imd* double mutants) or haemocytes and melanization ability (*domino Black cell* double mutants), then they become highly susceptible to bacterial infection (Braun *et al.*, 1998). Similarly, mutants incapable of producing anti-microbial peptides survive well if the haemocyte response is intact, but if haemocyte function is saturated by overloading the system then the mutant larvae become highly susceptible to bacterial challenge (Elrod-Erickson *et al.*, 2000). These experiments are strong evidence for redundancy and synergy in the insect immune system. We will look at each of the component parts of the immune system using, where possible, examples from blood-sucking insects.

To get inside the body of the insect, invading organisms must cross an epithelial barrier; there are several possible routes. The invader may directly attack the cuticle, perhaps simplifying the task by selecting the thinnest areas such as the intersegmental membranes or the trachea, but for insects, as for other highly organized Metazoa, the intestine is the main site of attack. There are two reasons for this. First, the intestine's role in digestive physiology requires that physically it is a relatively unprotected epithelial layer. Second, most insects ingest many potentially hostile organisms during the course of their feeding activities. Importantly, blood-sucking insects are in an unusual and privileged position here because blood taken directly from the host's circulatory system is largely a sterile food. For

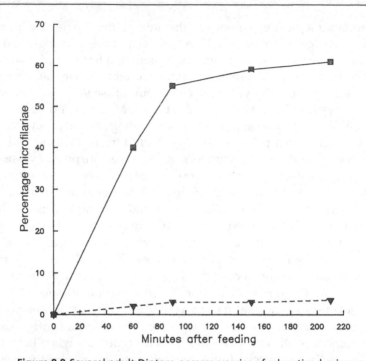

Figure 8.8 Several adult Diptera possess a series of sclerotized spines and teeth in the foregut that are capable of fatally damaging invading microfilariae in the blood meal. The percentage of *Wuchereria bancrofti* microfilariae that successfully migrate from the gut of *Aedes aegypti* (■, which lacks an armature) are compared with *Anopheles gambiae* species A (▼, which possesses a well-developed armature) (McGreevy *et al.*, 1978).

insects that feed exclusively on blood (e.g. tsetse flies, many lice and triatomine bugs) this means that the meal is only rarely a source of potential infection. This will presumably act to lessen the selective pressure ensuring efficient intestinal defences in blood-sucking insects. This may well be an important factor in permitting invasion of the vector by parasitic organisms.

The insect fore- and hindgut are of ectodermal origin and are lined by cuticle. Occasionally this cuticle may be arranged into armatures that help protect the insect against invading organisms (Fig. 8.8). The midgut epithelium is of endodermal origin and does not have a cuticular lining. In most insects the food (and any organisms ingested with it) is still separated from the midgut cells by an extracellular layer known as the peritrophic matrix (Lehane, 1997). The peritrophic matrix can function as a barrier preventing invading organisms ingested with the food coming into contact with the midgut epithelium, although this is not always the case. To appreciate why, we need to look at how the peritrophic matrix is formed.

Table 8.5 *Comparison of the rate of formation of the peritrophic matrix among various mosquito species.*

	Formation (hours)	
Species	First signs	Fully formed
Culex tarsalis	8–12	24
Culex nigripalpus	6	24
Culex pipiens pipiens	12	–
Aedes aegypti	–	4–6
	4–8	12
Aedes triseriatus	0.8	4
Anopheles stephensi	15–20	30
	12	48
Anopheles atroparvus	–	24

Two types of peritrophic matrix are recognized, based on their method of production. Type I peritrophic matrix is formed from secretions of cells along the complete length of the midgut. This is the most common method for the production of peritrophic matrix and is widespread among insects and many other animal groups (Peters, 1968). Type I peritrophic matrix is found in many haematophagous insects, including adult mosquitoes (Fig. 6.2), blackflies, sandflies and tabanids. Type I peritrophic matrix is not present in the hungry insect but is produced in response to the blood meal. Even though relatively few studies have been performed, it is clear that the thickness and rate of development of type I peritrophic matrix is species-specific (Table 8.5). For example, if we compare peritrophic matrix formation in the three temperate simuliids *Simulium equinum*, *S. ornatum* and *S. lineatum*, we see that at 24 hours after the blood meal the mean peritrophic matrix thicknesses are 9.03 mm, 11.7 mm and 18.95 mm, respectively. In *S. ornatum* a third of the final thickness (3.91 mm) is achieved in the first two minutes following the blood meal and by one hour it is 80 per cent (9.25 mm) of its final thickness. By comparison, in *S. equinum* the peritrophic matrix is only about 4.25 mm thick one hour after the meal and in *S. lineatum* only about 40 per cent of the total thickness (8.07 mm) is achieved in one hour after feeding (Reid and Lehane, 1984). Clearly such differences will be important in terms of the defensive barrier effect of type I peritrophic matrix against invaders in the blood meal.

Type II peritrophic matrix is found in all larval Diptera and in adult tsetse flies, hippoboscids and biting muscids. It is produced by an organ known commonly as the proventriculus, but more correctly as the cardium, which is situated at the junction of the fore- and midgut. The type II peritrophic matrix passes backwards down the length of the midgut, forming an unbroken cylinder that contains and separates the food from the midgut epithelium. Type II peritrophic matrix is continuously secreted and is fully formed on leaving the proventriculus (Fig. 6.2). Consequently, unlike the situation with type I peritrophic matrix, there is usually no time during the blood meal when the peritrophic matrix is absent or only partially formed. These fundamental differences in type I and type II peritrophic matrix, ensuring ingested parasites never have direct access to the midgut epithelium in insects with type II peritrophic matrix but are given access to the epithelium in insects with type I peritrophic matrix in the lag phase before the peritrophic matrix is produced following the blood meal, may be crucial in determining which insects are vectors and which are not. For example, biological vectors of arbovirus possess a type I peritrophic matrix that allows the virus to access midgut epithelial cells directly; they are not transmitted biologically by insects with a type II peritrophic matrix in which virus particles cannot directly contact the midgut epithelium.

Clearly, to permit digestion and subsequent absorption of the meal, while retaining the bulk of the meal within the endoperitrophic space, the peritrophic matrix must be a semi-permeable filter. On the basis of experimental results on isolated peritrophic matrix preparations, we estimate the pore size in the type II peritrophic matrix of the adult tsetse *Glossina morsitans morsitans*, to be about 9 nm, making the peritrophic matrix permeable to globular molecules of up to a molecular weight of about 150 kDa (Miller and Lehane, 1990). It has been suggested that 'pore' size in some types of type I peritrophic matrix may be much greater (200 nm in *Locusta* (Peters et al., 1973)).

So, there are two ways in which the peritrophic matrix may fail to be an effective physical barrier preventing invading organisms contacting midgut cells. The first occurs when the pores of the peritrophic matrix are larger than the invading organism. Clearly the 9 nm pores seen in the peritrophic matrix of the tsetse fly are too small to allow the free passage of any potential invaders, but the 200 nm pores reported in some types of type I peritrophic matrix would allow the free passage of arboviruses while denying entry to bacteria and eukaryote organisms. Second, as stated above, the type I peritrophic matrix may fail to be an effective physical barrier when it is not in position when the blood meal is taken. The more rapidly the peritrophic matrix is produced, the shorter this window of contact between the blood meal and the midgut epithelium will be. As noted above, the speed at which the peritrophic matrix is produced is species-specific and highly variable (Table 8.5).

Even when it is not an absolute barrier to infection, the peritrophic matrix can still be a factor limiting the intensity of infection. For example, microfilariae ingested along with the blood meal must penetrate the midgut of blackflies within about the first four hours, or they become trapped (Eichler, 1973; Laurence, 1966; Lewis, 1953). Less than 50 per cent of the microfilariae are normally successful in making this migration in simuliids and the rest die in the blood meal. Microfilariae ingested by mosquitoes are usually more successful and up to 90 per cent escape from the gut (Ewert, 1965; Zahedi, 1994). To achieve this they also leave the midgut quickly, beginning their migration within minutes of the blood meal being taken. Usually, over 50 per cent have left within two hours (Laurence and Pester, 1961; Wharton, 1957). It is widely speculated, but not yet clearly proven, that it is the developing type I peritrophic matrix that is the most significant barrier, although other factors such as the progressive gelling or clotting of the blood may also be important (Kartman, 1953; Sutherland et al., 1986).

The developing type I peritrophic matrix of susceptible mosquitoes may also be the factor limiting the numbers of successfully migrating malaria ookinetes, even though these ookinetes can secrete chitinase to help them to break down the peritrophic matrix (Shahabuddin, 1998; Sieber et al., 1991). Malaria gametocytes, ingested with the blood meal, have to undergo a maturation period before they develop into ookinetes capable of migrating through the midgut wall. During this maturation period, the mosquito's digestive cycle begins, including the development and thickening of the type I peritrophic matrix. If the gametocyte maturation period is side-stepped by the feeding of already matured *Plasmodium berghei* ookinetes to *Anopheles atroparvus*, then the number of parasites successfully invading the midgut increases greatly (Janse et al., 1985). The simplest explanation of these results is that more of these ookinetes penetrate the midgut because, at this earlier time, the peritrophic matrix is either absent or thinner. The view that the mosquito peritrophic matrix is a barrier to *Plasmodium* is supported by other experiments. Type I peritrophic matrix of *Aedes aegypti* can be artificially thickened, and this significantly reduces the number of *Plasmodium gallinaceum* ookinetes successfully penetrating the gut (Billingsley and Rudin, 1992; Ponnudurrai et al., 1988). Also supporting the idea that the peritrophic matrix is a barrier are the observations that *Plasmodium berghei* can infect *Anopheles atroparvus* but *Plasmodium falciparum* cannot. One explanation for this is found in the fact that *P. berghei* ookinetes develop more rapidly than those of *P. falciparum* and can enter the midgut before the peritrophic matrix of the mosquito forms. In contrast, both parasites can infect *Anopheles stephensi* because the peritrophic matrix of this mosquito takes much longer to be produced, by which time ookinetes of both species will have matured (Ponnudurai et al., 1988).

Organisms that are successful in crossing the cuticular/peritrophic matrix barrier and the outer epithelial barrier of an insect are then presented with the defensive mechanisms of the insect's blood system. Insect blood is contained in an open haemocoelic space that is not lined by endothelium, so the organs of the body are separated from the blood only by the basement membranes of their own cells. The key event in immunity is the ability to recognize foreignness; because of its position the basement membrane lining the haemocoel probably plays a key role in self / non-self recognition in insects. In innate immunity, foreignness is recognized through pattern recognition receptors (PRRs). These will bind to pathogen-associated molecular pattern (PAMP) molecules. The PAMPs clearly recognized to date are, not surprisingly, the exposed wall components of pathogens. PAMPs include lipopolysaccharide (LPS) and peptidoglycans from bacteria and β-1.3 glucans from fungi. Two types of PRR are believed to occur in insects, those that are soluble in the haemolymph and those that are associated with cell membranes; some PRR may exist in both forms. It is already known that the family of peptidoglycan recognition proteins (PGRPs) are insect PRR (Choe et al., 2002; Gottar et al., 2002; Kurata, 2004; Michel et al., 2001; Ramet et al., 2002b). PGRP–LC, PGRP-LA and PGRP-LE appear to be humoral receptors acting upstream of signalling pathways controlling expression of anti-microbial peptides (see below and Fig. 8.9). PGRP-LC is also involved in phagocytosis. PGRP-LE is particularly involved in epithelial immune responses and also in the upregulation of the prophenol oxidase cascade (see below) (Kurata, 2004). There are 12 and 7 PGRP genes in the Drosophila and Anopheles genomes, respectively. If all are PRR and have different selectivities, then these PRR alone could provide a broad spectrum of recognition capabilities. As might be predicted from their PRR function PGRP genes are expressed in the most appropriate sites from an immunity point of view, the haemocytes, fat body, midgut and cuticular epithelium (Werner et al., 2000). In addition to PRR the insect also contains opsonins, which are molecules capable of marking an object as foreign. Many molecules have been suggested to be potential opsonins, including lectins, hemolin and LPS-binding protein (Lavine and Strand, 2002; Lehane et al., 2004). The best evidence is available for the complement-like protein αTEP1, which binds to the surface of bacteria and is required for optimal phagocytosis of bacteria in vitro and in vivo (Levashina et al., 2001) (Levashina, personal communication, 2001). In addition, binding of the same molecule to Plasmodium mediates parasite killing in Anopheles mosquitoes (Blandin et al., submitted).

Recognition of foreignness will trigger a humoral and/or cellular immune response in the insect. Humoral responses in the insect include the production of a range of anti-microbial peptides (AMP). These are largely made by the fat body, but are also produced by epithelial surfaces such as the insect gut (Lehane et al., 1997; Tzou et al., 2000). In Drosophila there are seven families of anti-microbial peptide that can be grouped into three functional groups: (1) drosomycin and metchnikowin show antifungal

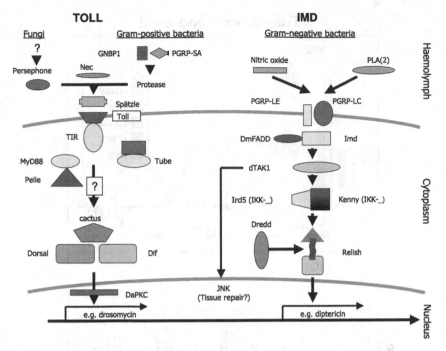

Figure 8.9 The molecular components involved in Toll and Imd, the two major signalling pathways in the *Drosophila* immune response (De Gregorio *et al.*, 2002), modified from Tzou *et al.* (2002) and Hoffmann (2003). As presented, the figure suggests an exclusive gram-negative and gram-positive response pathway; in reality the eventual picture is likely to be more complex (Hoffmann, 2003; Leulier *et al.*, 2003) with considerable crosstalk. In the Toll pathway fungal challenge leads to the activation of the serine protease persephone, which cleaves Spätzle (Ligoxygakis *et al.*, 2002). The pattern recognition receptor for fungi is unknown. The pattern recognition receptor for gram-positive bacteria is PGRP-SA (Michel *et al.*, 2001) and GNBP1 (Hoffmann, 2003), but the protease-cleaving Spätzle is unknown. Spätzle-cleaving proteases can be inhibited by the serpin Necrotic (Levashina *et al.*, 1999). Cleaved Spätzle binds Toll via a leucine-rich domain. Intracellularly a receptor–adaptor complex is formed comprising Toll, the kinase, Pelle and the death domain containing proteins MyD88 and Tube. Via an unknown kinase, this complex leads to the disassociation of the ankyrin domain protein Cactus from the NFκB class proteins Dorsal and Dif, which then enter the nucleus. The atypical kinase DaKPC works within the nucleus refining the transcriptional activity of these NFκB proteins (Avila *et al.*, 2002) that regulate transcription of a large number of genes (Irving *et al.*, 2001). In the Imd pathway the pattern recognition receptor can be PGRP-LC (Choe *et al.*, 2002; Gottar *et al.*, 2002; Ramet *et al.*, 2002b) or PGRP-LE (Takehana *et al.*, 2002). There is evidence that PGRP-LC is membrane-spanning. The response of the Imd pathway is influenced by nitric oxide (Foley and O'Farrell, 2003) and the phospholipase A(2)-generated fatty acid cascade (Yajima *et al.*, 2003). Unusually, Relish is endoproteolytically cleaved, probably by DREDD (Stoven *et al.*, 2003) under the influence of Ird5 and Kenny (Silverman *et al.*, 2000), to achieve its activation. JNK can be activated downstream of Imd, possibly regulating genes involved in tissue repair (Boutros *et al.*, 2002). Detailed comparison of *Drosophila* and *Anopheles* immunity genes has already been performed (Christophides *et al.*, 2002), and a comparison of the *Drosophila* genes and *Glossina morsitans morsitans* midgut EST has been published (Lehane *et al.*, 2004).

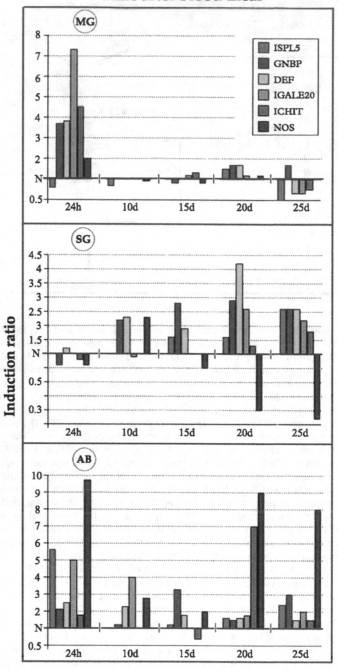

Figure 8.10 Estimated induction levels of immune marker genes in salivary glands (SG), midguts (MG) and abdominal wall tissues (containing ovaries and Malpighian tubules) (AB) of malaria-infected *Anopheles gambiae* 24 hours, 10, 15, 20 and 25 days after feeding on an infected mouse, as compared with mosquitoes fed on naive mice. It is clear that induction of immunity genes occurs as the particular tissues become infected with malaria parasites (Dimopoulos *et al.*, 1998).

activity; (2) diptercin, attacin, cecropin and drosocin are mainly active against gram-negative bacteria; (3) defensin, which is mainly active against gram-positive bacteria (Bulet et al., 1999). In Anopheles gambiae there are three families of AMP: defensins, cecropins and gambicin (Dimopoulos, 2003). The potential broad spectrum of activity of insect AMP is illustrated by the fact that one mosquito, An. gambiae, like Drosophila, uses a defensin family molecule against gram-positive bacteria, while another mosquito, Aedes aegypti, uses a defensin family molecule against gram-negative bacteria (Shin et al., 2003).

The recognition system in insects is sophisticated enough to ensure that upregulation of particular AMP can be specifically tailored to the particular pathogen challenge received. So a gram-negative bacterium infection in Drosophila results in upregulation of diptericin, attacin, cecropin and drosocin genes, while coating the fly with fungal spores specifically upregulates anti-fungal AMP (Lemaitre et al., 1997). Remarkably, the two signalling pathways, termed Toll and Imd (Fig. 8.9), which lead to upregulation of these AMP genes in Drosophila, bear a striking resemblance to pathways regulating innate immune responses in mammals and plants, indicating a very ancient evolutionary origin for these responses (Hoffmann and Reichhart, 2002).

AMP may play a role in anti-parasite activities in insects. AMP genes are upregulated in vector insects in response to parasite infection (Dimopoulos et al., 2002; Hao et al., 2001; Lowenberger et al., 1996; Richman et al., 1997). However, it is not always clear this is entirely a response to the parasite as many parasites, in entering the body of the insect, will cause physical injury and allow ingress of other pathogens (e.g. microfilaria crossing the insect midgut permit invasion of arboviruses (Turell et al., 1984b)), which may also upregulate immune responses. Some evidence suggests AMP may influence the outcome of the parasitic infection (Boulanger et al., 2002; Hao et al., 2001; Lowenberger et al., 1996; Shahabuddin et al., 1998), but not all AMP are necessarily involved, as gene knockout studies suggest that defensin 1 does not play a major anti-Plasmodium role in Anopheles gambiae (Blandin et al., 2002).

Other humoral factors are also important in the insect immune system. Lectins are proteins or glycoproteins that recognize and bind specific carbohydrate moieties. Because of this capability, they are able to agglutinate cells or precipitate complex carbohydrates. Lectins, with a range of carbohydrate specificities, are widespread in insects and are immune-inducible (Komano et al., 1980). It is unlikely, under most circumstances, that the number of invaders will be sufficiently high for lectins to exert defensive control by agglutinating them. Instead, it seems more probable that lectins act as opsonins, binding invader and receptors on the blood cell as a prelude to phagocytosis or encapsulation (Wilson et al., 1999; Yu and Kanost, 2000), or as inducers of apoptosis in the pathogen (Pearson *et al.*, 2000). Lectins are also found in the insect intestine in triatomine bugs (Pereira *et al.*, 1981) and tsetse flies (Ibrahim *et al.*, 1984). It has been suggested these gut-based lectins may also have a defensive function (Maudlin and Welburn, 1987; Peters *et al.*, 1983). Other humoral defence mechanisms are known to occur in the haemolymph, although their biochemical basis is not understood. For example, the cell-free tsetse fly haemolymph contains a non-inducible, temperature-sensitive factor that is capable of inactivating some trypanosomes (Croft *et al.*, 1982; East *et al.*, 1983). This factor shows species-specificity as it is active against trypanosomes that tsetse flies commonly encounter, such as *Trypanosoma brucei*, *T. vivax* and *T. congolense*, but it is not active against trypanosomes 'unusual' to tsetse flies, such as *T. dionisii* from bats. As such, it provides evidence for the evolutionary 'fine tuning' of the immune system to meet species-specific needs.

The insect haemolymph also contains a variable number of cells known collectively as haemocytes. As well as circulating haemocytes, insects may also have accumulations of fixed cells that take part in the defence responses (Hillyer and Christensen, 2002; Kaaya and Ratcliffe, 1982). No clear-cut classification of insect haemocytes is yet available, although attempts at generalization have been made (Brehelin, 1986). The blood cells of *Aedes aegypti* have been classified on the basis of morphology, lectin binding, enzyme activity and histochemistry into four types (Hillyer and Christensen, 2002). To illustrate the problems in the classification of blood cell types, these four cell types are compared in Box 8.1 to other insect blood cell types previously described. Attempts are now being made to classify blood cells on the basis of antibody binding and genetic markers (Gardiner and Strand, 1999; Lebestky *et al.*, 2000). The latter approach in particular holds much promise for sorting out the relationships of haemocytes in different taxa (Lavine and Strand, 2002).

The cells of the insect blood system are capable of a range of defensive responses. Phagocytosis of bacteria, yeasts, fungi, viruses, protozoa and apoptotic bodies occur in insects. In *Drosophila* the major cell type involved in phagocytosis is the plasmatocyte (Elrod-Erickson *et al.*, 2000; Meister and Lagueux, 2003). Phagocytosis occurs after the pathogen is bound either directly, or via an opsonin, to a receptor on the phagocyte surface. The complement-like protein αTEP1 of *Anopheles gambiae* is just such an opsonin, capable of enhancing phagocytosis of some gram-negative bacteria (Levashina *et al.*, 2001). After the pathogen-containing vacuole, the phagosome, forms it fuses with other vesicles that are believed to contain a variety of pathogen-killing molecules and a mature phagolysosome is formed. The signalling pathways involved in this process may be highly conserved in the animal kingdom (Soldatos *et al.*, 2003). In mammalian systems the killing factors in the phagolysosome are commonly reactive

Box 8.1 Four blood cell types characterized in *Aedes aegypti* are compared to haemocytes described in previous studies on a variety of insects (Hillyer and Christensen, 2002).

AEDES AEGYPTI haemocyte types	granulocyte	oenocytoid	thrombocytoid	adipohae-mocyte
Previous classifications				
granulocyte	(Butt and Shields, 1996) (Ochiai et al., 1992) (Akai and Sato, 1973)			
plasmatocyte	(Ashida et al., 1988) (Ksiazkiewicz-Ilijewa and Rosciszewska, 1979) (Wigglesworth, 1979) (Brehelin, 1982)			
oenocytoid		(Brehelin et al., 1978) (Ashida et al., 1988) (Butt and Shields, 1996) (Wigglesworth, 1979)		
thrombocytoid		(Ratcliffe and Rowley, 1979)	(Brehelin et al., 1978)	
coagulocyte	(Brehelin et al., 1978)			
nephrocyte			(Kaaya and Ratcliffe, 1982)	
proleukocyte		(Akai and Sato, 1973)		

oxygen intermediates (ROI) and reactive nitrogen intermediates (RNI). There is evidence for ROI and RNI use in insect cells (Nappi *et al.*, 2000; Whitten and Ratcliffe, 1999) and in haemolymph (but not specifically the blood cells) of mosquitoes (Luckhart and Rosenberg, 1999; Luckhart *et al.*, 1998).

Encapsulation is another possible immune response to pathogen entry. The capsules enclosing the pathogen are often formed by haemocytes, and these cellular capsules may eventually become melanized. Non-cellular melanized capsules also occur. A primary difference between cellular and non-cellular encapsulation is the speed at which they can occur. While cellular encapsulation may take anything from a few hours to several days to complete, humoral encapsulation may be achieved in 10–30 minutes (Gotz, 1986). In *Drosophila* the haemocytes involved in cellular encapsulation are lamellocytes, which are produced in large numbers from haemopoietic tissues following an appropriate challenge such as parasite invasion of the body (Meister and Lagueux, 2003). Cellular encapsulation occurs when haemocytes adhere to a pathogen in sufficient numbers to completely surround it. This normally involves a change in the 'stickiness' and 'spreading ability' of circulating haemocytes that can be brought about by cytokines such as the 25 amino acid peptide GBP and the 23 amino acid PSP (Lavine and Strand, 2002; Matsumoto *et al.*, 2003). These cytokines stimulate the export to the surface of the haemocyte of adhesive molecules previously held intracellularly in granules (Strand and Clark, 1999). There is circumstantial evidence for the involvement of integrins, well known in mammalian systems, in such a role in the adhesion process. The capsule stops increasing in size when a basement membrane-like covering appears over its surface and other haemocytes no longer attach to the surface of the capsule (Liu *et al.*, 1998). Presumably the basement membrane makes the surface of the capsule look like all the other surfaces bathed by the haemolymph; in other words, the basement membrane-covered capsule looks like 'self'. Inside the capsule the pathogen dies – possibly by asphyxiation, poisoning by its own waste products, through the actions of ROI, RNI or AMP, or through starvation (Chen and Chen, 1995). If the capsule becomes melanized, which is often the case, toxic quinones may also help kill the invader (Lavine and Strand, 2002). Filarial worms are commonly encapsulated in insects, either in cellular capsules (Liu *et al.*, 1998) or non-cellular capsules (Chen and Laurence, 1985; Chikilian *et al.*, 1994), both of which may be melanized. Similarly, *Plasmodium* oocysts can be enveloped in a non-cellular melanized capsule when they form on the outer surface of the midgut of a refractory strain of *Anopheles gambiae* (Paskewitz *et al.*, 1988).

Encapsulation is not limited to parasites in the haemocoel; it can also be used as a defence against intracellular parasites. The mosquito *Anopheles labranchiae atroparvus* is resistant to infection with the filarial nematode *Brugia pahangi*. Although the first-stage larvae can become established in the flight muscles of the mosquito, these intracellular parasites are soon enveloped in a melanizing capsule without the apparent involvement of haemocytes (Chikilian *et al.*, 1995; Lehane and Laurence, 1977). *Dirofilaria immitis* in Malpighian tubules can suffer a similar fate (Mahmood, 2000).

Melanization is a key component of many encapsulation reactions and is the end result of a multi-enzyme pathway known as the prophenol oxidase (proPO) cascade that results in the conversion of tyrosine to melanin. The central enzyme in this system is phenoloxidase, which oxidizes phenols to quinones, which then polymerize to melanin (Soderhall and Cerenius, 1998). The phenoloxidases are usually present in an inactive pro-form in the insect and are probably activated by a cascade of serine proteases following pathogen recognition by PRR or other appropriate challenges to the insect. This system is very tightly regulated because uncontrolled activation of the pathway would seriously damage or kill the insect, as a result of the toxic nature of some of the products of the cascade or of the danger of the wholesale melanization of the haemolymph space. Regulation is aided by the fact that components of the activated proPO cascade seem to form aggregates in discrete locations such as at the surface of a pathogen or wound site. For example, they seem to form on and adhere to parasite surfaces, hence the sequence of melanization seen in cellular encapsulation that proceeds outwards from the invader's surface. This may be a reflection of the activation of the cascade by components attached to PAMPs on the pathogen's surface. Another element of the tight control is seen in *Drosophila*, in which components of the proPO system are contained in crystal cells and only released from them on receipt of an appropriate stimulus (Meister and Lagueux, 2003).

There is a coagulation system in insects that depends upon humoral and haemocyte factors, and in addition haemocytes can help form a clot at wounding sites. Melanization is part of this wound-healing system and loss of crystal cells in mutant *Drosophila* impairs wound sealing and repair (Ramet *et al.*, 2002a; Royet *et al.*, 2003). Relatively little is known of these processes in insects, although similar systems have been very extensively studied in crustaceans and chelicerates (Iwanaga, 2002).

If a parasite is to develop successfully in a vector insect, it must either avoid triggering the insect's immune response (i.e. escape recognition) or be able to suppress or withstand its effects. It may escape recognition by having inherent surface properties that mimic those of the host, by acquiring material from the host that enables it to masquerade as host tissue ('self') or by becoming intracellular. *Aedes trivittatus* is refractory to infection with the filarial nematode *Brugia pahangi*, but, if the microfilariae are allowed to penetrate the midgut of a susceptible strain of *Aedes aegypti*

Table 8.6 *The melanization response to subsequent challenge of infected and uninfected* Aedes aegypti, *as shown by the intrathoracic injection of specific microfilariae (mff) that normally induce a strong melanization reaction.*

	No.	Challenge	Percentage mff melanized ± S.E.
Infected with *Brugia*	63	*Brugia pahangi* mff.	38.2 ± 6.3
pahangi	17	*Dirofilaria immitis* mff.	51.2 ± 6.6
Uninfected	68	*Brugia pahangi* mff.	64.5 ± 6.5
	20	*Dirofilaria immitis* mff.	71.3 ± 3.9

Christensen and LaFond, 1986.

before being intrathoracically injected into *Ae. trivittatus*, between 31 per cent to 43 per cent of them will avoid encapsulation and melanization in the haemocoel (Lafond *et al.*, 1985). Clearly something happens to change the properties of the parasite in its migration across the midgut of the susceptible mosquito. The first suggestion was that the parasite absorbs mosquito-derived material onto its surface and hides behind this coat, but attempts to demonstrate such adsorption, using indirect fluorescent antibody techniques, have failed. A second suggestion centres on the loss of the parasite's high electronegative surface charge as it migrates through the midgut of *Ae. aegypti* (Christensen *et al.*, 1987). This change in charge may account for the reduction in immune response observed, but conclusive evidence remains to be gathered. A parasitic strepsipteran insect is able to avoid the insect immune response by causing itself to be covered in a host-derived epithelial layer with the basement membrane of the epithelial layer facing the haemolymph of the host, which effectively makes the epithelium-enveloped parasite 'self' as far as the insect host's immune system is concerned (Kathirithamby *et al.*, 2003).

An alternative strategy is not to pose as self but to suppress the insect's immune response. Parasitoid wasps are known to be able to suppress the insect immune system (Strand and Pech, 1995). Suppression may be a sensible strategy for parasitoids, which do not require a fit host for their transmission. It would seem a more dangerous strategy for long-lived parasites that require a fit vector for their successful transmission because infections attendant on immunosuppression will reduce fitness (Lackie, 1986). Nevertheless, some parasites do depress the immune function of their vector insects. For example, suppression of the insect immune system is suggested by the decreased xenograft rejection seen in *Triatoma infestans* infected with *Trypanosoma cruzi* (Bitkowska *et al.*, 1982). Also, developing microfilariae and *Plasmodium* both actively suppress the melanizing immune response of the mosquito (Table 8.6) (Boete *et al.*, 2002; Christensen and LaFond, 1986). The effects of the immunosuppression are minimized in both these latter cases, being spatially limited in the case of the filaria and limited to the early stages of ookinete formation in the *Plasmodium* example. This compromise may avoid the dangers of more general immunosuppression mentioned above. From these data, it seems possible that immune suppression of the insect, over the short time period that parasites need for transmission, may be an acceptable risk for the parasite.

The genome-sequencing projects will greatly accelerate discoveries in insect innate immunity and in insect parasite interactions. For example, annotation of the *Anopheles gambiae* genome characterized 242 immune genes, including 18 gene families (Christophides *et al.*, 2002). In addition the use of microarrays identified 200-plus genes that are immune-responsive (Dimopoulos *et al.*, 2002). Immune systems in tsetse flies are also being investigated on a similar broad scale (Lehane *et al.*, 2003). Understanding the interrelated roles of these gene products in immune reactions in *Anopheles* and other vector insects is a major challenge, but one on which very rapid progress can be expected.

9

The blood-sucking insect groups

This section of the book gives an outline of the major groups of insect that feed on blood, concentrating on those groups that are habitual blood feeders. Detailed coverage of each blood-sucking group is not attempted; that would far exceed the space available in a book such as this. Instead, this chapter has been written as a quick reference section for those new to medical and veterinary entomology and for those who need a quick outline of a particular group in order to make most use of other parts of the book. A basic outline of each assemblage is given, covering the essential details of the group's medical and veterinary importance, morphology, life history and bionomics.

Insects are individuals, and even within a single species insects often do things in subtly different ways from each other; this is, after all, a prerequisite for evolution. Consequently, generalizing about a genus, family and whole class of insects is an imprecise business, but bearing in mind that there are many exceptions to these generalizations, this approach can still be a good way of introducing typical group characters. In discussing each group, quantitative values are often given for fecundity, duration of each stage in the life cycle, longevity, etc. These figures are good guidelines to expected values under optimum conditions, but in the field one can expect a good deal of variation.

Before looking at each group separately, we need to briefly look at how they fit into a general classification scheme and to explain some of the rules and conventions by which insects are named.

9.1   Insect classification

The Insecta are the dominant terrestrial class in the extremely successful phylum Arthropoda. The taxonomy of the group is the subject of much debate. A very traditional view of the group would divide the class Insecta into two subclasses: Apterygota (wingless) and Pterygota (winged). There are 29 orders of insects (McGavin, 2001) and most of these (including all blood-sucking species) fall into the Pterygota (Table 9.1). This subclass is further subdivided into two divisions: Exopterygota (Hemimetabola), in which the young (often called nymphs) resemble the parents except that

Table 9.1 *The groups of insect. Those groups containing blood-sucking insects are shown in bold.*

Phylum	Class	Order	Common name
Arthropoda	Insecta	Hymenoptera	Sawflies, wasps, bees, ants
		Trichoptera	Caddisflies
		Lepidoptera	Butterflies and moths
		Mecoptera	Scorpionflies
		Siphonaptera	Fleas
		Diptera	Flies, mosquitoes
		Coleoptera	Beetles
		Strepsiptera	Strepsipterans
		Neuroptera	Lacewings and antlions
		Megaloptera	Alderflies
		Raphidioptera	Snakeflies
		Thysanoptera	Thrips
		Hemiptera	Bugs, aphids
		Psocoptera	Booklice and barklice
		Phthiraptera	Lice
		Embioptera	Web spinners
		Plecoptera	Stoneflies
		Zoraptera	Angel insects
		Mantodea	Mantids
		Blattodea	Cockroaches
		Isoptera	Termites
		Dermaptera	Earwigs
		Grylloblattodea	Rock crawlers or ice crawlers
		Phasmatodea	Stick and leaf insects
		Orthoptera	Grasshoppers, crickets, locusts
		Odonata	Dragonflies and damselflies
		Ephemeroptera	Mayflies
		Thysanura	Siverfish
		Archaeognatha	Bristletails

they lack wings and are smaller; and Endopterygota (Holometabola), in which the young (usually called larvae or pupae) do not resemble the adults and in which the transformation from the larval configuration to the adult occurs in a quiescent stage, called the pupa.

Orders of insects may be subdivided into suborder, division, superfamily, family, subfamily, tribe, genus, subgenus, species and subspecies. In most everyday work, order, family, genus and species are the most common and useful groupings. To help with the complexities of taxonomy, the fourth edition of the *International Code of Zoological Nomenclature* (ICZN, 1999) lays down certain articles and opinions about the naming of these taxonomic compartments. Most orders of insects end in the term '-ptera', all superfamilies in the term '-oidea', all families in the term '-idae', all subfamilies in the term '-inae' and tribes in '-ini'. Species are indicated by the generic followed by the specific name, e.g. *Rhodnius pallescens* (usually abbreviated to the form *R. pallescens* after one full citation). In the scientific literature, especially when naming a little-known species, it is good practice on the first citation of a species to go further and to include the name of the author of the original taxonomic description, e.g. *Rhodnius pallescens* Barber. Rarely, the date of the original description is also included, so we get *Anopheles labranchiae* Falleroni, 1927. The appearance of the author in brackets after the specific name, such as *Triatoma infestans* (Klug), indicates the generic name has been changed since the naming and original description of the species by that author. Subspecies are identified by a trinomial, such as *Glossina morsitans centralis*. In the case of species complexes (see Section 3.2) such as the *Simulium damnosum* complex, another convention has been adopted. Because *Simulium damnosum* refers to a species as well as the complex as a whole, it is often necessary to distinguish between the use of the term in the two instances. This is done by appending the terms *sensu lato*, usually abbreviated to *s.l.*, when the name refers to the complex and *sensu stricto*, usually abbreviated to *s.s.* or *s.str.*, when the name refers to the species.

9.2   Phthiraptera

The order Phthiraptera contains the lice, of which about 4000 species have been described. The traditional classification divides Phthiraptera into three suborders, Mallophaga, Anoplura and Rhynchophthirina, and I will use this convenient classification here. Many authorities now raise Mallophaga and Anoplura to the status of order and then separate Mallophaga into three suborders, Amblycera, Ischnocera and Rhynchophthirina. Others suggest that Anoplura, Amblycera, Ischnocera and Rhynchophthirina be considered as four suborders of the Phthiraptera.

The Anoplura are the smallest suborder and they contain the sucking lice, all of which are blood-feeding parasites of eutherian mammals; most (67 per cent) are found on rodents. Although anoplurans are haematophagous, only a few Mallophaga regularly feed on blood. To be more specific, the Ischnocera do not as a rule feed on blood, although it may be taken if readily available on the skin's surface, the Amblycera may take blood as a regular component of the diet; and the two species in the third suborder, the Rhynchophthirina, are both blood feeders.

Three species of lice are regularly found on people: *Pediculus humanus* (= *P. humanus humanus* = *P. corporis*), the body louse; *P. capitis* (= *P. humanus capitis*), the head louse; and *Pthirus pubis*, the crab or pubic louse. There is some confusion concerning the status of the first two because they interbreed under laboratory conditions, but there is little evidence of this occurring naturally.

As well as the psychological damage humans may suffer as a result of the stigma associated with louse infestation, they also run the risk of physical damage. Individuals heavily infested with lice commonly show allergic responses to saliva injection or to the inhalation of louse faeces. Such individuals literally feel 'lousy' as a result of their lice, possibly because of sleep disturbance or long-term exposure to debilitating salivary antigens. There is evidence that such individuals can be educationally impaired (hence the term 'nitwit', with 'nits' being a colloquial term for louse eggs). The allergic responses to louse infestation may cause the skin of heavily infested individuals to become thickened and heavily pigmented. Itching and scratching may lead to secondary infection of scarified areas.

In livestock, such itching and scratching may lead to economically noticeable damage to the wool or hides, and the value of wool may be further impaired by staining from louse faeces. Cattle, particularly calves, also respond to lice by increased licking and this may lead to the formation of hair balls in the gut. Lice populations reach peak numbers during winter when the stock have longer and thicker coats. The irritation caused by the lice may disturb eating or sleeping habits to the extent that egg or milk yields may be impaired. Heavy infestations of *Haematopinus eurysternus*, particularly on adult beef cattle, may cause anaemia, lowering of food conversion efficiency, abortion, sterility and sometimes even death. The foot louse of sheep, *Linognathus pedalis*, may cause lameness.

The human body lice *Pediculus humanus* are the vectors of *Rickettsia prowazekii*, the causative agent of epidemic typhus; *Borrelia recurrentis*, the causative agent of epidemic relapsing fever; and, less commonly, *Bartonella quintana*, the causative agent of quintana (trench) fever. Louse-borne epidemic typhus has been a scourge of impoverished, overcrowded, undernourished communities throughout history, particularly in the temperate regions of the world. To take just one example, around the end of the First World War up to 30 million people are believed to have been afflicted by the disease in Europe, with a fatality rate of more than 10 per cent. Although the disease can be experimentally transmitted by all three human lice, it appears that in epidemics the disease is transmitted by the body louse. Typhus is not transmitted by the bite of the louse but rather in the louse's faeces or when the infected louse is crushed. The infective material may be scratched through the skin, or it can gain access to the body across mucous membranes. The rickettsiae are fatal to the louse, whose gut is damaged or ruptured by the invading and replicating organisms. There are two defined ways in which the organism persists between epidemics. In the short term the rickettsiae remain viable in dry faeces for over two months. In the longer term infected, but relatively unaffected, humans are the reservoir. As many as 16 per cent of these asymptomatic carriers, who have recovered from the primary attack of the disease, may subsequently fall ill with Brill Zinsser's disease and, if louse-infested, form the focus of another epidemic. There is some evidence that *Rickettsia prowazekii* is also found in domesticated and other animals and that transmission among these may occur through ticks, lice or fleas, but the relationship and importance of this to human disease is unknown.

Epidemic relapsing fever, caused by the spirochaete *Borrelia recurrentis*, occurs throughout the world. Mortalities as high as 50 per cent were associated with this much-feared disease before the advent of modern antibacterial agents. Transmission does not occur by the bite of the louse or in the faeces, instead the infected louse must be crushed to release the infective organisms, which enter the body through skin abrasions or across mucous membranes. Lice, unlike ticks, do not transovarially transmit the organisms to the next generation of potential vectors.

Adult lice vary in size from about 0.5 mm to about 8 mm depending on species. They are dorsoventrally flattened, wingless, brownish-grey insects with three to five segmented antennae, ocelli are absent and eyes are slight or absent (Fig. 9.1). In Amblycera the antennae are partly protected in grooves on the head. The Anoplura, which are blood feeders, have mouthparts that are highly modified for sucking. The mouthparts are housed in the stylet sac inside the head with only the opening of the sac visible at the anterior tip of the head. The Mallophaga, only some species of which are blood feeders, have chewing mouthparts that are used to cut their various foods. Incidentally in some cases, certainly by design in others (such as the Rhynchophthirina), this may lead to bleeding of the host and the mallophagan may then feed on the released blood. In fact, the Rhynchophthirina hold their mouthparts on an extended rostrum that is almost certainly an adaptation to facilitate blood feeding through the thick skin of their hosts (see Fig. 2.1). Lice have a tough, leathery cuticle capable of considerable

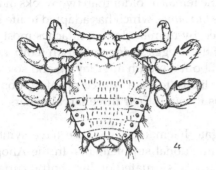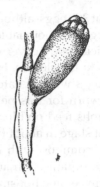

Figure 9.1 An adult crab louse, *Pthirus pubis*, and an egg of the same species firmly glued to a host hair. (Redrawn from Smith, 1973.)

expansion after a blood meal because of the provision of concertina-like folds.

In the Anoplura the three thoracic segments are fused. In the Mallophaga the prothorax is free, although the meso- and metathorax may be fused. Lice tend to have short, stout legs each bearing one (most mammal-infesting forms) or two (bird-infesting forms) pairs of terminal claws that are used for holding onto the host covering. Lice can move quite rapidly in the pelage. The crab louse, *Pthirus pubis*, can move a distance equivalent to the length of the torso in about 30 minutes (Burgess *et al.*, 1983) but mostly remains attached to the skin by its mouthparts. In contrast, the body louse, *Pediculus humanus*, can move much more rapidly, at rates of $0.5\,\mathrm{cm\,s^{-1}}$. In *P. humanus* the foreleg is modified for holding the hindleg of the female during copulation.

Lice are ectoparasitic at all stages of their lives, taking blood meals of up to a third of their own body weight every few hours. Lice will die within one or two days if deprived of blood. The newly emerged female usually feeds before copulation. The pregnant female louse seeks out a particular temperature zone within the host's covering in which to lay her eggs, gluing them to the host's covering using material from special cement-producing glands. *Pediculus capitis* glues eggs to the base of human hair and the sheep louse, *Damalinia ovis*, lays most of the eggs within 6 mm of the host's skin. Egg-laying sites are also determined by other factors, particularly the diameter of the host's hair. For example, *Haematopinus asini* lays its eggs on the coarser hairs of the horse's tail, mane, forelock and that zone above the hooves of some horses known as the feathers. Lice produce relatively few eggs in a batch: human lice produce up to six eggs per day; the cattle louse *Haematopinus eurysternus* produces about two eggs per day. Lice produce eggs every day because they feed several times a day. Adult female *Pediculus humanus* live for a month or so and in that time produce about 300 eggs, although when the female is older than two weeks many of the eggs are not viable. *Pediculus humanus*, which has adapted to life in the artificial body covering presented by human clothes, attaches most of its eggs to the clothes. Hatching of these eggs is dependent on temperature, but eggs will not hatch beyond about a month after laying and so clothes not worn for this period are unlikely to be a source of infestation. The nymphs feed on blood and pass through their three larval instars to the adult stage in about two weeks.

In common with other obligate haematophages, lice have symbiotic micro-organisms that provide nutritional supplements. In the Anoplura the mycetome housing the symbionts is situated on the ventral surface of the ventricular region of the midgut, and in adult females symbionts are also associated with the ovary, from which they are transovarially transmitted to the offspring. In the Mallophaga the mycetocytes are distributed among the cells of the fat body and, in females, symbionts are also associated with the ovary.

9.3   Hemiptera

The order Hemiptera contains those insects usually known as true bugs. Most hemipterans are either entomophagous or plant feeders, but three families contain several species of blood-sucking insects, which range from the small to the very large. The first of these families is the Cimicidae, all members of which are blood-feeding. Cimicids are now found throughout the world but are particularly well represented in the northern hemisphere. The majority are parasites on bats and/or birds, although a minority of species feed on larger mammals. These include the two species normally known as bedbugs, *Cimex lectularius* and *C. hemipterus*, which normally feed on humans. In addition the primarily bat-feeding form, *Leptocimex boueti*, also regularly feeds on humans. *Cimex lectularius* has been carried to all corners of the world in people's belongings but is encountered most often in temperate regions. *Cimex hemipterus* is largely a tropical species, while *L. boueti* is found only in tropical west Africa. Bedbugs bite at night, and heavy infestations can disturb the sleeping human. Some individuals develop marked responses to bedbug bites that may include oedema, inflammation, and an indurated ring at the site of the bite. Heavy infestation has been associated with chronic iron-deficiency anaemia. A role for cimicids in the transmission of disease has yet to be proven, although in the laboratory bedbugs have been shown to excrete hepatitis-B surface antigen for up to six weeks after an infected meal (Ogston and London, 1980). Cimicids are also important economic pests of poultry: prominent species include *Haematosiphon inodorus* in Central America and *Ornithocoris toledo*, formerly important in Brazil.

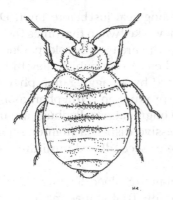

Figure 9.2 An adult bedbug. (Redrawn from Smith, 1973.)

Adult bedbugs are typical cimicids, being wingless, dorsoventrally flattened, brownish insects about 4–7 mm in length. Viewed from above they have a rather oval shape (Fig. 9.2); the short rudimentary wings, or hemelytra, are clearly seen and the broad head and well-developed eyes are clearly visible. When these insects are not feeding, the mouthparts are folded beneath the head and thorax. During feeding they are swung forwards in front of the head. The adult male carries a curved aedeagus at the tip of the 11-segmented abdomen. On the ventral surface of the female's fourth abdominal segment is the opening of the organ of Berlese (or Ribaga) used to store sperm. Mating in bedbugs is very unusual. The male penetrates the cuticle of the female and deposits his sperm into the organ of Berlese. The sperm eventually reach the ovaries by migrating through the haemolymph to the base of the oviducts, which they ascend to reach the unfertilized egg. The female lays about eight eggs a week and may produce 100 or more by the end of her long life. The eggs are cemented into cracks and crevices in which the adults congregate and hide between feeding forays, and they hatch in about seven to ten days. There are five nymphal instars, each of which requires one or two blood meals before moulting to the next stage. Under good conditions the egg-to-adult period may be as short as five weeks, but bedbugs are characterized by a marked ability to withstand starvation – in the laboratory adult bugs may go as long as 18 months between meals! Consequently adult lifespans and the length of time spent in each nymphal instar are particularly variable in these insects.

Nymphs and adults of both sexes feed on blood. As is normal in insects living entirely on blood, cimicids have symbiotic micro-organisms that provide supplementary nutrition. In bedbugs these are housed in a mycetome sited in the abdomen. Bedbugs are unusual in having two different symbiotic organisms in their mycetome, one pleiomorphic and one rod-shaped.

Bedbugs feed mainly at night, with a feeding peak just before dawn. During the ten-minute-or-so feeding period, they take two to five times their own body weight in blood from their hosts, who are usually asleep. Once they have fed, bugs return to the cracks and crevices in which they hide. Bug congregation in such places is influenced by an aggregation pheromone and then by thigmotaxis. The bugs also produce an alarm pheromone that can cause the rapid dispersal of these congregations. Bedbugs are unable to fly, so they are heavily reliant on passive dispersal from one home to another; movement is often in luggage or secondhand furniture – sometimes valuable antiques.

The second hemipteran family containing blood-feeding bugs is the Reduviidae, which includes the subfamily Triatominae made up of 138 species, in the main blood feeders. These bugs feed on a variety of vertebrates, including humans, and many are intimately associated with the habitual resting sites or nests of birds, mammals and other animals. Most triatomines (125 spp.) are confined to the Americas, but there are also one pantropical species (*Triatoma rubrofasciata*), six species in the Indian subcontinent and seven species in South-East Asia. Triatomines that are important from a human perspective are those found in the Americas, from 40 °N to 46 °S. These bugs have many local names, including cone nose bugs, kissing bugs or vinchucas. Some species, *Triatoma infestans* in particular and to a lesser extent *Rhodnius prolixus*, have become closely associated with humans' domestic and peridomestic environment. Sylvatic populations of *Triatoma infestans* are limited to the Cochabamba Valley of Bolivia, where they live among rocks and feed largely on wild guinea pig colonies. Throughout the rest of its range, which extends through parts of Argentina, Bolivia, Brazil, Chile, Paraguay, Peru and Uruguay, the insect is an entirely domestic and peridomestic species. The transition from a sylvatic to a domestic lifestyle is believed to have occurred in pre-Columbian times (Schofield, 1988). The bugs readily adapted to the unplastered, thatched-roof dwellings that are still so common in rural and unplanned urban areas of South America. Until the inception of the Southern Cone Project, an insecticide-based control programme aimed at reducing incidence of Chagas' disease (Dias *et al.*, 2002), *Triatoma infestans* was still extending its geographical range. These bugs will feed on humans and other available vertebrates, such as chickens, dogs, goats or guinea pigs. Houses become infested only if they have unplastered or cracked walls, thatched roofs or other places where bugs can easily hide. This species and others (Table 9.2) are the vectors of *Trypanosoma cruzi*, the causative agent of South American trypanosomiasis, or Chagas' disease. This is a zoonosis which, in its sylvatic cycle, is transmitted between a variety of animals, mainly rodents and marsupials, by sylvatic Triatominae. This is perhaps the classic parasitic disease of poverty because of the sharp economically based distinction

Table 9.2 *The geographical distribution of triatomine species, which have become highly adapted to the domestic-peridomestic environment of humans and so represent a particular threat as vectors of Chagas' disease.*

Vector species	Geographical distribution
Triatoma infestans	Widespread from Southern Argentina to north-east Brazil
Triatoma brasiliensis	Arid regions of north-eastern Brazil
Triatoma dimidiata	Humid coastal areas from Central America to Brazil
Panstrongylus megistus	Humid coastal regions of eastern and south-eastern Brazil
Rhodnius prolixus	Drier savannah areas from southern Mexico to northern South America

Information largely from Schofield (1988).

in vector insect distribution. Poor-quality houses in which infected people live can often be found next door to uninfested, higher-quality houses in which no one is infected. The high standard of living in California means that although 25 per cent of *Triatoma protract* in that state are infected with *Trypanosoma cruzi* (Wood, 1942) and will bite humans, cases of Chagas' disease are thankfully very rare indeed, with only nine autochthonous cases reported in the USA.

Adults of the smallest triatomine species, *Alberprosenia goyovargasi*, are only about 0.5 cm long, but those of the largest species, such as *Dipetalogaster maxima*, may be over 4.5 cm long. Triatomines are often distinctively marked with yellow or orange patches on the fringes of the abdomen, pronotum and bases of the forewings, which contrast against brown to black bodies. They all possess a characteristically elongated head bearing two prominent eyes and four-segmented antennae laterally inserted on the head (Fig. 9.3). As in the bedbugs, the non-feeding insect folds the straight, three-segmented elongated rostrum (proboscis) under the head. These mouthparts are swung forwards in front of the head for feeding. When viewed from above, the characteristic triangular shape of the pronotum is obvious. The proximal part of the forewings is relatively hard and rigid, gradually leading into the more membranous distal part. In the non-flying insects these forewings, or hemielytra, hide the completely membranous hindwings and cover most of the dorsal surface of the abdomen. In the unfed bug the edges of the flattened abdomen may form a raised rim around the wing-covered area.

The feeding adult bug takes up to three times its unfed body weight in blood and the abdomen swells to look like a ripe berry. In a typical house infestation, a person can expect to receive about 25 bites per night, and so the bug's attentions can contribute to chronic anaemia in the human population (Schofield, 1981). They normally feed every four to nine days, but

Figure 9.3 An adult triatomine bug viewed from above. (Redrawn from Faust *et al.*, 1977.)

bugs can withstand starvation for months, particularly the later nymphal instars.

Adult female *Triatoma infestans* produce small batches of eggs (10–30) subsequent to each blood meal; in her lifespan of six months to a year she will typically produce 100–300 eggs. The eggs of Triatominae are oval and about 2 mm long. They are white, yellow or pink, depending on species and stage of development. The eggs hatch in 10–30 days and there are five nymphal instars.

Triatomine bugs are obligate haematophages. Most feed at night, are stealthy, and have an almost painless bite, as evinced by *Triatoma infestans* which commonly feeds on the facial mucous membranes of its sleeping human hosts. In contrast, sylvatic bug species often have a very painful bite. Early instar nymphs may take up to 12 times their unfed body weight in blood, while adult bugs rarely take three times their unfed body weight. Each nymphal instar requires only one full blood meal to stimulate the moult into the next instar. Even so, most nymphs take more, smaller meals at each instar. Full development from egg to adult takes four months to a year to complete. In common with other insects feeding exclusively on blood, triatomine bugs have symbiotic micro-organisms. These are housed in the lumen of the anterior midgut and provide additional nutrients without which nymphs do not successfully develop into adulthood.

Triatoma infestans, which inhabits the human domestic environment, probably relies heavily on passive transfer in hand luggage, furniture, etc., for movement into previously uninfested regions. The adult insects are capable of strong flight, which is probably an important dispersal agent at a local level. Sylvatic species can use flight for dispersal, but it is also

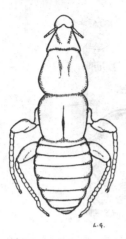

Figure 9.4 An adult polyctenid, *Eoctenes spasmae*. (Redrawn from Marshall, 1981.)

thought that the early-stage nymphs may disperse phoretically on their hosts.

The third family of blood-sucking bugs is the Polyctenidae (Fig. 9.4). All members of this group are permanently ectoparasitic, obligate haematophages living on microchiropteran bats of both the New and Old Worlds. They are of no economic importance. These eyeless, wingless insects are oviparous and there are only three postnatal nymphal instars.

9.4 Siphonaptera

The order Siphonaptera contains the fleas, of which about 2500 species and subspecies have been described. All adult fleas are ectoparasitic on warm-blooded hosts; about 94 per cent infest mammals (74 per cent live on rodents alone) and 6 per cent infest birds. Fleas are found throughout the world, with a concentration in temperate regions. There are about 20 flea species that will feed on humans. In the past *Pulex irritans* was a serious nuisance to humans, but it is now becoming rare in most industrialized countries. Meanwhile, the cat flea, *Ctenocephalides felis*, originally restricted to North Africa and the Middle East, has emerged to take its place, and this insect is now a serious household nuisance throughout Europe and North America.

Because of the stigma attached to flea infestation, affected humans may suffer more mentally than physically from the presence of fleas. Having said that, fleas can present a very serious disease risk. Heavy infestations of fleas are usually restricted to animals in poor condition, the health of which may deteriorate further as a result of infestation. Hosts show varying degrees of sensitivity to flea bites; some display severe reactions that may

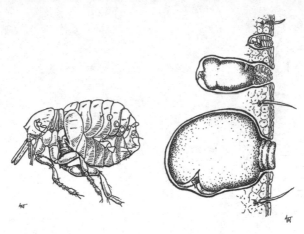

Figure 9.5 The embedding of the neosomic flea *Tunga calcigena*. (After Jordan 1962.)

seriously impair their health. For example, the dog and cat fleas (*Ctenocephalides canis* and *C. felis*, respectively) may cause moderate to severe pruritic reactions appearing as areas of moist dermatitis on their hosts. Secondary infection of these areas can occur, and local alopecia may result from excessive grooming by the affected animal.

As well as fleas, which move freely in the home and pelage of the host, there are those, like the tungids (Tungidae) and alakurts (Vermipsyllidae), that burrow into its skin and become neosomic (Fig. 9.5). There are also the sticktights, which attach to the host by their mouthparts for long periods, and some species display neosomy. Humans are afflicted by the chigoe, sand or jigger flea, *Tunga penetrans*, which causes a painful infestation by burrowing into the feet, often beneath the toenails. Originally a New World species, it has spread throughout the Neotropical and Afrotropical regions over the last 150 years. Heavy infestations can be crippling, and secondary infection of the ulcer, caused by the body's response to dead jiggers, can lead to other complications. The 'sticktight' flea, *Echidnophaga gallinacea*, is an important poultry pest in the tropics and subtropics. Young birds infested with this flea quickly die; older birds suffer anaemia, reduced egg output, and may also perish from heavy infestations.

Fleas are also the transmitters of disease, being most notorious for their involvement in the Black Death, the pandemic of plague that killed a quarter of the population of Europe (25 million people) in the fourteenth century. We know of two other plague pandemics. The first, in the sixth century AD, spread from Central Asia through the Middle East and into Africa. The last began in China in the 1890s and spread rapidly along trade routes to India, the Americas, Australia, South Africa, and on to many other parts of the world. It caused over 10 million deaths in India at the turn of the last century. This third pandemic is still lingering on, although the annual number of cases is now small. Plague is caused by the bacterium *Yersinia pestis*; it is a zoonotic disease usually cycled between rodents by their fleas. Humans become involved in the cycle when infected domestic or peridomestic rodents, such as the black or brown rat, die. The fleas then leave the dead rodents and look for another host, which might be human. The bacterium blocks the gut of infected fleas and is regurgitated into the wound when the flea attempts to feed. Mechanical transmission may also be of some importance during an epidemic of the disease. The two flea species most commonly involved in carrying the disease to humans are *Xenopsylla cheopis*, which is the major transmitter in urban areas of the tropics and subtropics, and *X. brasiliensis*, which is an important rural vector in Africa and India. In its pneumonic form, plague is directly transmitted between humans without the mediation of an insect vector.

Another bacterium, *Francisella tularensis*, is also transmitted by fleas and causes a plague-like disease called tularaemia in mammals. This disease occurs mainly in the northern hemisphere, possibly more commonly in the Old than the New World. The bacterium occurs in domesticated animals such as cattle, horses and sheep, in which the disease is particularly severe. There are many wild reservoir hosts, including rabbits, rodents and birds. The rodent populations are thought to be the major natural reservoir of the bacterium, with transmission among them by fleas and lice. The bacterium, which can infect humans through unbroken skin, can spread through a number of additional agencies. Mechanical transmission by insects, particularly tabanids, is well documented; it can also be airborne and can spread in contaminated food and water.

Fleas also transmit the causative agent of murine typhus, *Rickettsia typhi* (*mooseri*). The disease is a mild infection of wild rodents throughout the world and may cause significant mortality in affected human populations. It is transmitted between wild rodents by their fleas, lice and possibly their mites. Like plague, human infections are associated with domestic and peridomestic infestations of rats. The major vector is *X. cheopis* and, although other flea vectors may be involved, the evidence is inconclusive. The rickettsia is either released in the flea's faeces or when the flea is crushed on the skin, gaining entry to the body through skin abrasions or cuts, across mucous membranes or after inhalation. The flea's faeces can remain infective for several years. Rat salmonellosis, caused by *Salmonella enteritidis*, is also transmitted by flea bite and in flea faeces.

Although in most of its range myxomatosis is mechanically transmitted to rabbits largely by mosquitoes, in the British Isles the flea *Spilopsyllus cuniculi* is the major vector. This flea is a much more efficient vector of the virus than mosquitoes and has been artificially introduced into Australia as part of that country's long-standing campaign against rabbits. The flea *Echidnophaga myrmecobii* may also play a minor role in transmission of the disease in Australia.

Fleas are also the intermediate hosts of some helminths. The common dog and cat tapeworm, *Dipylidium caninum*, develops in the dog and cat fleas, *Ctenocephalides canis* and *C. felis*. The eggs are ingested by the flea larvae and the cysticercoid stage of the parasite develops in the haemo-coel of the adult flea. The mammal acquires the tapeworm when it ingests the infected flea (normally while grooming). This parasite also occasion-ally occurs in children. The tapeworm of rodents (and also humans), *Hymenolepis diminuta*, uses fleas as well as flour beetles and other insects as intermediate hosts, as may another tapeworm of rodents (and humans) *H. nana*. Fleas also transmit a filiarial worm of dogs, *Dipetalonema recondi-tum*. There is some evidence for the involvement of fleas in the transmission of many other bacterial, rickettsial and viral diseases. So, although the dis-eases outlined above are undoubtedly the most important transmitted by fleas, some surprises may still be in store.

Adult fleas are small (< 5 mm), brown, wingless, flattened insects (Fig. 9.6). Their characteristic laterally flattened shape is a feature they share only with some streblids. The flea's body is generally well covered with backward-projecting spines, and with combs. The rather arrow-shaped, immobile head bears the mouthparts, which project downwards, and the three-segmented antennae, which are protected in grooves on the head. The size of the mouthparts depends on the lifestyle of the flea, with stick-tight fleas having much longer mouthparts than more mobile forms. Fleas lack compound eyes, but most have lateral ocelli which are best developed in fleas infesting diurnally active hosts. The three-segmented thorax bears three pairs of well-developed legs. The hind pair are specialized for jump-ing and can propel the flea 20–30 cm in one jump. The abdomen of male fleas has an upturned appearance, which distinguishes it from the female, whose abdomen is more rounded.

Adult fleas show a spectrum of associations with the host. At one extreme are fleas that are more or less permanently attached to the host; examples are the jigger, *Tunga penetrans*, which is buried beneath the host's skin, or the sticktight fleas, such as *Echidnophaga* spp., which remain attached to the host by their mouthparts. In the middle of the spectrum we have those active forms that remain largely on the host, such as the bat fleas. At the other extreme, we have active forms such as the bird fleas that reside mostly in the nest, only visiting the host to feed. Many species fall between the last two categories: the adults are not permanent ectoparasites but spend a considerable part of their lives away from the host animal, com-monly in the host's nest. Adult fleas of both sexes feed on blood and will also feed on watery solutions. The frequency of feeding varies considerably

(a)

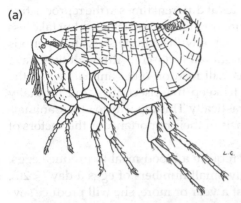

(b)

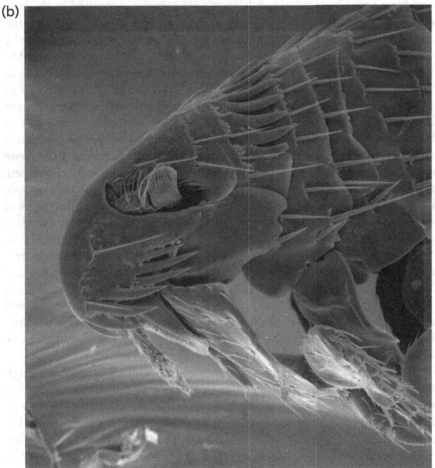

Figure 9.6 (a) External anatomy of the adult flea *Xenopsylla cheopis*. (Redrawn from Marshall, 1981.) (b) Scanning electron microscope view of the head of a flea (courtesy of Gregory S. Paulson).

among species and in the same species at different times in the reproductive cycle. In general, male fleas feed more frequently than females, but take less blood in the process. Fleas are capable of withstanding starvation periods of several months while their host animal is absent. Although they have distinct preferences for hosts, fleas will feed on other animals when the major host is absent. This serves to keep the insect alive, but normally reduces its fecundity, sometimes drastically. This capacity to utilize animals other than their usual host is the basis of their importance as the vectors of the zoonoses described above.

Fleas are anautogenous and so require a blood meal to produce eggs. Although the female produces only small numbers of eggs a day (< 20), during her lifetime, which may last a year or more, she will produce several hundred. The large, whitish eggs are normally deposited in the host animal's nest, but some fleas, like the vermipsyllids, deposit eggs that fall indiscriminately to the ground. The eggs of some species have a sticky coat that may attach them to, or coat them in, debris. The delay period to hatching, normally two to six days, is affected by humidity as well as temperature. The whitish, eruciform, apodous larva that hatches from the egg is very active and negatively phototropic and geotropic, so that it moves into the substrate. Like the eggs, the larval stages are susceptible to low humidities and move to zones of higher humidity. Most species have three larval instars, but some species have only two. The mature larva of some species is up to a centimetre long. Larvae feed on organic material which, because so many are nest dwellers, is usually plentiful. Much of the organic debris in the nest is derived from the host, but in addition adult fleas of some species do not completely digest their blood meals so that their faeces are commonly a food source for the larval stages. In the rabbit flea, *Spilopsyllus cuniculi*, the frequency of feeding and defecation increase dramatically during the reproductive period. As a consequence the burrow becomes well supplied with faeces, which the forthcoming larval states utilize. In species such as *Nosopsyllus fasciatus*, in which the larval as well as the adult diet consists largely of vertebrate blood, the fleas possibly harbour intestinal symbionts to supplement the diet. Species with a more catholic larval diet probably do not harbour such symbionts.

The larval stages are normally complete in under a month, although the larval span can be prolonged over many months, allowing the flea to survive through a difficult period. At pupation the larva empties its intestine and produces a silken cocoon in which it pupates over a period of several days. The pupa is sensitive to low humidity and low temperature. Under ideal conditions the pupal stage is completed within a week, but it can extend for a year or more and is used by the flea as a means of spanning periods of adverse conditions. After emergence, most adult fleas will commence feeding before mating; the exceptions are mostly bird fleas. Mating is influenced by species-specific, contact pheromones; more than one mating may be required to fertilize all the available eggs.

9.5 Diptera

The order Diptera contains many of the most important and familiar blood-sucking insects. They are of tremendous significance to man, from both an economic and health point of view, because of their role as parasite vectors. Because of the significance of this assemblage of insects, I will briefly outline the characteristics and taxonomy of the order before discussing the particular details of each group separately.

As their name implies, the obvious character separating adult Diptera from all other insects is that they possess only a single pair of wings. The second pair of wings (the hindmost) have been modified into short, knob-like structures called halteres, which are used as balance organs. Wing tracheation is reduced and is so remarkably constant that wing venation can often be used for taxonomic purposes. The wings are carried on a thorax in which the mesothorax is greatly enlarged and the prothorax and metathorax correspondingly reduced.

Most adult Diptera have large and highly mobile heads bearing well-developed eyes that are often larger in the male than the female, and mouthparts that lack mandibles and are suctorial. Many of the blood-sucking species show considerable morphological specialization of these sucking mouthparts to allow for penetration of the host's skin.

The Diptera are holometabolous (endopterygotes) and have four life stages: egg, larva, pupa and adult. Depending on species, the adult female may produce eggs or live young. The larvae, which are apodous and eruciform, prefer humid to wet habitats; in some species larvae are entirely aquatic and many terrestrial larval forms live in the humid surroundings of rotting or fermenting organic matter. Other larval forms are carnivorous, some living in the flesh or internal organs of vertebrate hosts and some, like the Congo floor maggot (*Auchmeromyia senegalensis*) feeding on blood. Some larval forms feed on live plant tissue. In the 'higher' forms the pupal stage is normally immobile, but in the 'lower' forms the pupal stage may exceptionally display various degrees of mobility, as is seen in mosquitoes. In 'higher' forms the pupa is commonly enclosed in the retained and hardened last larval skin and the structure arising is known as a puparium. The cyclorrhaphan flies (see below) have a special expandable bag called a ptilinum which is associated with the head. They use the ptilinum to force a cap from the end of the pupa at the time of eclosion. Adult Diptera are usually highly active and mobile.

THE BLOOD-SUCKING INSECT GROUPS

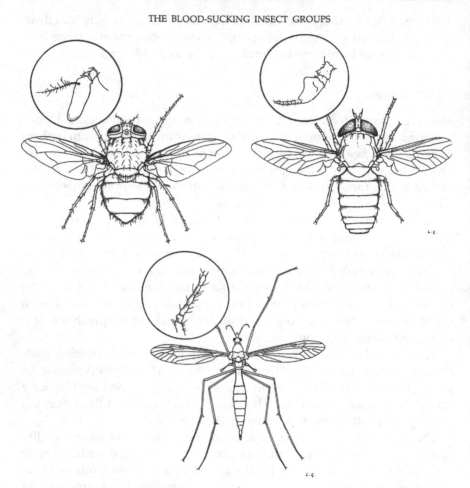

Figure 9.7 A typical example of a cyclorrhaphan (left), a brachyceran (right), and a nematoceran (bottom), emphasizing the antennal differences between the three groups. (Redrawn from Colyer and Hammond, 1968.)

The order Diptera is divided into three suborders. The first of these, the Nematocera, contains the soft-bodied flies such as the mosquitoes, sand-flies and blackflies (Fig. 9.7). The second suborder, the Brachycera, contains the tabanids and the blood-sucking rhagionids (Fig. 9.7). The third subor-der, the Cyclorrhapha, contains the insects most commonly termed 'flies' (Fig. 9.7). The taxonomy of the suborder Cyclorrhapha is complex. It is sub-divided into two series, the Ashiza and the Shizophora. The Shizophora is subdivided once more into three sections, Acalypterae, Calypterae and Pupipara. The families of insects that fit into each of these divisions are shown in Table 9.3.

Table 9.3 *The divisions of the order Diptera and the major families in each division. Families containing blood-sucking species are in bold type.*

Suborder	Series	Section	Family
Nematocera			Tipulidae
			Psychodidae
			Culicidae
			Chironomidae
			Ceratopogonidae
			Simuliidae
			Anisopodidae
			Bibionidae
			Mycetophilidae
			Sciaridae
Brachycera			Stratiomyidae
			Rhagionidae
			Tabanidae
			Asilidae
			Bombyliidae
			Empididae
			Dolichopodidae
Cyclorrhapha	Aschiza		Phoridae
			Syrphidae
	Schizophora	Acalypterae	Drosophilidae
			Chloropidae
			Sepsidae
			Gasterophilidae
			Piophilidae
		Calypterae	**Muscidae**
			Anthomyiidae
			Glossinidae
			Calliphoridae
			Sarcophagidae
			Oestridae
			Tachinidae
		Pupipara	**Hippoboscidae**
			Streblidae
			Nycteribiidae

9.5.1 Culicidae

Mosquitoes are perhaps the most familiar of all blood-sucking insects. They fall into the nematoceran family, the Culicidae, which is divided into 3 subfamilies and 41 genera. The Anophelinae and the Culicinae are blood feeders but the third subfamily, the Toxorhynchitinae, do not feed on blood and so do not concern us here. Anophelinae contain about 430 species in 3 genera, *Bironella, Chagasia* and, containing by far the largest number of species (416), *Anopheles*. Culicinae, containing about 3300 species, are a taxonomically more complex group. The most commonly encountered genera are *Culex, Aedes, Ochlerotatus, Sabethes, Mansonia, Culiseta, Psorophora, Wyeomyia, Coquillettidia, Haemagogus* and *Armigeres*.

Mosquitoes are found throughout the world, except the Antarctic. In many parts of their distribution, most particularly in tundra areas of the northern hemisphere, mosquito populations reach pest, and sometimes plague, proportions. However, their importance as pests is insignificant compared with their role as vectors, particularly of human diseases such as malaria. The World Health Organization estimated in 2001 that 273 million people were suffering from malaria annually and that 200 million of these were in sub-Sahelian Africa, where an estimated 1 million children a year die from the disease. They also estimated that over 2 billion people in 100 countries are at risk from malaria. Each of the four species of human malaria parasites are transmitted exclusively by *Anopheles* spp. Anopheline mosquitoes also transmit malaria parasites to other animals, for example *Plasmodium knowlesi* and *P. cynomolgi* to monkeys and *P. berghei* to rodents. Culicine mosquitoes transmit malaria parasites such as *P. gallinaceum* to wild and domesticated birds.

Mosquitoes also transmit filarial worms such as *Wuchereria bancrofti, Brugia malayi* and *B. timori*, which are parasitic in humans and cause elephantiasis. Over 120 million people are suffering from lymphatic filariasis in 80 countries, with 40 million seriously incapacitated and disfigured by the disease. These debilitating parasites are transmitted by many different mosquito species in various parts of their geographical distribution. To generalize, bancroftian filariasis is mainly vectored by *Anopheles* species and *Culex quinquefasciatus*, while brugian filariasis is principally transmitted by *Mansonia* and *Anopheles* species. The filarial worms of the dog, *Dirofilaria immitis* and *D. repens*, are also transmitted by mosquitoes, causing the disease known as dog heartworm.

Mosquitoes also transmit more than 200 arboviruses to humans and other animals. Culicine mosquitoes are most commonly involved as vectors, but anophelines do transmit a few viruses. The most important of the arboviruses transmitted are again infections of humans. These include dengue, which is largely an urban disease endemic in many parts of

South-East Asia and the western Pacific, also occurring in the Caribbean. The major vector is *Aedes aegypti*, with *Aedes albopictus* playing a subsidiary role in Asia. Yellow fever is a zoonotic disease of forest monkeys found in Africa, and Central and South America. It is transmitted between monkeys by tree-hole-breeding, forest-dwelling mosquitoes such as *Haemagogus* spp. and *Sabethes chloropterus* in Latin America and *Aedes africanus* in Africa. These insects bite their hosts high in the forest canopy. *Haemagogus* and *Sabethes* bite humans when they enter the forest, transferring the disease from monkey to humans; in Africa *Aedes bromeliae* (= *simpsoni*) transfers the virus from monkey to humans. Once the virus has infected humans, domesticated vectors such as *Ae. aegypti* are very efficient at spreading the disease among the urban population. Mosquitoes also transmit arboviruses causing encephalitis. To generalize, the vectors associated with particular viruses in the Americas are West Nile virus and Venezuelan equine encephalitis (mainly *Culex* spp.), eastern equine encephalitis (*Culiseta*, *Ochlerotatus*, *Culex* and *Coquillettidia* spp.), and western equine encephalitis (*Culex* and *Ochlerotatus* spp.). In Japan, South-East Asia and India, the major vector species transmitting Japanese encephalitis to humans is the rice-field-breeding *Culex tritaeniorhynchus*.

Adult mosquitoes are small (usually about 5 mm long), rather delicate insects, with slender bodies, long legs and elongated, forward-projecting mouthparts (Fig. 9.8). When resting, mosquitoes hold their single pair of wings over the abdomen like a pair of closed scissors. In most anophelines the wings have a dappled appearance because of alternating blocks of dark and light scales on the wings. In contrast, most culicines have wings that lack distinct markings. Characteristically, all mosquitoes have scales on their wing veins and the trailing edges of the wings. Males of important vector species are easily distinguished by their conspicuous, plumose antennae, which contrast with the pilose antennae of the female (Fig. 9.9). Adult members of the blood-sucking subfamilies can be distinguished in the field by their resting postures. Adult anophelines usually rest with the three divisions of the body in a straight line and the abdomen tilted up from the resting surface (Fig. 9.8). In contrast, the body of culicine adults forms an angle about the thorax and they tend to stand more or less in parallel with the resting surface (Fig. 9.8).

Adult mosquitoes of both sexes feed on sugary solutions, but only females take blood. Mosquitoes as a group feed on a tremendous range of vertebrates from fish and reptiles to birds and mammals, but each species typically has a narrow range of preferred hosts from which it normally feeds. Humans are a major host for most, but not all, important mosquito vectors of human disease. Many mosquitoes feed in a particular place; the canopy rather than the forest floor, or the reeds at the edge of the river rather than on birds resting on open water. This is important to humans

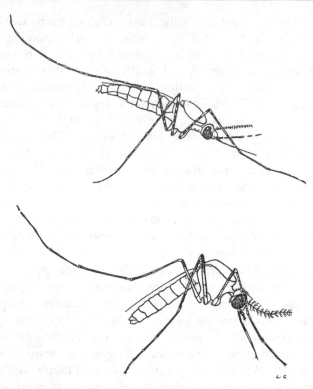

Figure 9.8 Adult anophelines (above) and culicines (below) commonly have these characteristic resting postures, which is a useful identification feature in the field. (Redrawn from Kettle, 1984.)

because, although most mosquitoes bite in the open, some of the more important vector species feed largely within dwellings (see below). Each species usually has a characteristic peak biting time or times, so *Anopheles gambiae* bites in the early hours of the morning (mainly 23.00 to 04.00 hours) and *Aedes aegypti* shows two biting peaks, one at dawn and another at dusk. To generalize, most anopheline species are night biters while the Culicinae contains both night- and day-biting species. In the tropics mosquitoes take a meal every two to four days. Once the meal has been taken, they find a quiet spot to digest it, and for some important vector species this is inside human dwellings, a behaviour that provides an opportunity for effective vector control of disease by spraying insecticides within these relatively accessible and limited spaces.

Although some mosquitoes need two blood meals to mature the first batch of eggs, most mosquitoes display gonotrophic concordance through most of their reproductive lives, each blood meal normally leading to the development of a batch of about 50–200 dark-brown to black ovoid eggs.

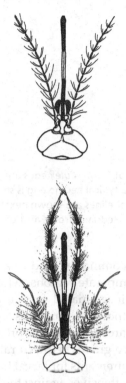

Figure 9.9 Female mosquitoes have pilose antennae, while the antennae of male mosquitoes are plumose. The examples shown here are *Culex annulirostris*. (Redrawn from Kettle, 1984.)

A few species, such as *Culex molestus* and *Wyeomyia smithii*, can develop one or more egg batches autogenously, but production of subsequent egg batches relies on blood feeding. *Anopheles*, most *Culex* and *Coquillettidia* lay their eggs directly onto the water surface, whereas *Mansonia* species lay their eggs in groups on the undersurface of the leaves of floating aquatic plants. Anopheline eggs are laid singly and usually possess floats (Fig. 9.10) that keep them on the water's surface. *Culex*, *Culiseta*, *Coquillettidia* and some *Uranotaenia* species lay their eggs in clusters arranged together into floating rafts (Fig. 9.10). In tropical regions these eggs hatch within three days of deposition. *Haemagogus*, *Psorophora*, *Aedes* and *Ochlerotatus* species lay their eggs on damp substrates just above the water's surface. These eggs, unlike those laid directly onto the water's surface, soon become resistant to desiccation and can remain viable for many months, delaying their hatch until they are immersed in water. While immersion is commonly the hatching trigger, these eggs are also capable of entering diapause so that hatching may be delayed

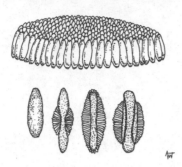

Figure 9.10 Mosquito eggs come in a variety of forms. *Culex* eggs are commonly produced as a floating raft (top), a typical aedine egg is shown bottom left and eggs of three species of anophelines are shown next to it to illustrate the egg floats typical of this genus. (Redrawn from Faust *et al.*, 1977.)

beyond the first immersion. Egg diapause is commonly used as an over-wintering device by aedine mosquitoes in temperate regions, while others overwinter as larvae or as hibernating, inseminated female adults. In tropical areas mosquitoes may survive the dry season as eggs or occasionally as aestivating adults, but in most areas mosquitoes breed continuously throughout the dry season, albeit at a greatly reduced rate.

Various mosquito species have specific requirements for larval habitats. Because mosquito control commonly involves offensives against the larvae, an understanding of their habitats is often central to the successful control of vector or nuisance populations. Almost all bodies of fresh or brackish water can be utilized as a larval habitat by one species or another. The exceptions are fast-flowing water (although backwaters or marshy areas along their edges may be used, especially by anophelines) and expanses of open water such as lakes, particularly if well stocked with larvivorous fish. Many species, particularly culicines, make use of natural container habitats such as broken coconut shells, rot holes in trees, bamboo stumps, leaf axils and pitcher plants. Some of these species have been able to make use of the tremendous number of man-made containers available. The classic example is *Aedes aegypti*, which has spread to urban and particularly to slum areas throughout the tropics and subtropics. Here it finds discarded tins and tyres, water storage vessels, flower pots in cemeteries and a host of other man-made containers suited to its breeding. Other species prefer water contaminated with organic matter. Slum areas throughout the tropics and subtropics, with their inadequate sanitation and rubbish disposal facilities, have provided polluted water that is an ideal breeding ground for one such mosquito, *Culex quinquefasciatus*. This mosquito has soared in numbers with the rapid, unplanned urbanization in such areas. Other mosquitoes, such as the important malaria vector *Anopheles gambiae*, need temporary sunlit pools, and farming (e.g. rice fields) and forestry

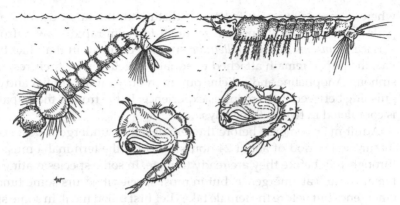

Figure 9.11 Culicine larvae (left) hang head down from the surface film and are easily distinguished from anopheline larvae (right) which lie with their bodies parallel to the water's surface. The comma-shaped pupae of culicines can be distinguished (with experience) from the very similar pupae of anophelines by the shape of the respiratory horns. (Redrawn from Faust *et al.*, 1977.)

activities in Africa have provided extensive additional breeding sites for these mosquitoes.

Each of the four larval stages of mosquitoes are aquatic, legless and have a thorax wider than the other body regions. A few larval mosquitoes are carnivorous (and in some cases cannibalistic), but the majority feed on micro-organisms using their mouth brushes to gather food particles. Most anopheline species are surface feeders; many other mosquito larvae feed on the bottom. Most culicine and anopheline larvae and pupae come to the surface to breathe atmospheric oxygen. Culicines bear a respiratory siphon on the penultimate, eighth abdominal segment and they use this to attach themselves to the water's surface film, from which they hang head down (Fig. 9.11). Anopheline larvae are easily distinguished from culicines because they do not have a respiratory siphon, although they are still air breathers and have paired spiracles on abdominal segment eight. When at the water's surface they lie with their bodies parallel to it (Fig. 9.11), suspended by a pair of thoracic palmate hairs and palmate hairs present on most of the abdominal segments. *Mansonia* and *Coquillettidia* spp. larvae breathe by plugging into the air vessels of plants using the modified valves of their conical respiratory siphons to cut their way into the plant tissues, to which they remain attached throughout their development.

In the hot tropics the larval stages can be completed within seven to ten days. The larvae then moult to become pupae, which are comma-shaped, non-feeding, air-breathing, aquatic stages in which the head and thorax are fused to form a cephalothorax. On the dorsal surface of the cephalothorax are paired breathing trumpets that attach the pupa to the surface film where it spends most of its time. If disturbed, the pupa uses the paired paddles on the last abdominal segment to swim rapidly away from the surface. *Mansonia* and *Coquillettidia* spp. are different in that, like the larvae, the pupae remain attached to aquatic vegetation by their respiratory siphons. Anopheline and culicine pupae are structurally similar and distinguishing between them requires experience. In the tropics the pupal stage is completed in two to three days.

Adult males emerge before the females; they undergo a post-eclosion maturation period of about 24 hours in which the terminalia must rotate through 180° before they are ready to mate. In some species mating of the female occurs at emergence, but in most species it occurs some time after emergence but before the female takes her first blood meal. In some species mating involves crepuscular swarming of the males above marker points (such as trees or prominent rocks) of species-specific height and setting. Females respond to the wing beat sounds of males and enter the swarm; coupling takes place and they leave the swarm and insemination occurs. Not all species need to swarm to mate. For example, *Aedes aegypti* use the host animal as a meeting point for the sexes, with male *Ae. aegypti* being attracted to host animals during the same period of the day as the females.

Male mosquitoes do not usually travel more than about 100 metres from the larval site. Females travel further to find both blood meals and new larval sites to colonize. There are considerable inter- and intraspecific differences in the distances females travel. Probably few actually fly more than one or two kilometres, but it is well known that individuals of some species may be carried considerable distances on the wind. For example, in the USA *Aedes vexans* has been observed to be displaced by over 320 kilometres on weather fronts, and the saltmarsh mosquito, *Ochlerotatus taeniorhynchus*, has often caused a nuisance in built-up areas many kilometres downwind of its breeding sites. The malaria vector *Anopheles pharoensis* has been recorded many tens of kilometres downwind of breeding sites in North Africa.

It is likely that under tropical field conditions male mosquitoes live for about seven to ten days and a few females for up to a month. In temperate regions females live two or more months, and hibernating adults survive for up to eight months.

9.5.2 Simuliidae

The nematoceran family Simuliidae contains some 1800 species in about 25 genera. Four genera, *Austrosimulium*, *Cnephia*, *Prosimulium* and *Simulium*, are of economic importance. These flies are known locally by a variety of names but most commonly as blackflies or buffalo gnats. In temperate and particularly sub-Arctic regions of the world, the flies can appear in huge swarms that can make life a misery for humans, their livestock and wildlife. Indeed, both humans and on some occasions thousands of livestock have died from the attacks of these huge swarms (16 474 domestic animals in southern Romania in 1923 (Baranov, 1935; Bradley, 1935)). Death probably occurs from a combination of effects, mainly toxaemia (or anaphylactic shock) from the insect's bite, blood loss and breathing problems from inhalation of large numbers of insects. Lower biting levels can cause economic losses in livestock from the worry they cause to the animal (Hunter and Moorhouse, 1976).

Blackflies are also important vectors of disease organisms. They are mechanical transmitters of myxomatosis, vectors of *Leucocytozoon* and avian trypanosomes of domestic and wild birds, and occasional vectors of the arbovirus Venezuelan equine encephalitis. But most importantly, blackflies are vectors of filarial worms, the most significant being *Onchocerca volvulus*, which causes the human disease river blindness in Africa, and Central and South America. Blindness can be so common in areas of intense transmission, such as the Volta and Niger river basins of west Africa, that it seriously interrupts the social organization of the population and thus impairs the development of the country (Senghor and Samba, 1988). Onchocerciasis in Africa is mainly transmitted by members of the *Simulium damnosum* complex, which contains about 40 cytotypes, many of which are species status. The Onchocerciasis Control Programme of the World Health Organization was instituted in 1974 to attempt to control the disease. This immense programme, which is based on the control of the vectors, demands that up to 50 000 kilometres of river are monitored or treated with insecticide each week. This successful programme has reduced prevalence from an original level of 25 to 30 per cent to below 5 per cent (Boatin, 2003). Because adult *Onchocerca volvulus* can live in a person for 15 years, the campaign must continue for at least this period to achieve any long-term success. The programme has been supplemented with drug-based control using ivermectin through the African Programme for Onchocerciasis Control (APOC) (Molyneux et al., 2003). *Simulium amazonicum* is the main vector of *Mansonella ozzardi* in Brazil, Venezuela, Colombia, Guyana and southern Panama, while *Culicoides* species transmit this infection in the Caribbean islands, Trinidad, Surinam and Argentina.

Adult buffalo gnats are small flies, 1 to 5 mm long. They have a single pair of clear, scaleless, broad wings between 1 and 6 mm long with large anal lobes and anterior veins that are usually prominent. When the fly is at rest the wings are folded over the body like a closed pair of scissors. The flies are usually blackish or greyish but often have silvery, orange or golden hairs patterning the thorax and/or legs. The eyes in the female fly are dichoptic; in the male the eyes are holoptic and the upper ommatidia are larger than the lower (Fig. 9.12). The male eye specializations are used for detecting flying females in the mating swarm. There are no ocelli. Antennae are cigar

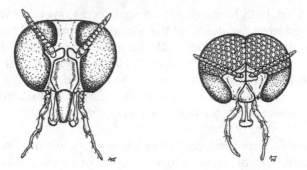

Figure 9.12 Male blackflies are holoptic (eyes meeting on top of the head) and females are dichoptic (two separate eyes). (Redrawn from Smith, 1973.)

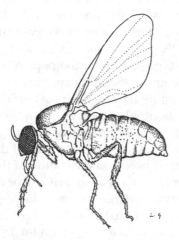

Figure 9.13 Side view of adult female *Simulium damnosum s.l.* (Redrawn from Smith, 1973.)

shaped, commonly 11-segmented and lack prominent hairs (Fig. 9.13). The thoracic scutum is well developed and gives the flies their characteristic hump-backed appearance (Fig. 9.13). Compared to other Nematocera, the legs of blackflies are distinctly short and stout (Fig. 9.13). Adults are diurnal and show an essentially bimodal behaviour pattern with maximum activity in the early morning and late afternoon.

Blackflies breed in water, commonly in rapid-flowing 'white' water. Each species has a preference for particular breeding sites. Thus, the African species *Simulium damnosum s.l.* breeds in the rapids of small to very large rivers, *S. ochraceum* in very small streams of South America and the east African species *S. neavei* in small streams and rivers attached in a phoretic association with freshwater crabs of *Potamonautes* spp. Most species lay eggs in the evening, in clusters containing 150–600 eggs on submerged

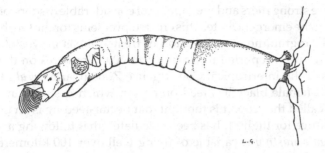

L·G

Figure 9.14 Blackfly larva are easily recognized from their cigar shape, mid-ventral proleg just behind the well-formed head and the circle of hooklets on the end of the body used to attach the larva to the substrate. When viewed in water the two large cephalic fans may also be seen. (Redrawn from Smith, 1973.)

vegetation or rocks, but in some species (e.g. *S. ochraceum*) the female flies over the water broadcasting eggs onto the surface. In many temperate species the egg may diapause; in others the larval stage may overwinter. In the tropics the egg and larval stages may be completed in about a week. The larvae are commonly encountered in water bodies and are easily recognized from their characteristic shape (Fig. 9.14). The larva spins a small pad of silk on the substrate from its salivary glands and attaches firmly to this by a ring of hooklets on its posterior end. The larva is sedentary throughout most of its life, but it has another, smaller circlet of hooks on the proleg (Fig. 9.14) and it can detach each circlet in turn and move using looping movements of the body. The larva can also produce silk from its salivary glands and can use these threads to relocate downstream and to move back upstream if the thread has been anchored. The larvae are filter feeders using their pair of cephalic fans to abstract fine particulate matter (10–100 rm) from the passing water flow. Larvae grow to about 12 mm long, undergoing up to eight larval moults to achieve this. The mature larva, or gill spot-larva (which is actually a prepupa or pharate pupa), pupates in a cocoon spun from salivary gland silk. During daylight, two to six days after pupation, the adult emerges rapidly from the pupa. It reaches the water surface by climbing up an available object, or by floating up in a gas bubble, and is capable of immediate flight.

Males form mating swarms near the emergence site or occasionally near the female's vertebrate host. Male flies grasp females in flight and insemination occurs once the pair have landed or fallen to the ground. In some species mating may occur in the absence of swarming, with males seeking out females on, or close to, the female's host animal. Insemination may take only seconds or nearly an hour, depending on species.

Blackflies are strong fliers and are capable of considerable dispersion and migration from the emergence site. This created problems for the Onchocerciasis Control Programme of re-invasion of treated areas in west Africa, where flies have been reported to travel up to 400 kilometres on the prevailing winds of the Intertropical Convergence Zone (Garms *et al.*, 1979). Both male and female blackflies feed on nectar, which is stored in a gut diverticulum called the crop. It is thought that nectar-feeding is an important energy source for flight. It has been calculated that following a sugar meal, *Simulium venustum* is capable of flying well over 100 kilometres in still air (Hocking, 1953).

Only female flies take blood meals. They are exophilic and exophagic and take blood meals during the day. Blackflies are pool feeders: the mouthparts are used to cut the skin and blood is taken from the pool that forms in the wound. The female completes the meal in a few minutes. Blackflies display host preferences. Most species, such as *Simulium euryadminiculum* and *S. rugglesi*, are ornithophilic, but there are many mammophilic species such as *S. venustum*, and others such as *S. damnosum s.l.* feed on both birds and mammals. Although some species of blackfly are autogenous, a blood meal is required by most females for egg batch maturation. The ovarian cycle takes 3 to 5 days in *S. damnosum* and about 500 eggs are matured in each cycle. The female has a maximum life expectancy of about 30 days and so can complete about 6 ovarian cycles. Other species, such as *Austrosimulium pestilens*, may complete the gonotrophic cycle within one day and other species have field life expectancies of at least 85 days. Fly numbers usually show seasonal peaks. In the tropics most flies appear during and immediately after the rainy season; in the more temperate regions of the world they appear in the summer months.

9.5.3   Ceratopogonidae

The nematoceran family Ceratopogonidae contains the biting midges known locally as punkies, no-see-ums or even sandflies (not to be confused with the phlebotomines – see Section 9.5.4). The blood-sucking midges are contained within four genera, *Forcipomyia*, *Leptoconops*, *Austroconops* and, by far the largest genus, *Culicoides*, with about 1400 described species (Beckenbach and Borkent, 2003). These tiny flies can make life a misery when they are present in large numbers, and the reaction to their bite is often intense. They are a serious threat to the tourist and leisure industries in several parts of the world, including parts of Scotland, Florida and the Caribbean. They are known vectors of several arboviruses of veterinary importance including bluetongue, bovine ephemeral fever, African horse sickness and Akabane virus, as well as the medically important Oropouche virus. The ongoing outbreaks of bluetongue virus in 15 Mediterranean countries have already resulted in the deaths of well over 600 000 sheep (Fu *et al.*, 1999). It is

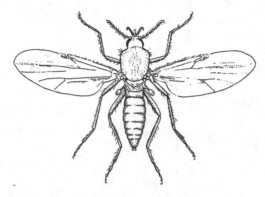

Figure 9.15 Adult *Culicoides nubeculosus*. (Redrawn from Smith, 1973.)

possible that the role of biting midges as vectors of arboviruses is considerably underestimated at present. Biting midges also transmit parasitic protozoa, including *Hepatocystis* spp., to mammals, some *Leucocytozoon* spp. and *Akiba* spp. to birds, and *Parahaemoproteus* spp. mainly to birds. They also transmit some filarial worms such as *Mansonella ozzardi*, *M. perstans* and *M. streptocerca* to humans and *Onchocerca gibsoni* to livestock.

The adult flies are very small, usually less than 1.5 mm long (Fig. 9.15). The black-and-white patterning typical of the single pair of wings of most species is the first and easiest clue for field identification. At rest the short, broad wings are folded scissor-like over the body. There is no distinct wing venation in the genus *Culicoides*, but it is more pronounced in the other genera. The mouthparts, which are used to cut the skin rather than pierce it, are held beneath the head. The long, usually 15-segmented antennae (12–14 in *Leptoconops* spp.) project well clear of the head and are plumose in the male and non-plumose in the female. The most important genus of biting midges, *Culicoides*, is distinguished by the presence of two depressions, known as humeral pits, placed anteriorly on the dorsal surface of the thorax.

In temperate parts of the world biting midges are largely a summer problem, in the tropics they may be present all year round. Both sexes feed on sugary solutions but only the females feed on blood. Ceratopogonids are exophilic, and exophagic, and mammophilic and ornithophilic species occur. Peak feeding time varies among species. Many are crepuscular, but perhaps the majority are active in the evening and first half of the night. Most activity takes place in still, warm, humid conditions; feeding is inhibited in winds of 1–$2\,\mathrm{m\,s^{-1}}$, probably by the inability of the fly to find hosts efficiently. Any exposed surface may be bitten, but when biting humans the scalp is often paid particular attention. Although autogeny is common in biting midges, most species require a blood meal for egg maturation

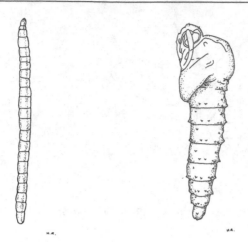

Figure 9.16 Larva and pupa for *Leptoconops spinosifrons*.

to occur. Mating usually occurs during swarming. The ovarian cycle take two to four days and females show gonotrophic concordance, maturing from 40–50 eggs in each cycle. These are laid in wet substrates, commonly marshy or sandy areas on the edge of water bodies (fresh or salt water). Some species use decaying organic matter as larval sites, breeding in rot holes in trees or in rotting banana stumps, the dung of herbivores, etc. Larval sites are very species-specific and adult flies often do not move far from their breeding sites. However, they can fly a few hundred metres and can travel in excess of 100 kilometres on the wind, a difficult issue for control campaigns.

We know little of the egg-laying behaviour in the biting midges, but females of *Leptoconops spinosifrons* burrow 3–6 cm into the sandy substrate to deposit their eggs (Duval *et al.*, 1974). In temperate areas biting midges may overwinter in the egg or fourth instar larval state, whereas in the tropics eggs usually hatch in under 10 days. There are four larval instars. The majority of ceratopogonid larvae are typically nematoceran, having 12 body segments lacking appendages or conspicuous setae; a dark, sclerotized head; and two retractable anal papillae, probably used in salt absorption (Fig. 9.16). The final instar larva may be up to 6 mm long. These larvae develop within their particular substrate and only rarely appear on the surface or swimming, in their characteristic serpentine fashion, in overlying water. It is possible that migration does occur at the wet fringe of expanding or contracting water bodies. (Larvae of the 50 or so species in the Forcipomyiinae do have anterior and posterior prolegs and have a more active, crawling lifestyle than most vermiform ceratopogonid larvae.) Although larval development of some species in the tropics may be complete in two weeks, larval development in biting midges is usually an extended process which, in the extreme case of some Arctic forms, may take two years. Larvae of some species are predatory, feeding on soil-dwelling metazoa such as nematodes; larvae of other species feed on micro-organisms. The short-lived pupae (2–3 days), which are culicid-like (Fig. 9.16), live in the surface layers of the substrate.

9.5.4 Psychodidae

The 600 or so species of phlebotomine sandflies are contained within the nematoceran subfamily the Phlebotominae, of the family Psychodidae. Two genera, *Lutzomyia* and *Phlebotomus*, contain the major vector species. Most phlebotomine sandfly species occur in the tropics, but some important disease vectors are found in temperate areas such as the Mediterranean. Two broad patterns of vector distribution are seen. In the Old World vector species are largely confined to the drier regions of the southern part of the temperate zone, but in the New World vector species are mainly found in wetter, forested areas.

Phlebotomine sandflies transmit the protozoan parasites causing leishmaniasis in humans and, in some cases, other animals. The epidemiology of leishmaniasis is complex (Ashford, 2001). To generalize in the briefest terms, in many if not most areas leishmaniasis is a zoonotic disease. Cutaneous leishmaniasis (also called dermal leishmaniasis or oriental sore) is largely an Old World disease occurring in the south of the Palaearctic region and transmitted principally by *Phlebotomus* spp. American dermal leishmaniasis (also called chiclero's ear, espundia and uta, and which includes mucocutaneous leishmaniasis) occurs largely in wet, forested areas of Central and South America from Mexico to Chile, where its vectors are *Lutzomyia* spp. Visceral leishmaniasis (also called kala-azar) is widely distributed through the drier areas of the southern Palaearctic region of the Old World, where its vectors are *Phlebotomus* spp.; it is also found in the drier parts of north-east Brazil, where the vectors are *Lutzomyia* spp.

In the Andean region of South America phlebotomine sandflies also transmit *Bartonella bacilliformis*, which is the causative agent of Carrion's disease of humans (also known as bartonellosis or Oroya fever). Sandflies also transmit viral diseases, including vesicular stomatitis virus, to cattle and horses, and sandfly fevers (also known as three-day fevers or pappataci fever) to humans. Although not important medically or economically, sandflies can also transmit trypanosomes (Anderson and Ayala, 1968), and probably reptilian malaria parasites (Ayala and Lee, 1970). Although their importance is primarily as vectors of disease, sandflies may occasionally be present in sufficient numbers to reach pest status because of the severe irritation caused by their bites.

Adult sandflies are small (1.5–4.0 mm long), slender-bodied, hairy, brownish insects with long, stilt-like legs. These blood-sucking flies are

Figure 9.17 Female of *Phlebotomus papatasi*. Note the erect wings. (Redrawn from Smith 1973.)

Figure 9.18 Phlebotomine wing to show that vein two (arrow) branches twice. (Redrawn from Smith, 1973.)

readily recognized at rest because they hold their single pair of narrow, lanceolate wings almost erect over the body (Fig. 9.17), whereas resting, non-biting psychodids fold their wings roof-like over the body. Closer examination shows that vein two of the wings of phlebotomines branches twice towards the middle or tip of the wing, whereas in non-biting psychodids branching is nearer the wing base (Fig. 9.18). The non-blood-feeding male sandfly is easily distinguished from the female by the pair of prominent claspers at the abdomen tip.

In temperate areas sandflies are present only in the summer months; in the tropics they may be present throughout the year. Mating is associated with the host rather than by swarming. The males jostle together and wait for the females in mating 'leks' near or on the host. Both sandfly sexes feed on sugars from plants or aphid honeydew, but only the female feeds on blood. Cold-blooded vertebrates and mammals are common hosts for various species, fewer species seem to be ornithophilic. Flies bite exposed parts of the skin either at twilight or after dark during periods of settled calm weather. Most are exophagic but some will rest and feed indoors; not surprisingly, these endophilic, endophagic forms contain some of the most important vector species. Autogenous forms are known (e.g. *Phlebotomus papatasi*), but most species require a blood meal for egg production to occur. Less than 100 tiny (< 0.5 mm), ovoid eggs are laid singly at each oviposition

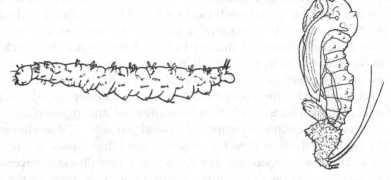

Figure 9.19 Phlebotomine larva showing matchstick hairs and pupa with last larval skin and caudal hairs. (Redrawn from Kettle, 1984.)

and darken to a brown or black colour within hours of being laid. Eggs, even of dry savannah dwellers, cannot withstand desiccation, and so oviposition sites need to be moist or wet. Oviposition sites are species-specific and range from termite mounds to cracked masonry, buttress roots of forest trees to leaf litter. In warm climates, eggs hatch in under 3 weeks and the 12-segmented larvae feed on organic debris. There are four larval instars. The larvae are grey with rather darker heads and are very difficult to locate, but if found are easily identifiable as phlebotomine larvae by the presence of so-called matchstick hairs on each segment (Fig. 9.19). In temperate areas, sandflies probably overwinter as mature larvae, in the tropics the larval period is usually less than two months. Pupae retain the skin of the final larval stage with its characteristic double pair of caudal bristles and can be recognized for this reason (Fig. 9.19). The pupal stage is completed in under two weeks. Like the egg stage, both larvae and pupae are very sensitive to drying out and require a humidity of more than 75 per cent. Because of their sensitivity to environmental conditions adult sandflies are often restricted to particular micro habitats (such as termitaria or mammal burrows) in an otherwise hostile landscape. Phlebotomine sandflies possess a short, hopping flight but can move up to 2 km over a period of several nights.

9.5.5 Tabanidae

The Brachyceran family Tabanidae consists of about 4300 species of medium to very large biting flies; indeed, it contains some of the largest blood-sucking insects. The three most important genera are *Chrysops*, *Haematopota* and *Tabanus*. The sheer size of tabanids and the pain their bite can cause mean they are not easily ignored. In consequence they have gained many local names, including clegs, mainly for *Haematopota* spp.; deerflies, mainly for *Chrysops* spp.; and horseflies and hippo-flies, mainly for *Tabanus* spp. Large accumulations of flies do occur in some areas, and with their painful bite tabanids can be pests of both livestock and humans. Livestock worry can reach a level at which economic losses occur.

Tabanids also transmit disease to both humans and livestock. Their painful bite results in the flies often being disturbed while feeding, but they are persistent and will commonly move directly to another animal to recommence feeding. Because of this, and the rather 'spongy' nature of the labella, which can hold up to a nanolitre of uncongealed blood, they are excellent mechanical vectors of several parasites. Tabanids mechanically transmit the flagellate *Trypanosoma evansi* that causes severe disease (surra) in horses, camels and dogs, and less severe illness in several other mammals, including cattle. Surra occurs over large areas of the tropics and subtropics. Tabanids are also the mechanical transmitters of other trypanosomes, including *T. vivax viennei* of cattle and sheep in South America; *T. equinum*, the causative agent of mal de Caderas of equines; *T. simiae* of pigs; and the African trypanosomes normally transmitted by tsetse flies. Tabanids are also the vectors of the posterior station (stercorarian) trypanosome *T. theileri* of cattle. In addition, tabanids are mechanical vectors of the bacterium *Francisella tularensis*, the causative agent of tularaemia in humans. This zoonotic disease is widespread throughout the Holarctic; the primary hosts are wild rodents. It is transmitted by a variety of agents, but in the USA the tabanid *Chrysops discalis* is a major vector. Tabanids also mechanically transmit other infectious agents, including anthrax, equine infectious anaemia virus, California encephalitis, western equine encephalitis, rinderpest and anaplasmosis.

Tabanids also transmit filarial worms, including *Loa loa* of humans; the closely related *L. loa papionis* of monkeys; the arterial worm of sheep, *Elaeophora schneideri*; and *Dirofilaria roemeri* of macropodid marsupials. Transmission of filarial worms is not mechanical, the tabanid is an obligatory stage in transmission, with the parasites undergoing development in the insect. The major vectors involved in the transmission of the human parasite *Loa loa* are *Chrysops dimidiatus* and *C. silaceus* in west Africa and *C. distinctipennis* in central Africa.

Adult tabanids are medium to very large (up to 25 mm), stocky insects (Fig. 9.20) variously coloured from black through brown to greens and yellows. The large, semi-lunar head bears two prominent eyes, which, in the live insects, are often patterned with iridescent reds and greens that usually fade after death. The eyes are commonly spotted in *Chrysops* spp., have zig-zag bands in *Haematopota* spp., and horizontal stripes and no patterning in *Tabanus* spp. In the male the eyes are holoptic; in the female, dichoptic, and this is an easy means of distinguishing the sexes. The thickset, porrect antennae consist of three components, the scape, pedicel and flagellum, and antennal shape is used to distinguish the three major genera,

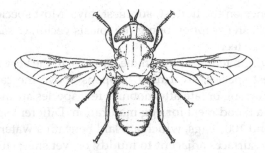

Figure 9.20 Adult female *Ancala africana*. (Redrawn from Smith, 1973.)

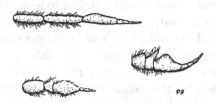

Figure 9.21 Antennal shape can help distinguish the three major genera of tabanids: *Chrysops* (top), *Tabanus* (middle) and *Haematopota* (bottom).

Tabanus, *Chrysops* and *Haematopota* (Fig. 9.21). Only the females suck blood; their mouthparts are adapted for cutting rather than piercing the skin and project below the head, not in front of it.

The thorax and abdomen are commonly patterned in various colours, which often provides a good guide for field identification. The single pair of wings may be very large (a span of up to 65 mm in the largest species), they may be clear (many *Tabanus* spp.) or may bear coloured patterns (usually mottled in *Haematopota* spp. and with a distinct stripe in *Chrysops*). At rest tabanids hold the wings either flat over the body like a pair of partially open scissors (*Chrysops* and *Tabanus* spp.), or they may hold the wings roof-like over the body (*Haematopota* spp.).

In temperate countries adult tabanids appear in summer. In the tropics they also commonly show a seasonal pattern, with the smallest numbers in the dry season. Both male and female flies feed on sugary solutions, which are important for their energy and water budgets. Adult flies survive three or four weeks in the field. The blood-feeding female utilizes a wide variety of mammals, particularly ungulates; some species also feed on reptiles and amphibians, but relatively few tabanids feed on birds. Most tabanids are restricted to certain habitats, and host choice is influenced by availability within the habitat. Most tabanids are woodland or forest dwellers, including *Chrysops* vectors of loiasis. Humans become infested only if they enter the forest habitat of these insects. Most species feed during full daylight and are often most in evidence on the hottest, sunniest days. Most species are exophilic and exophagic, an exception being the loiasis vector, *C. silaceus*, which will enter houses to feed.

Mating occurs after virgin females enter male swarms. In the very many canopy-inhabiting forest species, swarming happens above the canopy, normally in the early morning or late afternoon. A few species are autogenous, but most require a blood meal for egg maturation. Different species produce between 100 and 1000 eggs, which are laid beneath a waterproof layer on plants or other surfaces adjacent to muddy or wet sites; surfaces chosen for egg laying are often species-specific. The eggs hatch in one or two weeks and the larvae drop to the ground. In general, larval *Haematopota* spp. live in relatively dry soil, *Tabanus* spp. in wet soil close to water bodies, and the larvae of *Chrysops* spp. in wet mud, often in semi-submerged situations. There are between 4 and 11 larval instars, and larval development is prolonged and may take three years in some species (e.g. *Tabanus calens*). The larvae are greyish and have typically brachyceran reduced heads. Tabanid larvae may be identified by the raised tyre-like rings between body segments, by the six pseudopods that can often be found on segments 4–10, and by the presence of Graber's organ (a black pyriform sensory structure) on the dorsal surface in front of the spiracle-bearing terminal siphon. In general, *Chrysops* larvae are detritovores, whereas *Tabanus* and *Haematopota* larvae are carnivores (and cannibalistic). Pupation occurs in the soil at the drier fringes of the larval habitat and the pupae are capable of limited movement. Adults emerge in under three weeks. Although adult tabanids are strong fliers capable of considerable movement away from the breeding site, they do not normally become widely dispersed.

9.5.6   Rhagionidae

The Rhagionidae (= Leptidae) are a relatively little-studied family of brachyceran flies commonly known as snipe flies. Most are predatory upon other insects, but some species feed on blood. They are not known to be the vectors of any parasitic organisms. Because their bite is very painful and some humans react badly to it, these flies can be of nuisance value. The best known snipe flies occur in the genus *Symphoromyia*. These are found in the Holarctic region, where species such as *S. atripes* and *S. sackeni* are troublesome to both humans and other animals. *Spaniopsis* and *Austroleptis* spp. are nuisance flies in Australia, and *Atherix* spp. are found in the Nearctic and Neotropical regions.

Rhagionids are medium to large, elongated, rather bare, sombrely coloured flies (Fig. 9.22). The mouthparts are heavily sclerotized and are used for cutting and piercing the host's skin, and blood is then sucked up from the wound. Humans, horses, dogs, deer, cattle and frogs have all

Figure 9.22 Adult rhagionid, *Spaniopsis longicornis*. (Redrawn from Smith, 1973.)

been reported to be attacked by snipe flies. The adults also feed on sugary solutions. Little is known about the larvae, but larval sites include wet soil, leaf mould and rotting wood. Larvae of some species are predatory upon various insect larvae, earthworms and other soil fauna, but other species feed on mosses.

9.5.7 Muscidae

The cyclorrhaphan family Muscidae contains some 4200 species, only a few of which are haematophagous as adults. Some, like the stablefly *Stomoxys calcitrans*, have mouthparts that are well developed for penetrating vertebrate skin. Others, such as *Musca planiceps*, have strong, rasping prestomal teeth that dislodge scabs or scrape through thin skin. Other flies, such as the headfly *Hydrotaea irritans*, are facultative haematophages feeding from open wounds or sores. From an economic and health point of view, the most important genera are *Stomoxys* and *Haematobia*, and these two genera will be used as examples of biting flies.

The stablefly, *S. calcitrans*, is a pest species of worldwide distribution. It is particularly important as a worry to livestock and causes considerable annual losses to the agricultural industry. In some areas it may also be a direct nuisance to people. It can mechanically transmit trypanosomes and is especially important in the transmission of *Trypanosoma evansi*, which causes severe disease (surra) in horses, camels and dogs, and less severe illness in several other mammals, including cattle. It also transmits the Neotropical *T. equinum*, which causes mal de Caderas in equines, sheep, cattle and goats. Stableflies also play a minor role in the transmission of the

Figure 9.23 The characteristically forward-projecting proboscis of the stablefly, *Stomoxys calcitrans*. (Redrawn from Smith, 1973.)

African trypanosomes associated with tsetse fly. It is also the vector of the nematode *Habronema majus*, a stomach worm of equines. *Stomoxys calcitrans* has also been implicated in the transmission of other organisms such as polio virus, equine infectious anaemia, anthrax and fowl pox, although the regularity and hence importance of such events is unclear. *Stomoxys nigra*, *S. omega* and *S. inornata* are more localized pest and vector species found in tropical Africa and Asia.

The horn fly, *Haematobia irritans irritans*, is a major agricultural pest species throughout the northern hemisphere. Like the stablefly, it causes substantial economic losses in the agricultural industry through livestock worry. It is also a vector of *Stephanofilaria stilesi*, a skin parasite of cattle. The buffalo fly, *Haematobia irritans exigua*, is particularly important as an agricultural pest in the Australian region.

Other muscids, such as the sheep head fly *Hydrotaea irritans*, are facultative haematophages. Unable to penetrate the skin themselves, they gain access to blood by disturbing other insects that have penetrated the skin for them. Some *Hydrotaea* spp. may crowd around feeding insects such as tabanids, feeding concurrently with them or even crowding them off the wound. The head fly is a woodland species and in Europe causes damage to sheep, particularly lambs, both by animal annoyance, and through secondary bacterial infection of wounds and sores. It is also a vector of summer mastitis.

In temperate regions adult stableflies are commonly encountered throughout the summer and autumn basking on sunlit surfaces around stock pens but, despite the name, surprisingly rarely around stables. They are housefly-sized, but are easily distinguished from house flies by the forward-projecting proboscis (Fig. 9.23). Both male and female flies feed on blood as well as sugary solutions and a wide range of mammalian hosts are used. On sunny summer days they may feed twice, but one blood meal a day is probably normal. The bite is painful and often disturbs the host, leading to interrupted feeding and movement to another host – ideal conditions for the mechanical transmission of disease. Males congregate in sunlit patches that serve as mating leks.

The horn fly is much smaller than the stablefly, about half the size of a house fly. Its forward-projecting proboscis is squatter and more heavily built than the stablefly's. Both males and females take blood meals, mainly from bovids. The flies are almost permanently associated with their hosts. They commonly feed several times a day and defecate partially digested blood. Mating occurs on the host animal.

The female horn fly leaves the host animal to lay her brownish eggs. Up to 24 eggs are laid during any one cycle, in, under or close to the freshly deposited dung of the host animal, where they hatch in less than a day. The larvae feed in the dung and after three to eight days migrate to drier areas at the edge of the dung pat or in adjacent soil and pupariate. In temperate regions the flies overwinter as puparia, although in ideal conditions adults emerge from the puparium after three to eight days.

The anautogenous female stablefly lays her white eggs in rotting or fermenting vegetation, urine-soaked straw, etc., but rarely in dung alone. She lays up to 50 eggs during each ovarian cycle. These hatch 1 to 4 days after laying and the 3 larval stages are complete in 10–20 days. Pupation occurs in the drier parts of the medium and the adult emerges in under 10 days. Development time of the immature stages of these flies is often not readily correlated with ambient temperature because rotting and fermenting vegetation is warmer than its surroundings. In temperate regions the flies overwinter in the larval or pupal stage.

Both adult stableflies and horn flies are strong fliers and both species may disperse considerable distances from their emergence sites. Stableflies are largely inactive after dark, but it is thought that horn flies actively disperse at night.

9.5.8 Calliphoridae

No adult calliphorids are blood-sucking, but in some species the larvae feed on blood. The best-known example is the African species *Auchmeromyia senegalensis*, the Congo floor maggot. The adult fly is mainly coprophagous. The female lays batches of up to 60 eggs in dry sandy soil on the floor of huts, caves, etc. The three larval stages are extremely resistant to dry conditions and to starvation. The non-climbing larvae emerge from cracks and crevices to take nightly blood meals from hosts sitting or sleeping on the floor of the shelter. Recorded hosts are humans, suids, hyena, aardvark and dogs, which may be a reservoir for humans. The larva penetrates the skin using its mouth hooks and maxillary plates and completes the blood meal in about 20 minutes. The larval period can be completed in under two weeks given freely available food, but may take up to three months if food is scarce. Pupation occurs in protected situations and the pupal period is about two weeks. Several genera of calliphorids, including *Protocalliphora*, have larval stages that attack nesting birds.

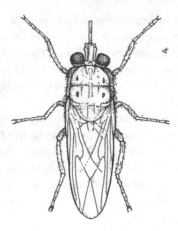

Figure 9.24 Plan view of a resting tsetse fly showing the folded wings. (Redrawn from Smith, 1973.)

9.5.9 Glossinidae

The cyclorrhaphan family Glossinidae contain medium to large insects commonly known as tsetse flies. Tsetse flies are important as vectors of trypanosomes to humans and to their domesticated animals. They are rarely, if ever, present in sufficient numbers to present a fly worry problem to livestock. As vectors of animal trypanosomiasis, or nagana, they preclude the use of up to ten million square kilometres of Africa for cattle-rearing and have also restricted development on that continent by limiting the use of draught and pack animals. The counter argument has also been put that the tsetse has prevented desertification of large areas of land from overgrazing and has been the saviour of Africa's game animals. These arguments have been clearly outlined by Jordan (1986). It is clear that nagana also slowed the conquest and colonization of Africa by preventing the free passage of the horse. The transmission of human sleeping sickness occurs on a more restricted scale, with only about 10 000 new cases reported each year throughout most of the twentieth century. However, an upsurge in disease has occurred at the end of the twentieth century, with 300 000–500 000 cases annually. The World Health Organization estimates that 45 million people are at risk. The danger from the disease was shown in the great epidemics of the colonial era. More than half a million people died of the disease in the Congo basin between 1896 and 1906, and as many as a quarter of a million in Uganda between 1900 and 1906.

Adult tsetse flies are brownish, rather slim, and elongated. The smallest are about 6 mm long and the largest are about 14 mm. They have a characteristic resting attitude in which the single pair of wings are folded like closed scissors over the fly's dorsal surface, their tips extending for a short distance beyond the rear of the abdomen (Fig. 9.24). The proboscis, which

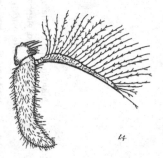

Figure 9.25 The branched hairs of the arista, a characteristic feature of tsetse flies.

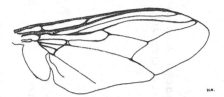

Figure 9.26 The hatchet shape of the discal cell of the wing, a characteristic feature of tsetse flies. (Redrawn from Kettle, 1984.)

has a distinct basal bulb, projects in front of the head and is flanked by palpi of almost equal length. Two characteristic features of tsetse flies are the branched arista of the antennae (Fig. 9.25) and the 'hatchet' shape of the discal cell of the wing (Fig. 9.26). Another distinctive feature of tsetse flies is the high-pitched buzzing sound that they produce when heating themselves endothermically (Howe and Lehane, 1986), and after which they have gained their onomatopoeic name. Tsetse flies are exophilic and exophagic, and live at low population densities estimated at an average of about 10 per hectare. They are diurnal insects that essentially show a bimodal pattern of activity (morning and evening), although there are species-specific differences. The pattern is modulated by environmental conditions and by the fly's physiological status (Brady, 1975). The tsetse is a viviparous insect, the female gives birth to a single, preposterously large, third instar larva about every nine days.

There are 23 species (of which 6 have one or more subspecies) and these are commonly arranged into 3 groups (Table 9.4): *fusca, palpalis* and *morsitans*. These groups are sometimes accorded the status of subgenera, when they are termed *Austenina, Nemorhina* and *Glossina*, respectively. Tsetse flies are now restricted to sub-Sahelian Africa (c. 4 °N to 29 °S, extending to 30° along the eastern coast) and two pockets in the Arabian peninsula. Fossil specimens have been found in the Oligocene shales of Florissant,

Table 9.4 *Characteristics of the three tsetse fly (Glossina) groups: fusca, palpalis and morsitans.*

Group	Species	Subspecies	Habitat	Geographical distribution	Vector status	Major hosts
fusca	G. haningtoni		lowland rainforest species	West and central Africa		river-hog, porcupine, ox, bushbuck
	G. nashi					
	G. tabaniformis					
	G. van hoofi					
	G. severini					
	G. nigrofusca	nigrofusca, hopkinsi				
	G. fusca	fusca, congolensis	edge of the rainforest, forest areas outside the lowland forest, forest areas along watercourses	West and central Africa and isolated areas of East Africa		bushbuck, river-hog, aardvark
	G. fuscipleuris		forest islands, often along watercourses, arid habitats			bushpig, hippo, buffalo, bushbuck, rhino, elephant, buffalo, giraffe, ostrich
	G. medicorum					
	G. schwetzi					
	G. brevipalpis			East Africa	locally important	
	G. longipennis			East Africa		
morsitans	G. morsitans			extensively in eastern and southern Africa	economically the most important vector	wart-hog, kudu, buffalo

Subgenus	Species	Subspecies	Habitat	Distribution	Importance	Hosts
morsitans		*morsitans*	savannah woodlands, in the drier areas restricted to mesophytic vegetation around waterways	Mozambique/Zimbabwe up to Tanzania		wart-hog, ox, humans, buffalo, kudu
		submorsitans		Ethiopia across to Senegal		wart-hog, humans, bushbuck, buffalo
		centralis		Botswana/Angola up to southern Uganda		buffalo
	G. *swynnertoni*			Northern Tanzania, southern Kenya	locally important	wart-hog, buffalo, giraffe, rhino
	G. *longipalpis*			Guinea to Cameroon	major economic importance	bushbuck, wart-hog, bushpig, buffalo
	G. *pallidipes*			Mozambique up to Ethiopia	locally important	bushbuck, buffalo
	G. *austeni*			East Coast from Mozambique to Somalia		bushpig, ox, duiker
palpalis	G. *palpalis*	*palpalis*	lowland rainforest, often associated with its waterways. Extending its range into savannah, following waterways	extensively in west and central Africa	important medically	humans, reptiles, bushbuck, ox
		gambiensis		Benin down to Angola		
		fuscipes		Sierra Leone to northern Benin extensively in west and central Africa	important medically	
		martinii		extensively in north zone of lowland forest		
		quanzensis		in south-east zone of lowland forest		
	G. *pallicera*	*pallicera*	rainforest	in south-west zone of lowland forest		
		newsteadi		forest areas of west Africa		
	G. *caliginea*		coastal mangrove swamp and rainforest	West African mangrove and adjoining forest		
	G. *tachinoides*		along watercourses in savannahs	West Africa and pockets as far east as Ethiopia	important medically	humans, ox, porcupine

(Information from several sources, notably Jordan, 1986 and Weitz, 1963.)

Colorado, North America (Cockerell, 1918), suggesting that in the past they had a more extensive range into the Nearctic. Tsetse flies currently compass more than 11 000 km² of Africa, being restricted in the north by the aridity of the deserts and in the south by low seasonal temperatures, although they are not evenly distributed within this range (Fig. 9.27).

The three tsetse groups are each closely associated with particular vegetation types. Eleven of the twelve species in the *fusca* group are forest species; some members (e.g. *Glossina nigrofusca*) are restricted to lowland rainforest, others (e.g. *G. fusca* and *G. brevipalpis*) are restricted to the edge of the rainforest and/or isolated areas of forest in the savannah, while one member, *G. longipennis*, is found in arid habitats. The *palpalis* group occurs in lowland forest, but many (e.g. *G. palpalis*) also extend into the drier savannahs along the woodlands of river banks and lake shores. Members of the *morsitans* group are restricted to woodlands, scrub and thicket in the savannah. Because of the relative intolerance of tsetse flies to low temperature (adults are inactive below *c.* 16 °C), they are not found in highland areas (above *c.* 1300–1800 m depending on latitude).

Male flies complete spermatogenesis before emergence from the puparium, but need several blood meals before they become fully fertile, at about seven days post-emergence. In contrast, the females become sexually receptive within a day of emergence. Because of the low density of tsetse fly populations, mating is most likely to occur when flies congregate around vertebrates to feed. The non-feeding, mature males that constitute the 'following swarm' found around moving hosts are probably there to mate with incoming, hungry, particularly virgin, females; the male locates the female visually. Mating requires the presence of a contact sex pheromone that is present in the cuticular waxes of the female. After mating the sperm migrates into the female's spermathecae, where it remains viable for the rest of her life. Despite laboratory evidence that females are willing to mate more than once, it seems that this rarely occurs in the field. Mating for more than 60 minutes is required to trigger the generation of hormones in the female that stimulate ovulation. The first ovulation occurs about nine days post-emergence and subsequently at nine- to ten-day intervals. Because female tsetse only produce a single offspring during each reproductive cycle, the female must be relatively long-lived to produce sufficient offspring for the perpetuation of the species. Many studies have determined the age of flies in the field from the structure of the ovary. Every female has two ovaries, each with two polytrophic ovarioles that ovulate in turn in a predictable order. At each ovulation structural changes occur in the ovarioles that can be recognized and quantified on dissection, allowing the estimation of the physiological age of field-caught female flies. In addition, age may be estimated using a biochemical technique (Lehane and Hargrove, 1988; Lehane and Mail, 1985; Msangi *et al.*, 1998). Using such

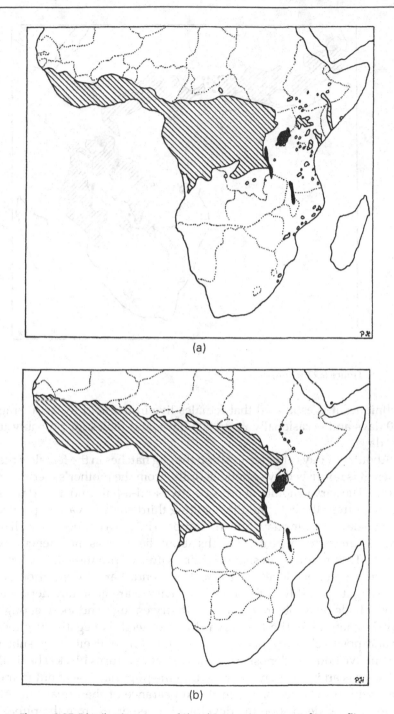

(a)

(b)

Figure 9.27 Distribution maps of the three major groups of tsetse flies: (a) *Glossina fusca* group; (b) *Glossina palpalis* group; (c) *Glossina morsitans* group. (Redrawn from Jordan, 1986.)

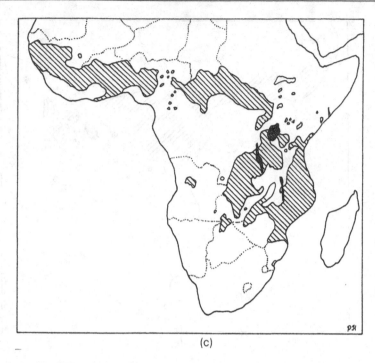

(c)

Figure 9.27 (*cont.*)

techniques it is estimated that females live in the field for an average of 100 days and occasionally up to 200 days, and males for probably up to 100 days.

About four days after fertilization the egg hatches in the female's uterus, where it is nourished by secretions (milk) from the mother's uterine (milk) gland. The larva moults about one, two-and-a-half and five days after hatching from the egg, and the resulting third-stage larva is deposited by the female about nine days after ovulation. The larva has been so well nourished in utero that at birth it weighs about the same as the unencumbered female fly. It is stubby in shape, white, with two prominent, black polypneustic respiratory lobes on its posterior end. Larval deposition occurs in daylight, probably in the afternoon. Larvae are generally deposited in dense shade on a substrate that is both dry enough and loose enough for larval penetration. The free larva does not feed, is negatively phototactic and positively thigmotactic, and rapidly buries itself in the substrate. Within five hours of deposition it pupariates and turns black. The final larval ecdysis and pupation occurs within the puparial case about four days later, with no visible change in the appearance of the puparium. About a month after deposition the developed fly emerges from the puparium; females normally emerge one or two days before the males. The emerging fly forces the cap from the puparium using its ptilinum and burrows up through the soil. It hardens off and is capable of flight about an hour after emergence. Over about the next nine days the fly's cuticle continues to develop, thicken and harden and the fly increases the mass of its thoracic flight musculature. This interval of post-emergence development is called the teneral period.

Both male and female flies feed exclusively on blood. Like other insects feeding in this way, tsetse flies maintain symbiotic micro-organisms to supplement their nutrition. These symbionts are held intracellularly in a specialized portion of the midgut called the mycetome. Because there is no feeding in the larval stage, the adult fly emerges from the puparium with few remaining reserves and it is critical that an early blood meal is obtained. Tsetse flies then feed every two to three days and take blood meals that average two to three times their unfed body weight, although, due to the size of the growing larva, the female takes progressively less blood as pregnancy proceeds. Tsetse flies are attracted to their food source by a mixture of olfactory and visual stimuli. There are five main patterns of host selection seen in *Glossina* spp. (Weitz, 1963):

(1)   Species with catholic tastes, such as *G. tachinoides*, *G. palpalis* and *G. fuscipes*, which feed on most available hosts.
(2)   Species such as *G. austeni*, *G. tabaniformis* and *G. swynnertoni*, which feed largely on suids.
(3)   Species such as *G. morsitans*, which feed on suids and bovids.
(4)   Species such as *G. pallidipes*, *G. fusca* and *G. longipalpis*, which feed largely on bovids.
(5)   Species such as *G. longipennis* and *G. brevipalpis*, which feed on other hosts such as rhinoceros and hippopotamus, respectively.

Disease problems arise only when the fly's host choice includes humans or their domesticated animals. One of the reasons why the vector status of the 23 species varies considerably (Table 9.4) is because the degree of overlap varies among fly species.

Tsetse flies are unusual in using proline as an energy source for flight (see Section 6.6). One consequence is they spend less then an hour a day flying. Resting sites are species-specific and an understanding of them is important in the planning of insecticide-based control campaigns. During the dry season flies tend to be restricted to certain primary foci. Appreciation of this fact permits the use of reduced quantities of insecticide in their control. Similarly, while *G. m. morsitans* may rest at heights of up to 12 m, *G. pallidipes* in the same habitat is rarely found resting more than 3 m above the ground. Clearly, appreciation of this difference means great savings can be made in the insecticide quantities used for treatments solely for *G. pallidipes* control. Day resting sites are usually the woody parts of the vegetation, and the higher the temperature the nearer the ground the fly tends to be. At night the fly moves to the upper surfaces of leaves.

9.5.10 Hippoboscidae

The Hippoboscidae are cyclorrhaphous flies that are exclusively ectoparasitic in the adult stage. They are often included in the artificial grouping the Pupipara. At present about 200 species are recognized. The flies are locally known as keds or forest flies. The best-known insect in the group is the sheep ked, *Melophagus ovinus*, which is found on sheep throughout the world (Fig. 9.28). The keds are irritating to the sheep, whose scratching, rubbing and biting damages the wool, which also becomes badly marked with the ked's faeces. The feeding activities of the keds also cause areas of damage to the sheep's skin known as 'cockle' which may make the hide unsaleable. Heavy infestations cause loss of condition in the sheep and can cause anaemia. Various species of hippoboscid will attack humans, and the bite of the pigeon fly, *Pseudolynchia canariensis*, is said to be painful. Apart from the sheep ked, hippoboscids are of little importance to humans. About 75 per cent of hippoboscids infest birds (the pigeon being the only domesticated species afflicted), the rest being found on mammals, particularly bovids and cervids. Hippoboscids transmit several relatively non-pathogenic parasites to these hosts. For example, *Trypanosoma melophagium* is transmitted to sheep by the sheep ked *T. theileri*, to cattle by *Hippobosca* spp., and the filarial worm *Dipetalonema dracunculoides* to cats and dogs by *H. longipennis*. *Pseudolynchia canariensis* is the vector of *Haemaproteus columbae* to pigeons.

Adult keds are highly adapted for life on the surface of the host animal. They are about 5–10 mm long, and are leathery insects; most have a single pair of rather tough wing membranes, all of which act as protection against abrasion by the host covering. Further protection is afforded by the partial sinking of the head into the thorax, the placing of the antennae in deep grooves, and having partially retractable mouthparts that are further protected by the rigid palpi. Like many other ectoparasites, keds are dorsoventrally flattened. Flattening may well allow increased freedom of movement among hair or feathers and allows adpression against the host or its covering. Both help the insect to evade the host's grooming activities. The legs are stout and bear paired claws in mammal-infesting forms and toothed claws in bird-infesting forms, and these are used to cling to the covering of the host. Adult hippoboscids present a spectrum of forms from permanently winged to wingless. About three-quarters of all Hippoboscidae are, like *Hippobosca* spp., winged throughout life (Fig. 9.28). *Allobosca* spp. lose only the wing tips on landing on a host, but in *Lipoptena* spp. the wings are lost at their base by abrasion, breakage or shedding once the host is located. *Melophagus* spp. do not have functional wings. Female

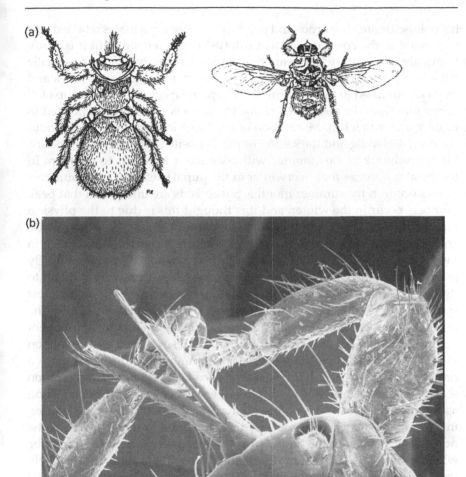

Figure 9.28 (a) Adult of the sheep ked, *Melophagus ovinus* (left), and the winged hippoboscid, *Hippobosca equina* (right). (b) Scanning electron micrograph of a dorsal view of a sheep ked head (courtesy of Gregory S. Paulson).

hippoboscids are viviparous and produce a single larva that is retained and nourished in the common oviduct (uterus) of the female until it is ready to pupate. The mature larvae are deposited in a variety of species-specific sites. The sheep ked, *M. ovinus*, is unique in that the deposited larva and pupa are attached to the fleece of the sheep. In *Lipoptena* spp. the pupa fall at random from the host. In *H. equina* the larva is commonly deposited in humus under bracken. Maturation of the larva in the female takes about a week, hardening and darkening of the deposited larva about six hours. Pupa produced in the summer will take about a month to develop. In temperate countries flies overwinter in the pupal stage and peak numbers of flies occur in the summer months. Sheep keds are unusual in that peak numbers occur in the winter, and it is thought this is due to the physical removal of keds at shearing, the development of resistance among ewes, and possibly the adverse effects of temperature increases in the fleece in the summer months. Adult flies emerging at a distance from the host fly immediately to it. Once on the host, hippoboscids are usually difficult to dislodge. Although in some species mating can occur on the wing, most species mate on the host. The first mating usually occurs soon after the female's emergence, but repeated matings are common, with subsequent matings occurring after larval deposition. Both sexes feed exclusively on blood. Once on the host *Hippobosca* spp. commonly feed several times a day, whereas the sheep ked feeds only once every 36 hours. In common with other insects in which all the life stages are dependent solely on blood as the nutrient source, hippoboscids have symbionts. These are housed in a mycetome on the intestine and are transferred to the offspring in the nutrients provided by the mother for her intra-uterine larva. As might be expected from the low fecundity of these flies, the adults are relatively long-lived, adult sheep keds living for about five months and *Hippobosca* spp. living for six to ten weeks.

9.5.11 Streblidae

The Streblidae are a relatively little-studied family of cyclorrhaphous flies in which the adults are exclusively ectoparasitic on bats. They are mainly found on colonial bats and many are strongly host-specific. Along with the Nycteribiidae, which are also exclusively ectoparasitic on bats, they are known as bat flies. They are often grouped with the Hippoboscidae and Nycteribiidae into the artificial grouping the Pupipara. Streblids are largely a New World group found in the tropics and subtropics where winter temperatures do not fall below 10 °C. Although they have occasionally been recorded as biting humans, they are of no medical or economic importance.

Streblids vary in length from about 1 mm up to about 6 mm (Fig. 9.29). Like all ectoparasites, streblids are exposed to the abrasive outer covering of the host. As a protective measure they are leathery and hold the

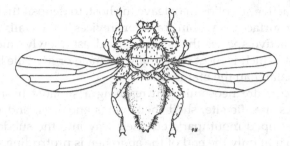

Figure 9.29 Adult streblid, *Trichobius lonchophylla*. (Redrawn from Marshall, 1981.)

appendages, such as the antennae, in grooves or pits. In common with many other ectoparasites their bodies are flattened, allowing increased freedom of movement through the host's covering and allowing adpression against the host, both of which help the fly avoid the host's grooming activities. Although most streblids are dorsoventrally flattened, interestingly the New World genus *Nycterophilia* are, like fleas, laterally compressed. Streblids show a range of forms from fully winged during adult life (about 80 per cent), through caducous, to apterous. Some streblids display a further modification in being able to fold away their wings beneath protective setae on the abdomen, which minimizes wing abrasion when the fly is on the host. Wing form correlates with leg form. Fully winged species have well-developed, rather long legs. Species with reduced or no wings have shorter legs that are modified in various ways for clinging to, or parting, the host's hair.

Both sexes of streblid feed on blood. It is probable that the pelage-dwelling forms such as *Megistopoda aranea*, which tend to remain more or less permanently on the host, feed at least once a day. Volant forms such as *Trichobius yunkeri*, which often leave the host, probably feed less often. Newly emerged females may mate on the host or, in forms such as *T. major* that commonly leave the host, mating may occur elsewhere in the roost. In common with other Pupipara, multiple matings are usual, with subsequent matings occurring after larval deposition. Unlike other Pupipara, it is believed that each mating is necessary for the production of further offspring. Streblids are viviparous and there is no free-feeding larval stage, so that all life stages are dependent upon blood for their nutrition. In consequence, it is virtually certain that streblids, like the tsetse flies, triatomine bugs and other insects in this position, have symbiotic micro-organisms providing nutritional supplements; however the site of the mycetome is unknown. The mature larvae are deposited by the female and immediately pupate. In *Trichobius* spp. the larva pupates within the mother, and it is the pupa not the larva that is deposited. Most streblid offspring are deposited in bat roosts. More active streblids may leave the host, to deposit their off-spring on particular surfaces or within cracks or crevices in the walls of the roost, in other less active species the larva is broadcast. Newly emerged flies seek out a host, which is most commonly a female or juvenile bat as these are more colonial than the mature males.

Female *Ascodipteron* are neosomic. On reaching a suitable host they migrate to a species-specific site, shed their wings and legs, and using their very well-developed mouthparts cut their way into the subdermal tissues of the bat so that only the end of the abdomen is protruding. They become anchored in this position by their mouthparts. The abdomen swells greatly, overgrowing the rest of the body so the head and thorax become embedded within it. In this position the fly feeds and matures its offspring which, as third-stage larvae, fall from the bat to the floor of the roost.

9.5.12 Nycteribiidae

The Nycteribiidae are a small family of relatively little-studied cyclorrhaphous flies in which the adults are exclusively ectoparasitic on bats. Along with the streblids, which are also exclusively ectoparasitic on bats, they are known as bat flies. They are often grouped with the Hippoboscidae and Streblidae into the artificial grouping the Pupipara. The various species of bat fly show a high degree of host specificity. They are largely an Old World group of insects, found in temperate as well as tropical areas and are of no medical or economic importance.

Nycteribiids are highly modified for their ectoparasitic existence, being rather leathery, dorsoventrally flattened, wingless insects in which the head is held protectively in a groove on the dorsal surface of the thorax (Fig. 9.30). Nycteribiids have body combs that may protect delicate areas of the body from abrasion and/or prevent removal of the insect during host grooming (see Section 7.2). Nycteribiids can be divided into two main types. The first group are small, like many *Nycteribii* spp., and tend to remain in the fur of the host's trunk, in which they are capable of 'swimming'. The second type is exemplified by several species of *Basilia* and *Penicillidia*. They tend to be larger and to live on the surface of their host's hair, into which they can escape if disturbed. Perhaps not surprisingly, these larger forms tend to congregate in areas protected from the host's grooming activities such as between the shoulder blades, under the chin or in the axilla of roosting hosts or, in flying hosts, in the tail region. The legs of nycteribiids are elongate and bear claws which allow the large forms to move rapidly over the host's surface. These large forms will often leave the host.

Adults of both sexes feed exclusively on blood. Nycteribiids are viviparous and there is no free-feeding larval stage, so that all life stages are dependent upon blood for their nutrition. In common with other insects in

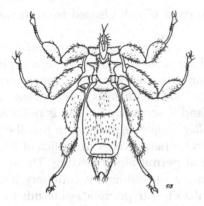

Figure 9.30 An adult nycteribiid. (Redrawn from Marshall, 1981.)

a similar position (e.g. tsetse flies and triatomine bugs), nycteribiids have symbiotic micro-organisms providing nutritional supplements. The mycetome housing these symbionts is sited in the dorsal part of the abdomen. Newly emerged females may mate before taking a blood meal. Multiple matings are common with subsequent matings occurring at the time of larval deposition. From egg production to larval deposition takes about nine days in *Basilia hispida*. Nycteribiids leave the host to deposit their offspring in a suitable crack or crevice, often on a vertical surface in the bat roost. The pupal stage lasts about a month. Newly emerged flies seek out a new host, most commonly a female or a juvenile bat as these are more colonial in their habits than the mature males. Nycteribiids produce between 3 and 16 offspring in a lifetime. Nycteribiids are associated with hibernating bats on which, although they continue to feed, reproduction is greatly slowed or ceases altogether, so population size falls to its lowest levels in the winter months.

9.6   Other groups

In addition to the insects already discussed there are a few instances of blood-sucking in other orders. In the main, these are probably reports of rare occurrences such as the account of blood feeding by a vespid wasp. In other cases, however, it is clear that blood-feeding is a regular occurrence.

In the order Lepidoptera there is at least one noctuid, *Calyptra* (= *Calpe*) *eustrigata*, that is known to feed on blood that it has obtained by piercing the skin of its forest-dwelling mammalian hosts (Banziger, 1971). This moth, which is found in South-East Asia, pierces the host's skin using its straight, lance-like proboscis. This insect probably evolved from species specialized for piercing fruits. Other geometrid and pyralid moths will feed on blood oozing from wounds, or from drops of blood released from the anus of feeding mosquitoes.

Species in which adults are ectoparasitic on mammalian hosts are found in the coleopteran groups Leptininae, Quediini, Amblyopinini and Languriidae. The best-known examples are in the four genera *Leptinus, Leptinillus, Silphosyllus* and *Platypsyllus*. These insects appear to feed mainly on ectodermis and its products, and it seems that blood is not a regular component of their diet. *Platypsyllus castoris* is unusual in that the larvae also live on their beaver hosts, and when present in sufficient numbers can cause superficial abrasions that permit blood feeding. These beetles appear to be progressing along the well-trodden evolutionary highway, from carnivore-detritovore nest inhabiters to phoretic epidermis feeders – which may eventually lead to fully fledged blood-feeding (see Section 2.1).

References

Aboul-Nasr, A. E. (1967) On the behaviour and sensory physiology of the bed bug (*Cimex lectularius*). I. Temperature reactions. *Bull. Soc. Ent. Egypte*, **51**, 43–54.

Adams, T. S. (1999) Hematophagy and hormone release. *Ann. Ent. Soc. Am.*, **92**, 1–13.

Adlington, D., Randolph, S. E. and Rogers, D. J. (1996) Flying to feed or flying to mate: gender differences in the flight activity of tsetse (*Glossina palpalis*). *Phys. Ent.*, **21**, 85–92.

Ahmadi, A. and McClelland, G. A. H. (1985) Mosquito-mediated attraction of female mosquitoes to a host. *Phys. Ent.*, **10**, 251–5.

Ahmed, A. M., Baggott, S. L., Maingon, R. and Hurd, H. (2002) The costs of mounting an immune response are reflected in the reproductive fitness of the mosquito *Anopheles gambiae*. *Oikos*, **97**, 371–7.

Ahmed, A. M., Maingon, R., Romans, P. and Hurd, H. (2001) Effects of malaria infection on vitellogenesis in *Anopheles gambiae* during two gonotrophic cycles. *Insect Mol. Biol.*, **10**, 347–56.

Ahmed, A M., Maingon, R. D., Taylor, P. J. and Hurd, H. (1999) The effects of infection with *Plasmodium yoelii* nigeriensis on the reproductive fitness of the mosquito *Anopheles gambiae*. *Invertebrate Reproduction and Development*, **36**, 217–22.

Aikawa, M., Suzuki, M. and Gutierez, Y. (1980) Pathology of malaria. In J. P. Kreier (ed.), *Malaria*. New York: Academic Press.

Akai, H. and Sato, S. (1973) Ultrastructure of the larval hemocytes of the silkworm, *Bombyx mori* L. (Lepidoptera: Bombycidae). *International Journal of Insect Morphology and Embryology*, **2**, 207–31.

Akman, L., Yamashita, A., Watanabe, H. et al. (2002) Genome sequence of the endocellular obligate symbiont of tsetse flies, *Wigglesworthia glossinidia*. *Nature Genetics*, **32**, 402–7.

Aksoy, S. (2000) Tsetse – a haven for microorganisms. *Parasitology Today*, **16**, 114–18.

Aksoy, S., Gibson, W. C. and Lehane, M. J. (2003) Perspectives on the interactions between tsetse and trypanosomes with implications for the control of trypanosomiasis. *Advances in Parasitology*, **53**, 1–84.

Albritton, A. (1952) *Standard Values in Blood*. Phiadelphia: Saunders.

Alekseev, A., Rasnityn, S. and Vitlin, L. (1977) On group behaviour of females of blood-sucking mosquitoes. Communication I. Discovery of the 'effect' of invitation. [In Russian] *Parazitologiyai Parazitarnye Bolezni*, **46**, 23–4.

Alekseyev, A. N., Abdullayev, I. T., Rasnitsyn, S. P. and Martsinovskiy, M. (1984) Comparison of flight capability of *Aedes aegypti* that are infected and not infected with plasmodia. *Med. Parazitol.*, **1**, 11–13.

Allan, S. A., Day, J. F. and Edman, J. D. (1987) Visual ecology of biting flies. *Ann. Rev. Ent.*, **32**, 297–316.

Allan, S. A. and Stoffolano, J. G. (1986a) Effects of background contrast on visual attraction and orientation of *Tabanus nigrovittatus* (Diptera: Tabanidae). *Environ. Entomol.*, **15**, 689–94.

(1986b) The effects of hue and intensity on visual attraction of adult *Tabanus nigrovittatus* (Diptera: Tabanidae). *J. Med. Ent.*, **23**, 83–91.

Altman, P. L. and Dittmer, D. S. (1971) Blood and other body fluids. In *Respiration and Circulation*. Bethesda, MDI: Federation of American Societies of Experimental Biology, p. 540.

Altner, H. and Loftus, R. (1985) Ultrastructure and function of insect thermo and hygroreceptors. *Ann. Rev. Ent.*, **30**, 273–95.

Amin, O. M. and Wagner, M. E. (1983) Further notes on the function of pronotal combs in fleas (Siphonaptera). *Ann. Ent. Soc. Am.*, **76**, 232–4.

Anderson, J. R. and Ayala, S. C. (1968) Trypanosome transmitted by a *Phlebotomus*: first report from the Americas. *Science*, **161**, 1023–5.

Anderson, J. R. and Hoy, J. B. (1972) Relationship between host attack rates and CO_2 baited insect flight trap catches of certain *Symphoromyia* spp. *J. Med. Ent.*, **9**, 373–92.

Anderson, R. A. and Brust, R. A. (1996) Blood feeding success of *Aedes aegypti* and *Culex nigripalpus* (Diptera: Culicidae) in relation to defensive behavior of Japanese quail (*Coturnix japonica*) in the laboratory. *Journal of Vector Ecology*, **21**, 94–104.

Anderson, R. A., Koella, J. C. and Hurd, H. (1999) The effect of *Plasmodium yoelii nigeriensis* infection on the feeding persistence of *Anopheles stephensi* Liston throughout the sporogonic cycle. *Proc. R. Soc. Lond. B Biol. Sci.*, **266**, 1729–33.

Anderson, R. C. (1957) The life cycles of Dipetalonematid nematodes (Filarioidea, Dipetalonematidae): the problem of their evolution. *J. Helminth.*, **31**, 203–24.

Anderson, R. M. (1986) Genetic variability in resistance to parasitic invasion: population implications for invertebrate host species. In A. M. Lackie (ed.), *Immune Mechanisms in Invertebrate Vectors*, Vol. 56, Oxford: Oxford University Press.

Anderson, R. M. and May, R. M. (1978) Regulation and stability of host–parasite population interactions. *Journal of Animal Ecology*, **47**, 219–47.

Ansell, J., Hamilton, K. A., Pinder, M., Walraven, G. E. L. and Lindsay, S. W. (2002) Short-range attractiveness of pregnant women to *Anopheles gambiae* mosquitoes. *Trans. R. Soc. of Trop. Med. and Hyg.*, **96**, 113–16.

Arlian, L. (2002) Arthropod allergens and human health. In *Annual Review of Entomology*. Annual Reviews Ic., Palo Alto, Vol. 47, 395–434.

Arrese, E. L., Canavoso, L. E., Jouni, Z. E. *et al.* (2001) Lipid storage and mobilization in insects: current status and future directions. *Insect Biochemistry and Molecular Biology*, **31**, 7–17.

Aschner, M. (1932) Experimentelle unter Suchungen über die Symbiose der Kleiderlaus. *Naturwiss.*, **20**, 501–5.

(1934) Studies on the symbiosis of the body louse. I. Elimination of the symbionts by centrifugation of the eggs. *Parasitol.*, **26**, 309–14.

(1946) The symbiosis of *Eucampsipoda aegyptica* Mcq. *Bull. Soc. Ent. Egypte*, **30**, 1–6.

Ashford, R. W. (2001) The *Leishmaniases*. In M. W. Service (ed.), *The Encyclopedia of Arthropod-transmitted Infections of Man and Domesticated Animals.* CABI, 269–79.

Ashida, M., Ochiai, M. and Niki, T. (1988) Immunolocalization of prophenoloxidase among hemocytes of the Silkworm, *Bombyx mori. Tissue and Cell*, **20**, 599–610.

Audy, J. R., Radovsky, F. J. and Vercammen-Grandjean, P. H. (1972) Neosomy: radical intrastadial metamorphosis associated with arthropod symbioses. *J. Med. Ent.*, **9**, 487–94.

Avila, A., Silverman, N., Diaz-Meco, M. T. and Moscat, J. (2002) The *Drosophila* atypical protein kinase C-ref(2)p complex constitutes a conserved module for signaling in the toll pathway. *Mol. Cell. Biol.* **22**, 8787–95.

Ayala, S. C. and Lee, D. (1970) Saurian malaria: development of sporozoites in two species of phlebotomine sandflies. *Science*, **167**, 891–2.

Bacot, A. M. and Martin, C. J. (1914) Observations on the mechanism of the transmission of the plague by fleas. *J. Hyg.*, **13**, 423–39.

Bailey, L. (1952) The action of the proventriculus of the worker honeybee. *J. Exp. Biol.*, **29**, 310–27.

Baker, J. R. (1965) The evolution of parasitic protozoa. In A. E. R. Taylor (ed.), *Symposium of the British Society for Parasitology*. Oxford: Blackwell, Vol. 3.

Baker, R. C. (1986) Pheromane-modulated movement of flying moths. In T. L. Payne, M. C. Birch and C. E. J. Kennedy (eds.), *Mechanisms in Insect Olfaction*. Oxford: Clarendon Press.

Baker, T. C. (1990) Upwind flight and casting flight: complementary phasic and tonic systems used for location of sex pheromone sources by male moths. In K. B. Doving (ed.), *International Symposium on Olfaction and Taste*, Oslo, Vol. X.

Balashov, Y. (1984) Interaction between blood-sucking arthropods and their hosts, and its influence on vector potential. *Ann. Rev. Ent.*, **29**, 137–56.

Ball, G. H. (1943) Parasitism and evolution. *Am. Nat.*, **77**, 345–64.

Banziger, H. (1971) Blood-sucking moths of Malaya. *Fauna*, **1**, 5–16.

Baranov, N. (1935) New information on the Golubatz fly, *S. columbaczense. Rev. Appl. Ent. B*, **23**, 275–6.

Barrera, A. (1966) Hallazgo de *Amblyopinus tiptoni* Barrera, 1966 en Costa Rica, A. C. (Col.: Staph.). *Acta zool. Mex.*, **8**, 1–3.

Barrera, A. and Machado-Allison, C. E. (1965) Coleopteros ectoparasiticos de Mamiferos. *Ciencia Mexico*, **23**, 201–8.

Bar-Zeev, M., Maibach, H. I. and Khan, A. A. (1977) Studies on the attraction of *Aedes aegypti* (Diptera: Culicidae) to man. *J. Med. Ent.*, **14**, 113–20.

Baudisch, K. (1958) Beiträge zur Zytologie und Embryologie einiger Insektensymbiosen. *Zeit. Morph. Okol. Tiere*, **47**, 436–88.

Baylis, M. and Mbwabi, A. L. (1995) Feeding-behavior of tsetse-flies (*Glossina pallidipes* Austen) on *Trypanosoma*-infected oxen in Kenya. *Parasitology*, **110**, 297–305.

Beach, R., Kiilu, G. and Leeuwenburg, J. (1985) Modification of sandfly biting behaviour by *Leishmania* leads to increased parasite transmission. *Am. J. Trop. Med. Hyg.*, **34**, 279–83.

Beard, C. B., Mason, P. W., Aksoy, S., Tesh, R. B. and Richards, F. F. (1992) Transformation of an insect symbiont and expression of a foreign gene in the Chagas-disease vector *Rhodnius prolixus. Am. J. Trop. Med. and Hyg.*, **46**, 195–200.

Beckenbach, A. T. and Borkent, A. (2003) Molecular analysis of the biting midges (Diptera: Ceratopogonidae), based on mitochondrial cytochrome oxidase subunit 2. *Mol. Phylogenet. Evol.* **27**, 21–35.

Beckett, E. B. and Macdonald, W. W. (1971) The development and survival of subperiodic *Brugia malayi* and *B. pahangi* larvae in a selected strain of *Aedes aegypti*. *Trans. R. Soc. Trop. Med. Hyg.*, **65**, 339–46.

Beenakkers, A. M. T., Van der Horst, D. J. and Van Marrewijk, W. J. A. (1984) Insect flight muscle metabolism. *Insect Biochemistry*, **14**, 243–60.

Beerntsen, B. T., James, A. A. and Christensen, B. M. (2000) Genetics of mosquito vector competence. *Microbiology and Molecular Biology Reviews*, **64**, 115–37.

Beerntsen, B. T., Severson, D. W., Klinkhammer, J. A., Kassner, V. A. and Christensen, B. M. (1995) *Aedes aegypti*: A quantitative trait locus (QTL) influencing filarial worm intensity is linked to QTL for susceptibility to other mosquito-borne pathogens. *Experimental Parasitology*, **81**, 355–62.

Beklemishev, V. N. (1957) Some general questions on the biology of bloodsucking lower flies. *Meditsinskaya parazitologiya i parazitarnye bolezni*, **5**, 562–6.

Belkaid, Y., Valenzuela, J. G., Kamhawi, S. *et al.* (2000) Delayed-type hypersensitivity to *Phlebotomus papatasi* sand fly bite: An adaptive response induced by the fly? *Proceedings of the National Academy of Sciences of the United States of America*, **97**, 6704–9.

Bell, J. F., Stewart, S. J. and Nelson, W. A. (1982) Transplant of acquired resistance to *Polyplax serrata* (Phthiraptera: Hoplopleuridae) in skin allografts to athymic mice. *J. Med. Ent.*, **19**, 164–8.

Belzer, W. R. (1978a) Factors conducive to increased protein feeding by the blowfly *Phormia regina*. *Phys. Ent.*. **3**, 251–7.

(1978b) Patterns of selective protein ingestion by the blowfly *Phormia regina*. *Phys. Ent.*, **3**, 169–257.

(1978c) Recurrent nerve inhibition of protein feeding in the blowfly *Phormia regina*. *Phys. Ent.*, **3**, 259–63.

(1979) Abdominal stretch receptors in the regulation of protein ingestion by the black blowfly, *Phormia regina*. *Phys. Ent.*, **4**, 7–14.

Bennet-Clark, H. C. (1963a) The control of meal size in the blood sucking bug, *Rhodnius prolixus*. *J. Exp. Biol.*, **40**, 741–50.

(1963b) Negative pressures produced in the pharyngeal pump of the bloodsucking bug, *Rhodnius prolixus*. *J. Exp. Biol.*, **40**, 223–9.

Bennett, G. F., Fallis, A. M. and Campbell, A. G. (1972) The response of *Simulium* (*EuSimulium*) *euryadminiculum* Davies (Diptera: Simuliidae) to some olfactory and visual stimuli. *Can J. Zool.*, **50**, 793–800.

Bequaert, J. C. (1953) The Hippoboscidae or house-flies (Diptera) of mammals and birds. Part 1. Structure, physiology and natural history. *Entomologia Americana*, **32**, 1–209.

Bergman, D. K. (1996) Mouthparts and feeding mechanisms of haematophagous arthropods. In S. K. Wikel, (ed.), *The Immunology of Host-Ectoparasitic Arthropod Relationships*. Wallingford: CAB International, 30–61.

Besansky, N. J. (1999) Complexities in the analysis of cryptic taxa within the genus *Anopheles*. *Parasitologia*, **41**, 97–100.

Bidlingmayer, W. L. (1994) How mosquitos see traps – role of visual responses. *J. Am. Mosq. Control Assoc.*, **10**, 272–9.

Bidlingmayer, W. L., Day, J. F. and Evans, D. G. (1995) Effect of wind velocity on suction trap catches of some Florida mosquitoes. *J. Am. Mosq. Control Assoc.*, **11**, 295–301.

Bidlingmayer, W. L. and Hem, D. G. (1979) Mosquito (Diptera: Culicidae) flight behaviour near conspicuous objects. *Bull. Ent. Res.*, **69**, 691–700.

(1980) The range of visual attraction and the effect of competitive visual attractants upon mosquito (Diptera: Culicidae) flight. *Bull. Ent. Res.*, **70**, 321–42.

Biessmann, H., Walter, M. F., Dimitratos, S. and Woods, D. (2002) Isolation of cDNA clones encoding putative odourant binding proteins from the antennae of the malaria-transmitting mosquito, *Anopheles gambiae*. *Insect Mol. Biol.*, **11**, 123–32.

Billingsley, P. B. (1990) The midgut ultrastructure of haematophagous arthropods. *Ann. Rev. Ent.*, **35**, 219–48.

Billingsley, P. F. and Downe, A. E. R. (1983) Ultrastructural changes in posterior midgut cells associated with blood-feeding in adult female *Rhodnius prolixus* Stal (Hemiptera: Reduviidae). *Can J. Zool.*, **61**, 1175–87.

(1985) Cellular localisation of aminopeptidase in the midgut of *Rhodnius prolixus* Stal (Hemiptera: Reduviidae) during blood digestion. *Cell and Tissue Research*, **241**, 421–8.

(1986) The surface morphology of the midgut cells of *Rhodnius prolixus* Stal (Hemiptera: Reduviidae) during blood digestion. *Acta Trop.*, **43**, 355–66.

(1988) Ultrastructural localisation of cathepsin B in the midgut of *Rhodnius prolixus* Stal (Hemiptera: Reduviidae) during blood digestion. *International Journal of Insect Morphology and Embryology*, **17**, 295–302.

(1989) Changes in the anterior midgut cells of adult female *Rhodnius prolixus* Stal. (Hemiptera: Reduviidae) after feeding. *J. Med. Ent.*, **26**, 104–8.

Billingsley, P. F. and Rudin, W. (1992) The role of the mosquito peritrophic membrane in bloodmeal digestion and infectivity of *Plasmodium* species. *J. Parasit.*, **78**, 430–40.

Bissonnette, E. Y., Rossignol, P. A. and Befus, A. D. (1993) Extracts of mosquito salivary gland inhibit tumour necrosis factor alpha. *Parasite Immunology*, **15**, 27–33.

Bitkowska, E., Dzbenski, T. H., Szadziewska, M. and Wegner, Z. (1982) Inhibition of xenograft rejection reaction in the bug *Triatoma infestans* during infection with a protozoan, Tryapnosoma *cruzi*. *J. Invert. Path.*, **40**, 186–9.

Bize, P., Roulin, A. and Richner, H. (2003) Adoption as an offspring strategy to reduce ectoparasite exposure. *Proc. R. Soc. Lond. B. Biol. Sci.*, **270** Suppl 1, 114–16.

Blackwell, A. (2000) Scottish biting midges: tourist attraction or deterrent? *Antenna*, **24**, 144–50.

Blackwell, A. and Page, S. (2003) Managing tourist health and Safety in the new millennium: global perspectives. In *Managing Tourist Health and Safety in the New Millennium*. Pergamon. 177–96.

Blandin, S., Moita, L. F., Kocher, T. *et al.* (2002) Reverse genetics in the mosquito *Anopheles gambiae*: targeted disruption of the Defensin gene. *EMBO Report*, **3**, 852–6.

Boatin, B. A. (2003) The current state of the Onchocerciasis Control Programme in West Africa. *Trop. Doct.*, **33**, 209–14.

Boete, C., Paul, R. E. L. and Koella, J. C. (2002) Reduced efficacy of the immune melanization response in mosquitoes infected by malaria parasites. *Parasitology*, **125**, 93–8.

Bos, H. J. and Laarman, J. J. (1975) Guinea pig lysine, cadaverine, and estradiol as attractants for the malaria mosquito *Anopheles stephensi*. *Ent. Exp. Appl.*, **18**, 161–72.

Bosch, O. J., Geier, M. and Boeckh, J. (2000) Contribution of fatty acids to olfactory host finding of female *Aedes aegypti*. *Chemical Senses*, **25**, 323–30.

Bosio, C. F., Beaty, B. J. and Black, W. C. (1998) Quantitative genetics of vector competence for Dengue-2 virus in *Aedes aegypti*. *Am. J. Trop. Med. Hyg.*, **59**, 965–70.

Bosio, C. F., Fulton, R. E., Salasek, M. L., Beaty, B. J. and Black, W. C. T. (2000) Quantitative trait loci that control vector competence for dengue-2 virus in the mosquito *Aedes aegypti*. *Genetics*, **156**, 687–98.

Bossard, R. L. (2002) Speed and Reynolds number of jumping cat fleas (Siphonaptera: Pulicidae). *Journal of the Kansas Entomological Society*, **75**, 52–4.

Boulanger, N., Munks, R. J., Hamilton, J. V. *et al.* (2002) Epithelial innate immunity. A novel antimicrobial peptide with antiparasitic activity in the blood-sucking insect *Stomoxys calcitrans*. *J. Biol. Chem.*, **277**, 49921–6.

Boutros, M., Agaisse, H. and Perrimon, N. (2002) Sequential activation of signaling pathways during innate immune responses in *Drosophila*. *Dev. Cell*, **3**, 711–22.

Bozza, M., Soares, M. B., Bozza, P. T. *et al.* (1998) The PACAP-type I receptor agonist maxadilan from sand fly saliva protects mice against lethal endotoxemia by a mechanism partially dependent on IL-10. *Eur. J. Immunol.*, **28**, 3120–7.

Bracken, G. K., Hanec, W. and Thorsteinson, A. J. (1962) The orientation of horseflies and deerflies (Tabanidae: Diptera). II. The role of some visual factors in the attractiveness of decoy silhouettes. *Can. J. Zool.*, **40**, 685–95.

Bracken, G. K. and Thorsteinson, A. J. (1965) The orientation behaviour of horse flies and deer flies (Tabanidae: Diptera). IV. The influence of some physical modifications of visual decoys on orientation of horse flies. *Ent. Exp. Appl.*, **8**, 314–18.

Bradbury, W. C. and Bennett, G. F. (1974) Behaviour of adult Simuliidae (Diptera). I. Response to colour and shape. *Can. J. Zool.*, **52**, 251–9.

Bradley, G. H. (1935) Notes on the southern buffalo gnat *Eusimulium pecuarum* (Riley) (Diptera: Simuliidae). *Proc. Ent. Soc. Washington*, **37**, 60–4.

Bradshaw, W. E. (1980) Blood-feeding and capacity for increase in the pitcher plant mosquito, *Wyeomyia smithii*. *Environ. Entomol*, **9**, 86–9.

Bradshaw, W. E. and Holtzapfel, C. M. (1983) Life cycle strategies in *Wyeomyia smithii*: seasonal and geographic adaptations. In V. K. Brown and I. Hodek (eds.), *Diapause and Life Cycle Strategies in Insects*. The Hague: Junk.

Brady, J. (1972) The visual responsiveness of the tsetse fly *Glossina morsitans* Westw. (Glossinidae) to moving objects: the effects of hunger, sex, host odour and stimulus characteristics. *Bull. Ent. Res.*, **62**, 257–79.

(1973) Changes in the probing responsiveness of starving tsetse flies (*Glossina morsitans* Westw.) (Diptera, Glossinidae). *Bull. Ent. Res.*, **63**, 247–55.

(1975) 'Hunger' in the tsetse fly: the nutritional correlates of behaviour. *J. Insect Physiol.*, **21**, 807–29.

Brady, J., Costantini, C., Sagnon, N., Gibson, G. and Coluzzi, M. (1997) The role of body odours in the relative attractiveness of different men to malarial vectors in Burkina Faso. *Ann. Trop. Med. Parasit.*, **91**, S121–S122.

Brady, J., Gibson, G. and Packer, M. J. (1989) Odour movement, wind direction and the problem of host finding by tsetse flies. *Phys. Ent.*, **14**, 369–380.

Brady, J., Griffiths, N. and Paynter, Q. (1995) Wind speed effects on odour source location by tsetse flies (*Glossina*). *Phys. Ent.*, **20**, 293–302.

Brady, J. and Shereni, A. (1988) Landing responses of the tsetse fly *Glossina morsitans morsitans* Weidemann and the stablefly *Stomoxys calcitrans* (L.). (Diptera: Glossinidae & Muscidae) to black and white patterns: a laboratory study. *Bull. Ent. Res.*, **78**, 301–11.

Braks, M. A. H., Scholte, E. J., Takken, W. and Dekker, T. (2000) Microbial growth enhances the attractiveness of human sweat for the malaria mosquito, *Anopheles gambiae sensu stricto* (Diptera: Culicidae). *Chemoecology*, **10**, 129–34.

Braun, A., Hoffmann, J. A. and Meister, M. (1998) Analysis of the *Drosophila* host defense in domino mutant larvae, which are devoid of hemocytes. *Proceedings of the National Academy of Sciences of the United States of America*, **95**, 14337–42.

Bray, R. S. (1963) The exo-erythrocytic phase of malaria parasites. *Int. Rev. Trop. Med.*, **2**, 41.

Bray, R. S., McCrae, A. W. R. and Smalley, M. E. (1976) Lack of a circadian rhythm in the ability of the gametocytes of *Plasmodium falciparum* to infect *Anopheles gambiae*. *Int. J. Parasit.*, **6**, 399–401.

Brecher, G. and Wigglesworth, V. B. (1944) The transmission of *Actinomyces rhodnii* Erkison in *Rhodnius prolixus* Stal. (Hemiptera) as its influence on the growth of the host. *Parasitol.*, **35**, 220–4.

Breev, K. V. (1950) The behaviour of blood-sucking Diptera and warble flies when attacking reindeer and the responsive reactions of reindeer. [In Russian] *Parasitologicheskii Sbornik*, **12**, 167–89.

Brehelin, M. (1982) Comparative study of structure and function of blood cells from two *Drosophila* species. *Cell and Tissue Research*, **221**, 607–15.

(1986) Insect haemocytes: a new classification to rule out the controversy. In M. Brehelin (ed.), *Immunity in Invertebrates*. Berlin, Heidelberg, New York, Tokyo: Springer Verlag Press, 36–49.

Brehelin, M., Zachary, D. and Hoffmann, J. A. (1978) A comparative ultrastructural study of blood cells from nine insect orders. *Cell and Tissue Research*, **195**, 45–57.

Briegel, H., Hefti, A. and DiMarco, E. (2002) Lipid metabolism during sequential gonotrophic cycles in large and small female *Aedes aegypti*. *J. Insect Physiol.*, **48**, 547–54.

Briegel, H., Knusel, I. and Timmermann, S. E. (2001) *Aedes aegypti*: size, reserves, survival, and flight potential. *Journal of Vector Ecology*, **26**, 21–31.

Brook, M. L. (1985) The effect of allopreening on tick burdens of moulting eudyptid penguins. *Auk*, **102**, 893.

Browne, S. M. and Bennett, G. F. (1980) Colour and shape as mediators of host-seeking responses of simuliids and tabanids (Diptera) in the Tantramar marshes, New Brunswick, Canada. *Can. J. Med. Entomol.*, **17**, 58–62.

(1981) Response of mosquitoes (Diptera: Culicidae) to visual stimuli. *J. Med. Ent.*, **6**, 505–21.

Bruce, D. (1895) *Preliminary Report on the Tsetse Fly Disease or Nagana, in Zululand*. Durban: Bennett and Davis.

Bruce, D. and Nabarro, D. (1903) Progress report on sleeping sickness in Uganda.

Buchner, P. (1965) *Endosymbiosis of Animals with Plant Microorganisms.* New York: Wiley.

Budd, L. T. (1999) *DFID-Funded Tsetse and Trypanosomiasis Research and Development since 1980 (V. 2. Economic Analysis).* London: Department for International Development.

Bulet, P., Hetru, C., Dimarcq, J. L. and Hoffmann, D. (1999) Antimicrobial peptides in insects; structure and function. *Developmental and Comparative Immunology,* **23**, 329–44.

Bungener, W. and Muller, G. (1976) Adharenz-Phänomene bei *Trypanosoma congolense. Tropenmed. Parasitol.,* **27**, 307–71.

Burg, J. G., Knapp, F. W. and Silapanuntakul, S. (1993) Feeding *Haematobia irritans* (Diptera, Muscidae) adults through a nylon-reinforced silicone membrane. *J. Med. Ent.,* **30**, 462–6.

Burgess, I., Maunder, J. W. and Myint, T. T. (1983) Maintenance of the crab louse, *Pthirus pubis,* in the laboratory and behavioural studies using volunteers. *Community Med.,* **5**, 238–41.

Burkett, D. A., Butler, J. F. and Kline, D. L. (1998) Field evaluation of colored light-emitting diodes as attractants for woodland mosquitoes and other diptera in north central Florida. *J. Am. Mosq. Control Assoc.,* **14**, 186–95.

Burkhart, C. N., Stankiewicz, B. A., Pchalek, I., Kruge, M. A. and Burkhart, C. G. (1999) Molecular composition of the louse sheath. *J. Parasit.,* **85**, 559–61.

Burkot, T. R. (1988) Non-random host selection by anopheline mosquitoes. *Parasitology Today,* **4**, 156–62.

Burkot, T. R., Narara, A., Paru, R., Graves, P. M. and Garner, P. (1989) Human host selection by anophelines: no evidence for preferential selection of malaria or microfilariae-infected individuals in a hyperendemic area. *Parasitology,* **98**, 337–42.

Bursell, E. (1961) The behaviour of tsetse flies (*Glossina swynnertoni* Austen) in relation to problems of sampling. *Proc. R. Ent. Soc. Lond. (A),* **36**, 9–20.

  (1975) Substrates of oxidative metabolism in dipteran flight muscle. *Comp. Biochem. Physiol.,* **52**, 235–8.

  (1977) Synthesis of proline by fat body of the tsetse fly (*Glossina morsitans*). Metabolic pathways. *Insect Biochem.,* **7**, 427–34.

  (1981) Energetics of haematophagous arthropods: influence of parasites. *Parasitology,* **82**, 107–10.

  (1984) Effects of host odour on the behaviour of tsetse. *Insect Sci. Applic.,* **5**, 345–9.

  (1987) The effect of wind-borne odours on the direction of flight in tsetse flies, *Glossina* spp. *Phys. Ent.,* **12**, 149–56.

Bursell, E. and Taylor, P. (1980) An energy budget for *Glossina* (Diptera: Glossinidae). *Bull. Ent. Res.,* **70**, 187–96.

Burton, R. (1860) *The Lake Regions of Central Africa.* London: Longman.

Butt, T. M. and Shields, K. S. (1996) The structure and behavior of gypsy moth (*Lymantria dispar*) hemocytes. *J. Invert. Path.,* **68**, 1–14.

Buxton, P. A. (1930) The biology of a blood-sucking bug, *Rhodnius prolixus. Trans. R. Soc. Trop. Med. Hyg.,* **78**, 227–36.

  (1947) *The Louse.* London: Edward Arnold.

  (1948) Experiments with lice and fleas. I. The baby mouse. *Parasitol.,* **39**, 119–24.

Canyon, D. V., Hii, J. L. K. and Muller, R. (1998) Multiple host-feeding and biting persistence of *Aedes aegypti*. *Ann. Trop. Med. Parasit.*, **92**, 311–16.

Cappello, M., Li, S., Chen, X. O. *et al.* (1998) Tsetse thrombin inhibitor: bloodmeal-induced expression of an anticoagulant in salivary glands and gut tissue of *Glossina morsitans morsitans*. *Proceedings of the National Academy of Sciences of the United States of America*, **95**, 14290–5.

Carde, R. T. (1996) Odour plumes and odour-mediated flight in insects. In G. R. Bock and G. Cardew (eds.), *Olfaction in Mosquito-Host Interactions*. Chichester: Wiley Ciba Foundation Symposium 200, 54–70.

Carwardine, S. L. and Hurd, H. (1997) Effects of *Plasmodium yoelii* nigeriensis infection on *Anopheles stephensi* egg development and resorption. *Med. Vet. Entomol*, **11**, 265–9.

Castanera, M. B., Aparicio, J. P. and Gurtler, R. E. (2003) A stage-structured stochastic model of the population dynamics of *Triatoma infestans*, the main vector of Chagas disease. *Ecological Modelling*, **162**, 33–53.

Caterino, M. S., Cho, S. and Sperling, F. A. H. (2000) The current state of insect molecular systematics: a thriving Tower of Babel. *Ann. Rev. Ent.*, **45**, 1–54.

Cavanaugh, D. C. (1971) Specific effect of temperature upon transmission of the plague bacillus by the oriental rat flea, *Xenopsylla cheopis*. *Am. J. Trop. Med. Hyg.*, **20**, 264–73.

Chadee, D. D. and Beier, J. C. (1997) Factors influencing the duration of blood-feeding by laboratory-reared and wild *Aedes aegypti* (Diptera: Culicidae) from Trinidad, West Indies. *Ann. of Trop. Med. Parasit.*, **91**, 199–207.

Chadee, D. D., Beier, J. C. and Martinez, R. (1996) The effect of the cibarial armature on blood meal haemolysis of four anopheline mosquitoes. *Bull. Ent. Res.*, **86**, 351–4.

Chagas, C. (1909) Ueber eine neue Trypanosomiasis des Menschen. *Memorias Institut Oswaldo Cruz*, **1**, 159–218.

Challier, A., Eyraud, M., Lafaye, A. and Laveissiere, C. (1977) Amelioration due rendement du piege biconique pour glossines (Diptera, Glossinidae) par l'emploi d'un cone inférieur bleu. *Cah. ORSTOM, Ser. Entomol. Med. Parasitol.*, **15**, 283–6.

Champagne, D. E., Nussenzveig, R. H. and Ribeiro, J. M. C. (1995a) purification, partial characterization, and cloning of nitric oxide-carrying heme-proteins (nitrophorins) from salivary-glands of the bloodsucking insect *Rhodnius prolixus*. *J. Biol. Chem.*, **270**, 8691–5.

Champagne, D. E. and Ribeiro, J. M. C. (1994) Sialokinin-I and Sialokinin-Ii – Vasodilatory Tachykinins from the Yellow-Fever Mosquito *Aedes aegypti*. *Proceedings of the National Academy of Sciences of the United States of America*, **91**, 138–42.

Champagne, D. E., Smartt, C. T., Ribeiro, J. M. C. and James, A. A. (1995b) The salivary gland-specific apyrase of the mosquito *Aedes aegypti* is a member of the 5'-nucleotidase family. *Proceedings of the National Academy of Sciences of the United States of America*, **92**, 694–8.

Chang, Y. H. and Judson, C. L. (1977) The role of isoleucine in differential egg production by the mosquito *Aedes aegypti* Linnaeus (Diptera: Culicidae) following feeding on human or guinea pig blood. *Comp. Biochem. Physiol. A.*, **57**, 23–8.

Chapman, R. F. (1961) Some experiments to determine the methods used in host finding by tsetse flies, *Glossina medicorum*. *Bull. Ent. Res.*, **52**, 83–97.

(1982) Chemoreception: the significance of receptor numbers. *Adv. Insect Physiol.*, **16**, 247–356.

Charlab, R., Rowton, E. D. and Ribeiro, J. M. C. (2000) The salivary adenosine deaminase from the sand fly *Lutzomyia longipalpis*. *Experimental Parasitology*, **95**, 45–53.

Charlwood, J. D., Billingsley, P. F. and Hoc, T. Q. (1995a) Mosquito-mediated attraction of female European but not African mosquitos to hosts. *Ann. Trop. Med. Parasit.*, **89**, 327–9.

Charlwood, J. D., Smith, T., Kihonda, J. (1995b) Density-independent feeding success of malaria vectors (Diptera, Culicidae) in Tanzania. *Bull. Ent. Res.*, **85**, 29–35.

Chen, C. C. and Chen, C. S. (1995) *Brugia pahangi*: effects of melanization on the uptake of nutrients by microfilariae in vitro. *Experimental Parasitology*, **81**, 72–8.

Chen, C. C. and Laurence, B. R. (1985) An ultrastructural study on the encapsulation of microfilariae of *Brugia pahangi* in the haemocoel of *Anopheles quadrimaculatus*. *Int. J. Parasitol*, **15**, 421–8.

Chikilian, M. L., Bradley, T. J., Nayar, J. K., Cashclark, C. E. and Knight, J. W. (1995) Ultrastructure of the intracellular melanization of *Brugia malayi* (Buckley) (Nematoda, Filarioidea) in the thoracic muscles of *Anopheles quadrimaculatus* (Say) (Diptera, Culicidae). *International Journal of Insect Morphology and Embryology*, **24**, 83–92.

Chikilian, M. L., Bradley, T. J., Nayar, J. K. and Knight, J. W. (1994) Ultrastructural comparison of extracellular and intracellular encapsulation of *Brugia malayi* in *Anopheles quadrimaculatus*. *Journal of Parasitology*, **80**, 133–40.

Choe, K. M., Werner, T., Stoven, S., Hultmark, D. and Anderson, K. V. (2002) Requirement for a peptidoglycan recognition protein (PGRP) in relish activation and antibacterial immune responses in *Drosophila*. *Science*, **296**, 359–62.

Christensen, B. M. (1978) *Dirofilaria immitis*: effects on the longevity of *Aedes trivittatus*. *Experimental Parasitology*, **44**, 116–23.

Christensen, B. M., Forton, K. F., Lafond, M. M. and Grieve, R. B. (1987) Surface changes on *Brugia pahangi* microfilariae and their association with immune evasion in *Aedes aegypti*. *J. Invert. Pathol.*, **49**, 14–18.

Christensen, B. M. and LaFond, M. M. (1986) Parasite induced suppression of the immune response in *Aedes aegypti* by *Brugia pahangi*. *J. Parasit.*, **72**, 216–19.

Christophides, G. K., Zdobnov, E., Barillas-Mury, C. *et al.* (2002) Immunity-related genes and gene families in *Anopheles gambiae*. *Science*, **298**, 159–65.

Ciurea, I. and Dinulescu, G. (1924) Ravages causes par la mouchede Goloubatz en Roumanie; ses attaques contre les animaux et contre l'homme. *Ann. Trop. Med. Parasit.*, **18**, 323–42.

Clay, T. (1963) A new species of *Haematomyzus* Piaget (Phthiraptera, Insecta). *Proc. Zool. Soc. Lond.*, **141**, 153–61.

Clifford, C. M., Bell, J. F., Moore, G. J. and Raymond, C. (1967) Effects of limb disability of lousiness in mice. IV. Evidence of genetic factors in susceptibility to *Polyplax serrata*. *Experimental Parasitology*, **20**, 56–67.

Coatney, G. R., Collins, W. E., McWilson, W. and Contacos, P. G. (1971) *The Primate Malarias*. Bethesda, MD: NIAID.

Cockerell, T. D. A. (1918) New species of North American fossil beetles, cockroaches and tsetse flies. *Proc. U. S. Natn. Mus.*, **54**, 301–11.

Coetzee, M., Craig, M. and le Sueur, D. (2000) Distribution of African malaria mosquitoes belonging to the *Anopheles gambiae* complex. *Parasitology Today*, **16**, 74–7.

Colless, D. H. and Chellapah, W. T. (1960) Effects of body weight and size of blood meal upon egg production in *Aedes aegypti* (Linnaeus) (Diptera, Culicidae). *Ann. Trop. Med. Parasit.*, **54**, 475–82.

Collins, F. H., Sakai, R. K., Vernick, K. D. *et al.* (1986) Genetic selection of a refractory strain of the malaria vector *Anopheles gambiae*. *Science*, **234**, 607–10.

Colman, R. W. (2001) *Hemostasis and Thrombosis: Basic Principles and Clinical Practice*. London: Lippincott, Williams and Wilkins.

Coluzzi, M., Concetti, A. and Ascoli, F. (1982) Effect of cibarial armature of mosquitoes (Diptera: Culicidae) on blood-meal haemolysis. *J. Insect Physiol.*, **28**, 885–8.

Coluzzi, M., Sabatini, A., della Torre, A., Di Deco, M. A. and Petrarca, V. (2002) A polygene chromosome analysis of the *Anopheles gambiae* species complex. *Science*, **298**, 1415–18.

Colvin, J., Brady, J. and Gibson, G. (1989) Visually-guided, upwind turning behaviour of free-flying tsetse flies in odour-laden wind: a wind-tunnel study. *Phys. Ent.*, **14**, 31–9.

Colyer, C. N. and Hammond, C. O. (1968) *Flies of the British Isles*. London: Frederick Warne.

Compton Knox, P. and Hayes, K. L. (1972) Attraction of *Tabanus* spp. (Diptera: Tabanidae) to traps baited with carbon dioxide and other chemicals. *Environ. Entomal.*, **1**, 323–6.

Cook, S. P. and McCleskey, E. W. (2002) Cell damage excites nociceptors through release of cytosolic ATP. *Pain*, **95**, 41–7.

Cornford, E. M., Freeman, B. J. and MacInnis, A. J. (1976) Physiological relationships and circadian periodicities in rodent trypanosomes. *Trans. R. Soc. Trop. Med. Hyg.*, **70**, 238–43.

Croft, S. L., East, J. S. and Molyneux, D. H. (1982) Antitrypanosomal factor in the haemolymph of *Glossina*. *Acta Trop.*, **39**, 293–302.

Cross, M. L., Cupp, E. W. and Enriquez, F. J. (1994) Modulation of murine cellular immune-responses and cytokines by salivary-gland extract of the black fly *Simulium vittatum*. *Tropical Medicine and Parasitology*, **45**, 119–124.

Cupp, E. W., Cupp, M. S., Ribeiro, J. M. C. and Kunz, S. E. (1998a) Blood-feeding strategy of *Haematobia irritans* (Diptera: Muscidae). *J. Med. Ent.*, **35**, 591–5.

Cupp, E. W. and Stokes, G. M. (1976) Feeding patterns of *Culex salinarius* Coquillett in Jefferson parish, Louisiana. *Mosq. News*, **36**, 332–5.

Cupp, M. S., Ribeiro, J. M. C., Champagne, D. E. and Cupp, E. W. (1998b) Analyses of cDNA and recombinant protein for a potent vasoactive protein in saliva of a blood-feeding black fly, *Simulium vittatum*. *J. Exp. Biol.*, **201**, 1553–61.

Dale, C., Young, S. A., Haydon, D. T. and Welburn, S. C. (2001) The insect endosymbiont *Sodalis glossinidius* utilizes a type III secretion system for cell invasion.

Proceedings of the National Academy of Sciences of the United States of America, **98**, 1883–8.

Dan, A., Pereira, M. H., Pesquero, J. L., Diotaiuti, L. and Beirao, P. S. L. (1999) Action of the saliva of *Triatoma infestans* (Heteroptera: Reduviidae) on sodium channels. *J. Med. Entomol.*, **36**, 875–9.

Daniel, T. L. and Kingsolver, J. G. (1983) Feeding strategy and the mechanics of blood sucking in insects. *J. Theor. Biol.*, **105**, 661–72.

David, C. T., Kennedy, J. S., Ludlow, A. R., Perry, J. N. and Wall, C. (1982) A reappraisal of insect flight towards a distant point source of wind-borne odor. *J. Chem. Ecol.*, **8**, 1207–15.

Davidson, G. and Draper, C. C. (1953) Field studies of some of the basic factors concerned in the transmission of malaria. *Trans. R. Soc. Trop. Med. Hyg.*, **47**, 522–35.

Davis, E. E. (1984) Regulation of sensitivity in the peripheral chemoreceptor systems for host-seeking behaviour by a haemolymph-borne factor in *Aedes aegypti*. *J. Insect Physiol.*, **30**, 179–83.

Davis, E. E. and Sokolove, P. G. (1975) Temperature response of the antennal receptors in the mosquito, *Aedes aegypti*. *J. Comp. Physiol.*, **96**, 223–36.

Day, J. F. and Edman, J. D. (1983) Malaria renders mice susceptible to mosquito feeding when gametocytes are most infective. *J. Parasitol.*, **69**, 163–70.

(1984a) The importance of disease induced changes in mammalian body temperature to mosquito blood feeding. *Comp. Biochem. Physiol.*, **77**, 447–52.

(1984b) Mosquito engorgement on normally defensive hosts depends on host activity patterns. *J. Med. Ent.*, **21**, 732–40.

De Azambuja, P., Guimares, J. A. and Garcia, E. S. (1983) Haemolytic factor from the crop of *Rhodnius prolixus*: evidence and partial characterisation. *J. Insect Physiol.*, **29**, 833–7.

De Gregorio, E., Spellman, P. T., Tzou, P., Rubin, G. M. and Lemaitre, B. (2002) The Toll and Imd pathways are the major regulators of the immune response in *Drosophila*. *EMBO J.*, **21**, 2568–79.

De Jong, R. and Knols, B. G. J. (1995) Selection of biting sites on man by 2 malaria mosquito species. *Experientia*, **51**, 80–4.

DeFoliart, G. R., Grimstad, P. R. and Watts, D. M. (1987) Advances in mosquito-borne arbovirus/vector research. *Ann. Rev. Ent.*, **32**, 479–505.

Dei Cas, E., Maurois, P., Landau, I. *et al.* (1980) Morphologie et infectivite des gametocytes de *Plasmodium inui*. *Annales Parasit.*, **55**, 621–33.

Dekker, T. and Takken, W. (1998) Differential responses of mosquito sibling species *Anopheles arabiensis* and *An. quadriannulatus* to carbon dioxide, a man or a calf. *Med. Vet. Entomol.*, **12**, 136–40.

Dekker, T., Takken, W. and Braks, M. A. H. (2001) Innate preference for host-odor blends modulates degree of anthropophagy of *Anopheles gambiae* sensu lato (Diptera: Culicidae). *J. Med. Ent.*, **38**, 868–71.

Dekker, T., Takken, W., Knols, B. G. J. *et al.* (1998) Selection of biting sites on a human host by *Anopheles gambiae* s. s., *An. arabiensis* and *An. quadriannulatus*. *Entomologia Experimentalis et Applicata*, **87**, 295–300.

Denotter, C. J., Tchicaya, T. and Schutte, A. M. (1991) Effects of age, sex and hunger on the antennal olfactory sensitivity of tsetse-flies. *Phys. Ent.*, **16**, 173–82.

Desquesnes, M. and Dia, M. L. (2003) *Trypanosoma vivax*: mechanical transmission in cattle by one of the most common African tabanids, *Atylotus agrestis*. *Experimental Parasitology*, **103**, 35–43.

Dethier, V. G. (1954) Notes on the biting response of tsetse flies. *Am. J. Trop. Med. Hyg.*, **3**, 160–71.

Detinova, T. S. (1962) *Age Grouping Methods in Diptera of Medical Importance*. Geneva: World Health Organization.

Dias, J. C., Silveira, A. C. and Schofield, C. J. (2002) The impact of Chagas disease control in Latin America: a review. *Mem. Inst. Oswaldo Cruz*, **97**, 603–12.

Diatta, M., Spiegel, A., Lochouarn, L. and Fontenille, D. (1998) Similar feeding preferences of *Anopheles gambiae* and *An. arabiensis* in Senegal. *Trans. R. Soc. Trop. Med. Hyg.*, **92**, 270–2.

Dickerson, G. and Lavoipierre, M. M. J. (1959) Studies on the methods of feeding blood-sucking arthropods. II. The method of feeding adopted by the bed-bug (*Cimex lectularius*) when obtaining a blood meal from the mammalian host. *Ann. Trop. Med. Parasit.*, **53**, 347–57.

Dimopoulos, G. (2003) Insect immunity and its implication in mosquito-malaria interactions. *Cell Microbiol.*, **5**, 3–14.

Dimopoulos, G., Christophides, G. K., Meister, S. *et al.* (2002) Genome expression analysis of *Anopheles gambiae*: responses to injury, bacterial challenge, and malaria infection. *Proceedings of the National Academy of Sciences of the United States of America*, **99**, 8814–19.

Dimopoulos, G., Seeley, D., Wolf, A. and Kafatos, F. C. (1998) Malaria infection of the mosquito *Anopheles gambiae* activates immune-responsive genes during critical transition stages of the parasite life cycle. *EMBO J.*, **17**, 6115 23.

Dobson, A. P. (1988) Parasite-induced changes in host behaviour, *Q. Rev. Biol.*, **63**(4), 140–65.

Downes, J. A. (1970) The ecology of blood-sucking diptera: an evolutionary perspective. In A. M. Fallis (ed.), *Ecology and Physiology of Parasites*. Toronto: University of Toronto Press.

Downs, C. M., Theberge, J. B. and Smith, S. M. (1986) The influence of insects on the distribution, microhabitat choice, and behaviour of the Burwash caribou herd. *Can. J. Zool.*, **64**, 622–9.

Dryden, M. W. (1989) Host association, on-host longevity and egg production of *Ctenocephalides felis felis*. *Vet Parasitol.*, **34**, 117–22.

Dujardin, J. C., Banuls, A. L., Llanos-Cuentas, A. *et al.* (1995) Putative *Leishmania* hybrids in the Eastern Andean valley of Huanuco, Peru. *Acta Trop.*, **59**, 293–307.

Duncan, P. and Vigne, N. (1979) The effect of group size in horses on the rate of attacks by blood-sucking flies. *Animal Behaviour*, **27**, 623–5.

Dunnet, G. M. (1970) Siphonaptera (Fleas). In CSIRO (ed.) *The Insects of Australia*. Melbourne: Melbourne University Press.

Durvasula, R. V., Gumbs, A., Panackal, A. *et al.* (1997) Prevention of insect-borne disease: an approach using transgenic symbiotic bacteria. *Proceedings of the National Academy of Sciences of the United States of America*, **94**, 3274–8.

Duval, J., Rajaonarivelo, E. and Rabenirainy, L. (1974) Ecologie de *Styloconops spinosifrons* (Carter, 1921) (Diptera, Ceratopogonidae) sur les plages de la côte Est de Madagascar. *Cahiers ORSTOM*, **12**, 245–58.

Dye, C. (1992) The analysis of parasite transmission by bloodsucking insects. *Ann. Rev. Ent.*, **37**, 1–19.

East, J., Molyneux, D. H., Maudlin, I. and Dukes, P. (1983) Effect of *Glossina* haemolymph on salivarian trypanosomes in vitro. *Annals of Tropical Medicine and Parasitology*, **77**, 97–9.

Edman, J. D. (1974) Host-feeding patterns of Florida mosquitoes III *Culex* (*Culex*) and *Culex* (*NeoCulex*). *J. Med. Ent.*, **11**, 95–104.

Edman, J. D., Day, J. F. and Walker, E. (1985) Vector-host interplay: factors affecting disease transmission. In L. P. Lounibos, R. Rey and J. H. Frank (eds.), *Ecology of Mosquitoes: Proceedings of a Workshop*. New Beach, Florida: Florida Medical Entomology Laboratory.

Edman, J. D. and Kale, H. W., II (1971) Host behaviour: its influence on the feeding success of mosquitoes. *Ann. Ent. Soc. Am.*, **64**, 513–16.

Edman, J. D. and Spielman, A. (1988) Blood feeding by vectors: physiology, ecology, behaviour and vertebrate defence. In T. P. Monath (ed.), *The Arboviruses: Epidemiology and Ecology*. Baton Rouge: C. R. C. Press, Vol. 1.

Edman, J. D. and Taylor, D. J. (1968) *Culex nigripalpus*: seasonal shift in the bird-mammal feeding ratio in a mosquito vector of human encephalitis. *Science*, **161**, 67–8.

Edman, J. D., Webber, L. A. and Kale, H. W. (1972) Effect of mosquito density on the interrelationship of host behavior and mosquito feeding success. *Am. J. Trop. Med. Hyg.*, **21**, 487–91.

Eichler, D. A. (1973) Studies on *Onchocerca gutterosa* (Neumann, 1910) and its development in *Simulium ornatum* (Meigen, 1818). 3. Factors affecting the development of the parasite in its vector. *J. Helm.*, **47**, 73–88.

Eiras, A. E. and Jepson, P. C. (1994) Responses of female *Aedes aegypti* (Diptera, Culicidae) to host odors and convection currents using an olfactometer bioassay. *Bull. Ent. Res.*, **84**, 207–11.

Elkinton, J. S., Schal, C., Ono, T. and Carde, R. T. (1987) Pheromone puff trajectory and upwind flight of male gypsy moths in a forest. *Phys. Ent.*, **12**, 399–406.

Ellis, D. S. and Evans, D. A. (1977) Passage of *Trypanosoma brucei rhodesiense* through the peritrophic membrane of *Glossina morsitans morsitans*. *Nature*, **267**, 834–5.

Elrod-Erickson, M., Mishra, S. and Schneider, D. (2000) Interactions between the cellular and humoral immune responses in *Drosophila*. *Current Biology*, **10**, 781–4.

Elsen, P., Amoudi, M. A. and Leclercq, M. (1990) 1st record of *Glossina fuscipes fuscipes* Newstead, 1910 and *Glossina morsitans submorsitans* Newstead, 1910 in Southwestern Saudi-Arabia. *Annales De La Societe Belge De Medecine Tropicale*, **70**, 281–7.

Emmerson, K. C., Kim, K. C. and Price, R. D. (1973) Lice. In R. J. Flynn (ed.), *Parasites of Laboratory Animals*. Ames, Iowa: Iowa State University Press.

Escalante, A. A. and Ayala, F. J. (1995) Evolutionary origin of *Plasmodium* and other apicomplexa based on ribosomal-RNA genes. *Proceedings of the National Academy of Sciences of the United States of America*, **92**, 5793–7.

Esseghir, S., Ready, P. D., KillickKendrick, R. and BenIsmail, R. (1997) Mitochondrial haplotypes and phylogeography of *Phlebotomus* vectors of *Leishmania* major. *Insect Mol. Biol.*, **6**, 211–25.

Evans, G. O. (1950) Studies on the bionomics of the sheep ked, *Melophagus ovinus* L., in West Wales. *Bull. Ent. Res.*, **40**, 459–78.

Ewert, A. (1965) Comparative migration of microfilariae and development of *Brugia pahangi* in various mosquitoes. *Am. J. Trop. Med. Hyg.*, **14**, 254–9.

Falleroni, D. (1927) Per la soluzione del problema malarico italiano. *Riv. Malariol.*, **6**, 344–409.

Fallis, A. M., Bennett, G. F., Griggs, G. and Allen, T. (1967) Collecting *Simulium venustum* female in fan traps and on silhouettes with the aid of carbon dioxide. *Can. J. Zool.*, **45**, 1011–17.

Fallis, A. M. and Raybould, J. N. (1975) Response of two African simuliids to silhouettes and carbon dioxide. *J. Med. Entomol.*, **12**, 349–51.

Fallis, A. M. and Smith, S. M. (1964) Ether extract from birds and CO_2 as attractants for some ornithophilic simuliids. *Can. J. Zool.*, **42**, 723–30.

Farkas, S. R. and Shorey, H. H. (1972) Chemical trail-following by flying insects: a mechanism for orientation to a distant odor source. *Science (Washington, D. C.)*, **178**, 67–8.

Farmer, J., Maddrell, S. H. P. and Spring, J. H. (1981) Absorption of fluid by the midgut of *Rhodnius*. *J. Exp. Biol.*, **94**, 301–16.

Faust, E. C., Russel, P. F. and Jung, R. C. (1977) *Craig and Fausts Clinical Parasitology*. Philadelphia: Lea and Febiger.

Favia, G., dellaTorre, A., Bagayoko, M. *et al.* (1997) Molecular identification of sympatric chromosomal forms of *Anopheles gambiae* and further evidence of their reproductive isolation. *Insect Mol. Biol.*, **6**, 377–83.

Feingold, B. F. and Benjamini, E. (1961) Allergy to flea bites: clinical and experimental observations. *Ann. Allerg.*, **19**, 1274–89.

Ferdig, M. T., Beerntsen, B. T., Spray, F. J., Li, J. and Christensen, B. M. (1993) Reproductive costs associated with resistance in a mosquito-filarial worm system. *Am. J. Trop. Med. Hyg.*, **49**, 756–62.

Ferguson, H. M. and Read, A. F. (2002a) Genetic and environmental determinants of malaria parasite virulence in mosquitoes. *Proc. R. Soc. Lond. B Biol. Sciences*, **269**, 1217–24.

(2002b) Why is the effect of malaria parasites on mosquito survival still unresolved? *Trends in Parasitology*, **18**, 256–61.

Ferrari, J., Muller, C. B., Kraaijeveld, A. R. and Godfray, H. C. (2001) Clonal variation and covariation in aphid resistance to parasitoids and a pathogen. *Evolution*, **55**, 1805–1814.

Ferris, G. R. (1931) The louse of elephants *Haematomyzus elephantis* Piaget (Mallophaga: Haematomyzidae). *Parasitol.*, **23**, 112–27.

Flores, G. B. and Lazzari, C. R. (1996) The role of the antennae in *Triatoma infestans*: Orientation towards thermal sources. *J. Insect Physiol.*, **42**, 433–40.

Foley, E. and O'Farrell, P. H. (2003) Nitric oxide contributes to induction of innate immune responses to gram-negative bacteria in *Drosophila*. *Genes Dev.*, **17**, 115–25.

Foster, W. A. (1976) Male sexual maturation of the tsetse flies *Glossina morsitans* Westwood and *G. austeni* Newstead (Diptera: Glossinidae) in relation to feeding. *Bull. Ent. Res.*, **66**, 389–99.

Fox, A. N., Pitts, R. J., Robertson, H. M., Carlson, J. R. and Zwiebel, L. J. (2001) Candidate odorant receptors from the malaria vector mosquito *Anopheles gambiae* and evidence of down-regulation in response to blood feeding.

Proceedings of the National Academy of Sciences of the United States of America, **98**, 14693–7.

Francischetti, I. M. B., Ribeiro, J. M. C., Champagne, D. and Andersen, J. (2000) Purification, cloning, expression, and mechanism of action of a novel platelet aggregation inhibitor from the salivary gland of the blood-sucking bug, *Rhodnius prolixus*. *J. Biol. Chem.*, **275**, 12639–50.

Francischetti, I. M. B., Valenzuela, J. G. and Ribeiro, J. M. C. (1999) Anophelin: kinetics and mechanism of thrombin inhibition. *Biochemistry*, **38**, 16678–85.

Fredeen, F. J. H. (1961) A trap for studying the attacking behaviour of black flies *Simulium articum* Mall. *Can. Entomol.*, **93**, 73–8.

Freier, J. E. and Friedman, S. (1976) Effect of host infection with *Plasmodium gallinaceum* on the reproductive capacity of *Aedes aegypti*. *J. Invert. Pathol.*, **28**, 161–6.

Friend, W. G. (1978) Physical factors affecting the feeding responses of *Culiseta inornata* to ATP, sucrose and blood. *Ann. Ent. Soc. Am.*, **71**, 935–40.

Friend, W. G. and Smith, J. J. B. (1971) Feeding in *Rhodnius prolixus*: mouthpart activity and salivation and their correlation with changes of electrical resistance. *J. Insect Physiol.*, **17**, 233–43.

——— (1975) Feeding in *Rhodnius prolixus*: increasing sensitivity to ATP during prolonged food deprivation. *J. Insect Physiol.*, **21**, 1081–4.

——— (1977) Factors affecting feeding by blood-sucking insects. *Ann. Rev. Ent.*, **22**, 309–31.

Friend, W. G. and Stoffolano, J. G. (1984) Feeding responses of the horsefly, *Tabanus nigrovittatus*, to physical factors, ATP analogues and blood fractions. *Phys. Ent.*, **9**, 395–402.

Fu, H., Leake, C. J., Mertens, P. P. and Mellor, P. S. (1999) The barriers to bluetongue virus infection, dissemination and transmission in the vector, *Culicoides varipennis* (Diptera: Ceratopogonidae). *Arch. Virol.*, **144**, 747–61.

Gad, A. M., Maier, W. A. and Piekorski, G. (1979) Pathology of *Anopheles stephensi* after infection with *Plasmodium berghei berghei*. I. Mortality rate. *Z. Parasit.*, **60**, 249–61.

Gade, G. and Auerswald, L. (2002) Beetles' choice: proline for energy output: control by AKHs. *Comp. Biochem. Physiol. B.*, **132**, 117–29.

Galun, R. (1966) Feeding stimulants of the rat flea *Xenopsylla cheopis* Roth. *Life Sci.*, **5**, 1335–42.

——— (1986) Diversity of phagostimulants used for recognition of blood meal by haematophagous arthropods. In D. Borovsky and A. Spielman (eds.), *Host-Regulated Development Mechanisms in Vector Arthropods*. Florida: IFAS, University of Florida.

——— (1987) The evolution of purinergic receptors involved in recognition of a blood meal by haematophagous insects. *Mem. Inst. Oswaldo Cruz*, **82**, 5–9.

Galun, R., Avidor, Y. and Bar-Zeev, M. (1963) Feeding response in *Aedes aegypti*: stimulation by adnosine triphosphate. *Science*, **124**, 1674–5.

Galun, R., Friend, W. G. and Nudelman, S. (1988) Purinergic reception by culicine mosquitoes. *J. Comp. Physiol. A.*, **163**, 665–70.

Galun, R. and Kabayo, J. P. (1988) Gorging response of *Glossina palpalis palpalis* to ATP analogues. *Phys. Ent.*, **13**, 419–23.

Galun, R., Koontz, L. C. and Gwadz, R. W. (1985) Engorgement response of anopheline mosquitoes to blood fractions and artificial solutions. *Phys. Ent.*, **10**, 145–9.

Galun, R., Vardimonfriedman, H. and Frankenburg, S. (1993) Gorging response of culicine mosquitos (Diptera, Culicidae) to blood fractions. *J. Med. Ent.*, **30**, 513–17.

Garcia, R. and Radovsky, F. J. (1962) Haematophagy by two non-biting myscid flies and its relationship to tabanid feeding. *Can. Entomol.*, **94**, 1110–16.

Gardiner, E. M. and Strand, M. R. (1999) Monoclonal antibodies bind distinct classes of hemocytes in the moth *Pseudoplusia includens*. *J Insect Physiol.*, **45**, 113–26.

Garms, R., Walsh, J. F. and Davies, J. B. (1979) Studies on the reinvasion of the Onchocerciasis Control Programme in the Volta River Basin by *Simulium damnosum* s.l. with emphasis on the south-western areas. *Tropenmed. Parasitol.*, **30**, 345–62.

Garrett-Jones, C. and Shidrawi, G. R. (1969) Malaria vectorial capacity of a population of *Anopheles gambiae*. *Bulletin of the World Health Organization*, **40**, 531–45.

Gaston, K. A. and Randolph, S. E. (1993) Reproductive under-performance of tsetse-flies in the laboratory, related to feeding frequency. *Phys. Ent.*, **18**, 130–6.

Gatehouse, A. G. (1970) The probing response of *Stomoxys calcitrans* to certain physical and olfactory stimuli. *J. Insect Physiol.*, **16**, 61–74.

(1972) Some responses of tsetse flies to visual and olfactory stimuli. *Nature New Biol.*, **236**, 63–4.

Gaunt, M. W. and Miles, M. A. (2002) An insect molecular clock dates the origin of the insects and accords with palaeontological and biogeographic landmarks. *Molecular Biology and Evolution*, **19**, 748–61.

Gaunt, M. W., Yeo, M., Frame, I. A. (2003) Mechanism of genetic exchange in American trypanosomes. *Nature*, **421**, 936–9.

Gautret, P. (2001) *Plasmodium falciparum* gametocyte periodicity. *Acta Trop.*, **78**, 1–2.

Gautret, P. and Motard, A. (1999) Periodic infectivity of *Plasmodium* gametocytes to the vector: a review. *Parasite-Journal De La Societe Francaise De Parasitologie*, **6**, 103–11.

Geden, C. J. and Hogsette, J. A. (1994) Research and extension needs for integrated pest management for arthropods of veterinary importance. Proceedings of a Workshop in Lincoln, Nebraska, 12–14 April, Lincoln, Nebraska.

Gee, J. C. (1975) Diuresis in the tsetse fly *Glossina austeni*. *J. Exp. Biol.*, **63**, 381–90.

Geier, M., Bosch, O. J. and Boeckh, J. (1999) Ammonia as an attractive component of host odour for the yellow fever mosquito, *Aedes aegypti*. *Chemical Senses*, **24**, 647–53.

Gentile, G., Della Torre, A., Maegga, B., Powell, J. R. and Caccone, A. (2002) Genetic differentiation in the African malaria vector, *Anopheles gambiae* s.s., and the problem of taxonomic status. *Genetics*, **161**, 1561–78.

Ghosh, K. N. and Mukhopadhyay, J. (1998) The effect of anti-sandfly saliva antibodies on *Phlebotomus argentipes* and *Leishmania donovani*. *Int. J. Parasit.*, **28**, 275–81.

Gibson, G. (1992) Do tsetse-flies see zebras: a field-study of the visual response of tsetse to striped targets. *Phys. Ent.*, **17**, 141–7.

Gibson, G. and Brady, J. (1985) Anemotactic flight paths of tsetse flies in relation to host odour: preliminary video study in nature. *Phys. Ent.*, **10**, 395–406.

(1988) Flight behaviour of tsetse flies in host odour plumes: the initial response to leaving or entering odour. *Phys. Ent.*, **13**, 29–42.

Gibson, G. and Torr, S. J. (1999) Visual and olfactory responses of haematophagous Diptera to host stimuli. *Med. Vet. Entomol.*, **13**, 2–23.

Gibson, G. and Young, S. (1991) The optics of tsetse-fly eyes in relation to their behavior and ecology. *Phys. Ent.*, **16**, 273–82.

Gikonyo, N. K., Hassanali, A., Njagi, P. G. N. and Saini, R. K. (2000) Behaviour of *Glossina morsitans morsitans* Westwood (Diptera: Glossinidae) on waterbuck *Kobus defassa* Ruppel and feeding membranes smeared with waterbuck sebum indicates the presence of allomones. *Acta Trop.*, **77**, 295–303.

Gillespie, R. D., Mbow, M. L. and Titus, R. G. (2000) The immunomodulatory factors of bloodfeeding arthropod saliva. *Parasite Immunology*, **22**, 319–31.

Gillett, J. D. (1967) Natural selection and feeding speed in a blood sucking insect. *Proc. R. Soc. London Ser. B.*, **167**, 316–29.

Gillett, J. D. and Connor, J. (1976) Host temperature and the transmission of arboviruses by mosquitoes. *Mosq. News*, **36**, 472–7.

Gillies, M. T. (1980) The role of carbon dioxide in host-finding by mosquitoes (Diptera: Culicidae): a review. *Bull. Ent. Res.*, **70**, 525–32.

Gillies, M. T. and Wilkes, T. J. (1969) A comparison of the range of attraction of animal baits and carbon dioxide for some West African mosquitoes. *Bull. Ent. Res.*, **59**, 441–56.

(1970) The range of attraction of single baits for some West African mosquitoes. *Bull. Ent. Res.*, **60**, 225–35.

(1972) The range of attraction of animal baits and carbon dioxide for mosquitoes. Studies in a freshwater area of West Africa. *Bull. Ent. Res.*, **61**, 389–404.

Glasgow, J. P. (1961) The feeding habits of *Glossina swynnertoni*. *J. An. Ecol.*, **30**, 77–85.

Goodchild, A. J. P. (1955) Some observations on growth and egg production of the blood-sucking Reduviids, *Rhodnius prolixus* and *Triatoma infestans*. *Proc. R. Ent. Soc.*, **30**, 137–44.

Gooding, R. H. (1968) A note on the relationship between feeding and insemination in *Pediculus humanus*. *J. Med. Ent.*, **5**, 265–6.

(1972) Digestive processes of haematophagous insects. I. A literature review. *Quaest. Ent.*, **8**, 5–60.

(1974) Digestive processes in haematophagous insects. Control of trypsin secretion in *Glossina morsitans morsitans*. *J. Insect Physiol.*, **20**, 957–64.

(1975) Inhibition of diuresis in the tsetse fly (*Glossina morsitans*) by ouabain or acetazolamide. *Experientia*, **31**, 938–9.

(1977) Digestive processes of haematophagous insects. XIV Haemolytic activity in the midgut of *Glossina morsitans morsitans* Westwood (Diptera: Glossinidae). *Can. J. Zool.*, **55**, 1899–1905.

Gordon, R. M., Crewe, W. and Willett, K. C. (1956) Studies on the deposition, migration and development to the blood forms of tryapnosomes belonging to the *Trypanosoma brucei* group. I. An account of the process of feeding adopted by the tsetse fly when obtaining a blood meal from the mammalian host, with special reference to the ejection of saliva and the relationship of the feeding process to the deposition of the metacyclic trypanosomes. *Ann. Trop. Med.*, **50**, 426–37.

Gore, T. C. and Pittman-Noblet, G. (1978) The effect of photoperiod on the deep body temperature of domestic turkeys and its relationship to the diurnal periodicity of *Leucocytozoon smithi* gametocytes in the peripheral blood of turkeys. *Poultry Science*, **57**, 603–7.

Gorman, M. J., Cornel, A. J., Collins, F. H. and Paskewitz, S. M. (1996) A shared genetic mechanism for melanotic encapsulation of CM- Sephadex beads and a malaria parasite, *Plasmodium cynomolgi* B, in the mosquito, *Anopheles gambiae*. *Experimental Parasitology*, **84**, 380–6.

Gottar, M., Gobert, V., Michel, T. *et al.* (2002) The *Drosophila* immune response against Gram-negative bacteria is mediated by a peptidoglycan recognition protein. *Nature*, **416**, 640–4.

Gotz, P. (1986) Encapsulation in arthropods. In M. Brehelin (ed.), *Immunity in Invertebrates*. Springer Verlag, Berlin Heidelberg, New York, Tokyo, 153–70.

Graf, R., Raikhel, A. S., Brown, M. R., Lea, A. O. and Briegel, H. (1986) Mosquito trypsin: immunocytochemical localisation in the midgut of blood-fed *Aedes aegypti* (L.). *Cell*, **245**, 19–27.

Graham, H. (1902) Dengue: a study of its mode of propagation and pathology. *Medical Record*, **61**, 204–7.

Grant, A. J., Wigton, B. E., Aghajanian, J. G. and O' Connell, R. J. (1995) Electrophysiological responses of receptor neurons in mosquito maxillary palp sensilla to carbon-dioxide. *J. Comp. Physiol.*, **177**, 389–96.

Grassi, B., Bignami, A. E. and Bastianelli, G. (1899) Ciclo evolutivo delle semilune nell' *Anopheles claviger* ed altri studi sulla malaria dall' ottobre 1898 all maggio 1899. *Atti. Soc. Studi Malaria*, **1**, 143–27.

Green, C. H. (1986) effects of colors and synthetic odors on the attraction of *Glossina pallidipes* and *Glossina morsitans morsitans* to traps and screens. *Phys. Ent.*, **11**, 411–21.

  (1989) The use of two-coloured screens for catching *Glossina palpalis palpalis* (Robineau-Desvoidy) (Diptera: Glossinidae). *Bull. Ent. Res.*, **79**, 81–93.

Green, C. H. and Cosens, D. (1983) Spectral responses of the tsetse fly, *Glossina morsitans morsitans*. *J. Insect Physiol.*, **29**, 795–800.

Griffiths, N. and Brady, J. (1995) Wind structure in relation to odour plumes in tsetse fly habitats. *Phys. Ent.*, **20**, 286–92.

Griffiths, N., Paynter, Q. and Brady, J. (1995) Rates of progress up odour trails by tsetse flies: a mark-release video study of the timing of odour source location by *Glossina pallidipes*. *Phys. Ent.*, **20**, 100–8.

Grimstad, P. R., Paulson, S. L. and Craig, G. B., Jr (1985) Vector competence of *Aedes hendersoni* (Diptera: Culicidae) for La Crosse virus and evidence of a salivary gland escape barrier. *J. Med. Ent.*, **22**, 447–53.

Grimstad, P. R., Ross, Q. E. and Craig, G. B., Jr (1980) *Aedes triseriatus* (Diptera: Culicidae) and La Crosse virus. II. Modification of mosquito feeding behaviour by virus infection. *J. Med. Ent.*, **17**, 1–7.

Grossman, G. L. and Pappas, L. G. (1991) Human skin temperature and mosquito (Diptera, Culicidae) blood feeding rate. *J. Med. Ent.*, **28**, 456–60.

Grubhoffer, L., Hypsa, V. and Volf, P. (1997) Lectins (hemagglutinins) in the gut of the important disease vectors. *Parasite-Journal De La Societe Francaise De Parasitologie*, **4**, 203–16.

Guarneri, A. A., Diotaiuti, L., Gontijo, N. F., Gontijo, A. F. and Pereira, M. H. (2000) Comparison of feeding behaviour of *Triatoma infestans*, *Triatoma brasiliensis* and *Triatoma pseudomaculata* in different hosts by electronic monitoring of the cibarial pump. *J. Insect Physiol.*, **46**, 1121–7.

Guerenstein, P. G. and Nunez, J. A. (1994) Feeding response of the hematophagous bugs *Rhodnius prolixus* and *Triatoma infestans* to saline solutions: a comparative-study. *J. Insect Physiol.*, **40**, 747–52.

Gwadz, R. W. (1969) Regulation of blood meal size in the mosquito. *J. Insect Physiol.*, **15**, 2039–44.

Haarlov, N. (1964) Life cycle and distribution pattern of *Lipoptena cervi* (Dipt., Hippobosc.) on Danish deer. *Oikos*, **15**, 93–129.

Hacker, C. S. and Kilama, W. L. (1974) The relationship between *Plasmodium gallinaceum* density and the fecundity of *Aedes aegypti*. *J. Invert. Pathol.*, **23**, 101–5.

Hackett, L. W. and Missiroli, A. (1931) The natural disappearence of malaria in certain regions of Europe. *Am. J. Hyg.*, **13**, 57–78.

Hafner, M. S., Sudman, P. D., Villablanca, F. X. *et al.* (1994) Disparate rates of molecular evolution in cospeciating hosts and parasites. *Science*, **265**, 1087–90.

Hailman, J. P. (1979) Environmental light and conspicuous colours. In E. H. Burtt (ed.), *The Behavioural Significance of Colour*. New York: Garland STPM Press.

Hall, D. R., Beevor, P. S., Cork, A., Nesbitt, B. F. and Vale, G. A. (1984) 1-Octen-3-ol: a potent olfactory stimulant and attractant for tsetse isolated from cattle odours. *Insect Sci. Applic.*, **5**, 335–9.

Hall, L. R. and Titus, R. G. (1995) Sand fly vector saliva selectively modulates macrophage functions that inhibit killing of *Leishmania major* and nitric oxide production. *Journal of Immunology*, **155**, 3501–6.

Halstead, S. B. (1990) Dengue. In K. S. Warren and A. A. F. Mahmoud (eds.), *Tropical and Geographical Medicine*. New York: McGraw Hill, 675–85.

Handman, E. (2000) Cell biology of *Leishmania*. In *Advances in Parasitology*, Vol. 44, 1–39.

Hansens, E. J., Bosler, E. M. and Robinson, J. W. (1971) Use of traps for study and control of salt marsh flies. *J. Econ. Entomol.*, **64**, 1481–6.

Hao, Z., Kasumba, I., Lehane, M. J. *et al.* (2001) Tsetse immune responses and trypanosome transmission: implications for the development of tsetse-based strategies to reduce trypanosomiasis. *Proc. Natl. Acad. Sci. USA*, **98**, 12648–53.

Haque, A. and Capron, A. (1982) Transplacental transfer of rodent microfilariae induces antigen-specific tolerance in rats. *Nature*, **299**, 361–3.

Hargrove, J. W. (1980a) The effect of ambient-temperature on the flight performance of the mature male tsetse-fly, *Glossina morsitans*. *Phys. Ent.*, **5**, 397–400.

(1980b) The importance of model size and ox odour on the alighting response of *Glossina morsitans* Westwood and *Glossina pallidipes* Austen (Diptera: Glossinidae). *Bull. Ent. Res.*, **70**, 229–34.

Hargrove, J. W., Holloway, M. T. P., Vale, G. A., Gough, A. J. E. and Hall, D. R. (1995) Catches of tsetse (*Glossina* spp) (Diptera, Glossinidae) from traps and targets baited with large doses of natural and synthetic host odor. *Bull. Ent. Res.*, **85**, 215–27.

Hargrove, J. W. and Williams, B. G. (1995) A cost-benefit-analysis of feeding in female tsetse. *Med. Vet. Ent.*, **9**, 109–19.

Harrington, L. C., Edman, J. D. and Scott, T. W. (2001) Why do female *Aedes aegypti* (Diptera: Culicidae) feed preferentially and frequently on human blood? *J. Med. Ent.*, **38**, 411–22.

Harris, J. A., Hillerton, J. E. and Morant, S. V. (1987) Effect on milk production of controlling muscid flies, and reducing fly-avoidance behaviour by the use of Fenvalerate ear tags during the dry period. *Journal of Dairy Research*, **54**, 165–71.

Harris, P., Riordan, D. F. and Cooke, D. (1969) Mosquitoes feeding on insect larvae. *Science*, **164**, 184–5.

Harrison, G. (1978) *Mosquitoes, Malaria and Man: A History of the Hostilities since 1880*. London: Murray.

Hart, B. L. and Hart, L. A. (1994) Fly switching by Asian elephants: tool use to control parasites. *Animal Behaviour*, **48**, 35–45.

Hawking, F. (1962) Microfilaria infestation as an instance of periodic phenomena seen in host-parasite relationships. *Ann. N. Y. Acad. Sci.*, **98**, 940–53.

(1976) Circadian rhythms in *Trypanosoma congolense*. *Trans. R. Soc. Trop. Med. Hyg.*, **70**, 170.

Hawking, F., Gammage, K. and Worms, M. J. (1972) The asexual and sexual circadian rhythms of *Plasmodium vinckei chabaudi*, *P. berghei* and *P. gallinaceum*. *Parasitology*, **65**, 189–201.

Hawking, F., Wilson, M. E. and Gammage, K. (1971) Guidance for cyclic development and short-lived maturity in in the gametocytes of *Plasmodium falciparum*. *Parasitology*, **65**, 549–59.

Hawking, F., Worms, M. J. and Gammage, K. (1968) 24- and 48-hour cycles of malaria parasites in the blood: their purpose, production and control. *Trans. R. Soc. Trop. Med. Hyg.*, **62**, 731–60.

Hawking, F., Worms, M. J., Gammage, K. and Goddard, P. A. (1966) The biological purpose of the blood-cycle of the malaria parasite *Plasmodium cynomolgi*. *Lancet*, **2**, 422–4.

Hecker, H. and Rudin, W. (1981) Morphometric parameters of the midgut cells of *Aedes aegypti* L. (Insecta, Diptera) under various conditions. *Cell*, **219**, 619–27.

Helle, T. and Aspi, J. (1983) Does herd formation reduce insect harassment among reindeer? A field experiment with animal traps. *Acta Zool. Fernica*, **175**, 129–31.

Hendry, G. and Godwin, G. (1988) Biting midges in Scottish forestry: a costly irritant or a trivial nuisance? *Scottish Forestry*, **42**, 113–19.

Henry, V. G. and Conley, R. H. (1970) Some parasites of European wild hogs in the southern Appalachians. *J. Wildl. Mgmt.*, **34**, 913–17.

Hill, C. A., Fox, A. N., Pitts, R. J. *et al.* (2002) G protein coupled receptors in *Anopheles gambiae*. *Science*, **298**, 176–8.

Hill, P., Saunders, D. S. and Campbell, J. A. (1973) The production of 'symbiont-free' *Glossina morsitans* and an associated loss of female fertility. *Trans. R. Soc. Trop. Med. Hyg.*, **67**, 727–8.

Hillyer, J. F. and Christensen, B. M. (2002) Characterization of hemocytes from the yellow fever mosquito, *Aedes aegypti*. *Histochemistry and Cell Biology*, **117**, 431–40.

Hinnebusch, B. J., Fischer, E. R. and Schwan, T. G. (1998) Evaluation of the role of the *Yersinia pestis* plasminogen activator and other plasmid-encoded factors in temperature-dependent blockage of the flea. *Journal of Infectious Diseases*, **178**, 1406–15.

Hinton, H. E. (1958) The phylogeny of the panorpoid orders. *Ann. Rev. Ent.*, **3**, 181–206.

Hoc, T. Q. and Schaub, G. A. (1996) Improvement of techniques for age grading hematophagous insects: ovarian oil-injection and ovariolar separation techniques. *J. Med. Ent.*, **33**, 286–9.

Hocking, B. (1953) The intrinsic range and speed of flight of insects. *Trans. R. Soc. Lond.*, **104**, 223–345.

(1957) Louse control through textile fibre size. *Bull. Ent. Res.*, **48**, 507–14.

(1971) Blood-sucking behaviour of terrestrial arthropods. *Ann. Rev. Ent.*, **16**, 1–26.

Hockmeyer, W. T., Schieffer, B. A., Redington, B. C. and Eldridge, B. V. (1975) *Brugia pahangi* effects upon the flight capability of *Aedes aegypti*. *Experimental Parasitology*, **38**, 1–5.

Hoffmann, J. A. (2003) The immune response of *Drosophila*. *Nature*, **426**, 33–8.

Hoffmann, J. A. and Reichhart, J. M. (2002) *Drosophila* innate immunity: an evolutionary perspective. *Nature Immunology*, **3**, 121–6.

Hogg, J. C. and Hurd, H. (1995) *Plasmodium yoelli nigeriensis* – the effect of high and low intensity of infection upon the egg production and bloodmeal size of *Anopheles stephensi* during three gonotrophic cycles. *Parasitol.*, **111**, 555–62.

Holloway, M. T. P. and Phelps, R. J. (1991) The responses of *Stomoxys* spp. (Diptera, Muscidae) to traps and artificial host odors in the field. *Bull. Ent. Res.*, **81**, 51–5.

Hooke, R. (1664) *Micrographia*.

Hopkins, F. H. E. (1949) The host associations of the lice of mammals. *Proc. Zool. Soc. Lond.*, **119**, 387–604.

Hopwood, J. A., Ahmed, A. M., Polwart, A., Williams, G. T. and Hurd, H. (2001) Malaria-induced apoptosis in mosquito ovaries: a mechanism to control vector egg production. *J. Exp. Biol.*, **204**, 2773–2780.

Hosoi, T. (1958) Adenosine 5′ phosphates as the stimulating agent in blood for inducing gorging of the mosquito. *Nature*, **181**, 1664–5.

(1959) Identification of blood components which induce gorging in the mosquito. *J. Insect Phys.*, **3**, 191–218.

Houseman, J. G., Downe, A. E. R. and Morrison, P. E. (1985a) Similarities in digestive proteinase production in *Rhodnius prolixus* (Hemiptera: Reduviidae) and *Stomoxys calcitrans* (Diptera: Muscidae). *Insect Biochem.*, **15**, 471–4.

Houseman, J. G., Morrison, P. E. and Downe, A. E. R. (1985b) Cathepsin B and aminopeptidase in the posterior midgut of *Euschistus euschistoides* (Hemiptera: Phymatidae). *Can. J. Zool.*, **63**, 1288–91.

Howe, M. A. and Lehane, M. J. (1986) Post-feed buzzing in the tsetse, *Glossina morsitans morsitans*, is an endothermic mechanism. *Phys. Ent.*, **11**, 279–86.

Huang, C. T. (1971) Vertebrate serum inhibitors of *Aedes aegypti* trypsin. *Insect Biochem.*, **1**, 27–38.

Hudson, A. (1970) Notes on the piercing mouthparts of three species of mosquitoes (Diptera: Culicidae) viewed with the scanning electron miscroscope. *Can. Entomal.*, **102**, 501–9.

Hudson, B. W., Feingold, B. F. and Kartman, L. (1960) Allergy to flea bites. II. Investigations of flea bite sensitivity in humans. *Experimental Parasitology*, **9**, 264–70.

Huff, C. (1929) Ovulation requirements of *Culex pipiens* Linn. *Biol. Bull. (Woods Hole)*, **56**, 347–50.

Huff, C. (1931) A proposed classification of disease transmission by arthropods. *Science*, **74**, 456–7.

Hughes, A. L. and Piontkivska, H. (2003) Phylogeny of trypanosomatidae and bodonidae (Kinetoplastida) based on 18S rRNA: evidence for paraphyly of *Trypanosoma* and six other genera. *Molecular Biology and Evolution*, **20**, 644–52.

Humphries, D. A. (1967) Function of combs in ectoparasites. *Nature*, **215**, 319.

Hunter, D. M. and Moorhouse, D. W. (1976) The effects of *Austrosimulium pestilens* on the milk production of dairy cattle. *Aust. Vet. J.*, **52**, 97–9.

Huq, M. (1961) African horse sickness. *Veterinary Record*, **73**, 123.

Hurd, H. (1998) Parasite manipulation of insect reproduction: who benefits? *Parasitology*, **116**, S13–21.

(2003) Manipulation of medically important insect vectors by their parasites. *Ann. Rev. Ent.*, **48**, 141–61.

Hurd, H., Hogg, J. C. and Renshaw, M. (1995) Interactions between bloodfeeding, fecundity and infection in mosquitoes. *Parasitology Today*, **11**, 411–6.

Hursey, B. S. (2001) The programme against African trypanosomiasis: aims, objectives and achievements. *Trends in Parasitology*, **17**, 2–3.

Ibrahim, E. A. R., Ingram, G. A. and Molyneux, D. H. (1984) Haemagglutinins and parasite agglutinins in haemolymph and gut of *Glossina*. *Tropenmed. Parasit.*, **35**, 151–6.

ICZN (1999) *International Code of Zoological Nomenclature*. London: ICZN.

Irving, P., Troxler, L., Heuer, T. S. *et al.* (2001) A genome-wide analysis of immune responses in *Drosophila*. *Proceedings of the National Academy of Sciences of the United States of America*, **98**, 15119–24.

Isawa, H., Yuda, M., Orito, Y. and Chinzei, Y. (2002) A mosquito salivary protein inhibits activation of the plasma contact system by binding to factor XII and high molecular weight kininogen. *J. Biol. Chem.*, **277**, 27651 8.

Isawa, H., Yuda, M., Yoneda, K. and Chinzei, Y. (2000) The insect salivary protein, prolixin-S, inhibits factor IXa generation and Xase complex formation in the blood coagulation pathway. *J. Biol. Chem.*, **275**, 6636–41.

Iwanaga, S. (2002) The molecular basis of innate immunity in the horseshoe crab. *Current Opinion in Immunology*, **14**, 87–95.

James, M. T. and Harwood, R. F. (1969) *Herm's Medical Entomology*. London: Macmillan.

Janeway, C. A., Travers, P., Walport, M. and Schlomchik, M. (2001) *Immunology*. Edinburgh: Churchill Livingstone.

Janse, C. J., Rouwenhorst, R. J., van der Klooster, P. F. J., van der Kaay, H. J. and Overdulve, J. P. (1985) Development of *Plasmodium berghei* ookinetes in the midgut of *Anopheles* atroparvus mosquitoes and in vitro. *Parasit.*, **91**, 219–25.

Jefferies, D. (1984) Transmission of disease by haematophagous arthropods. Unpublished Ph.D. thesis, University of Salford.

Jenkins, D. W. (1964) Advances in medical entomology using radio-isotopes. *Experimental Parasitology*, **3**, 474–90.

Jenni, L., Molyneux, D. H., Livesey, J. L. and Galun, R. (1980) Feeding behaviour of tsetse flies infected with salivarian trypanosomes. *Nature*, **283**, 383–5.

Jobling, B. (1976) On the fascicle of blood-sucking Diptera. In addition a description of the maxillary glands in *Phlebotomus papatasi*, together with the musculature of the labium and pulsatory organ of both the latter species and also of some other Diptera. *J. Nat. Hist.*, **10**(4), 457–61.

Johnson, K. P., Adams, R. J. and Clayton, D. H. (2002a) The phylogeny of the louse genus *Brueelia* does not reflect host phylogeny. *Biological Journal of the Linnaean Society*, **77**, 233–47.

Johnson, K. P., Weckstein, J. D., Witt, C. C., Faucett, R. C. and Moyle, R. G. (2002b) The perils of using host relationships in parasite taxonomy: phylogeny of the *Degeeriella* complex. *Mol. Phylogenet. Evol.*, **23**, 150–7.

Jones, C. J. (1996) Immune responses to fleas, bugs and sucking lice. In S. K. Wikel (ed.), *The Immunology of Host–Ectoparasitic Arthropod Interactions*. Wallingford: CAB International, 150–74.

Jones, J. C. and Pillitt, D. R. (1973) Blood-feeding behavior of adult *Aedes aegypti* mosquitoes. *Biol. Bull.*, **145**, 127–39.

Jordan, A. M. (1974) Recent development in the ecology and methods of control of tsetse flies (*Glossina* spp.), a review. *Bull. Ent. Res.*, **63**(4), 361–99.

Jordan, A. M. (1986) *Trypanosomiasis Control and African Rural Development*. London: Longman.

Jordan, A. M. and Curtis, C. F. (1968) The performance of *Glossina austeni* when fed on lop-eared rabbits and goats. *Trans. R. Soc. Trop. Med. Hyg.*, **62**, 123–4.

  (1972) Productivity of *Glossina morsitans morsitans* Westwood maintained in the laboratory, with particular reference to the sterile-insect release method. *Bull. W. H. O.*, **46**, 33–8.

Jordan, K. (1962) Notes on the *Tunga caecigena* (Siphonaptera: Tungidae). *Bull. Br. Mus. Nat. Hist. (Ent.)*, **12**(4), 353–64.

Julius, D. and Basbaum, A. I. (2001) Molecular mechanisms of nociception. *Nature*, **413**, 203–10.

Kaaya, G. P. and Ratcliffe, N. A. (1982) Comparative study of haemocytes and associated cells of some medically important dipterans. *J. Morph.*, **173**, 351–65.

Kamhawi, S. (2000) The biological and immunomodulatory properties of sand fly saliva and its role in the establishment of *Leishmania* infections. *Microbes Infect*, **2**, 1765–73.

Kamhawi, S., Belkaid, Y., Modi, G., Rowton, E. and Sacks, D. (2000) Protection against cutaneous Leishmaniasis resulting from bites of uninfected sand flies. *Science*, **290**, 1351–4.

Kangwangye, T. N. (1977) Reactions of large mammals to biting flies in Rwenzori National Park, Uganda. In C. P. F. Lima (ed.), *Proceedings of the First East African Conference on Entomological Pest Control*.

Kartman, L. (1953) Factors influencing infection of the mosquito with *Dirofilaria immitis* (Leidy, 1856). *Experimental Parasitology*, **2**, 27–78.

Kathirithamby, J., Ross, L. D. and Johnston, J. S. (2003) Masquerading as self? Endoparasitic Strepsiptera (Insecta) enclose themselves in host-derived epidermal bag. *Proc. Natl. Acad. Sci. USA*, **100**, 7655–9.

Katz, O., Waitumbi, J. N., Zer, R. and Warburg, A. (2000) Adenosine, AMP, and protein phosphatase activity in sandfly saliva. *Am. J. Trop. Med. Hyg.*, **62**, 145–50.

Kavaliers, M., Choleris, E. and Colwell, D. D. (2001) Learning from others to cope with biting flies: social learning of fear-induced conditioned analgesia and active avoidance. *Behavioral Neuroscience*, **115**, 661–74.

Keiper, R. R. and Berger, J. (1982) Refuge-seeking and pest avoidance by feral horses in desert and island environments. *Applied Animal Ethology*, **9**, 111–20.

Kellogg, F. E. (1970) Water vapour and carbon dioxide receptors in *Aedes aegypti*. *J. Insect Physiol.*, **16**, 99–108.

Kellogg, F. E. and Wright, R. H. (1962) The guidance of flying insects. V. Mosquito attraction. *Can. Entomol.*, **94**, 1009–16.

Kelly, D. W. (2001) Why are some people bitten more than others? *Trends in Parasitology*, **17**, 578–81.

Kelly, D. W., Mustafa, Z. and Dye, C. (1996) Density-dependent feeding success in a field population of the sandfly, *Lutzomyia longipalpis*. *Journal of Animal Ecology*, **65**, 517–27.

Kelly, D. W. and Thompson, C. E. (2000) Epidemiology and optimal foraging: modelling the ideal free distribution of insect vectors. *Parasitology*, **120**, 319–27.

Kennedy, J. S. (1940) The visual responses of flying mosquitoes. *Proc. Zool. Soc. Lond.*, **109**, 221–42.

(1983) Zigzagging and casting as a programmed response to wind-borne odour: a review. *Phys. Ent.*, **8**, 109–20.

Kettle, D. S. (1984) *Medical and Veterinary Entomology*. London: Croom Helm.

Khan, A. A. and Maibach, H. I. (1970) A study of the probing response of *Aedes aegypti*. I. Effect of nutrition on probing. *J. Econ. Ent.*, **63**, 974–6.

(1971) A study of the probing response of *Aedes aegypti*. 2. Effect of desiccation and blood feeding on probing to skin and an artificial target. *J. Econ. Entomol.*, **64**, 439–42.

Kilama, W. L. and Craig, G. B. (1969) Monofactorial inheritance of susceptibility to *Plasmodium gallinaceum* in *Aedes aegypti*. *Ann. Trop. Med. Parasit.*, **63**, 419–32.

Killeen, G. F., McKenzie, F. E., Foy, B. D., Bogh, C. and Beier, J. C. (2001) The availability of potential hosts as a determinant of feeding behaviours and malaria transmission by African mosquito populations. *Trans. R. Soc. Trop. Med. Hyg.*, **95**, 469–76.

Kim, K. C. (1985) Evolution and host association of Anoplura. In K. C. Kim (ed.), *Coevolution of Parasitic Arthropods and Mammals*. New York: Wiley.

Kim, K. D. and Adler, P. H. (1985) Evolution and host association of Anoplura. In K. C. Kim (ed.), *Coevolution of Parasitic Arthropods and Mammals*. New York: Wiley.

Kingsolver, J. G. (1987) Mosquito host choice and the epidemiology of malaria. *Am. Nat.*, **130**, 811–27.

Kirch, H. J., Spates, G., Droleskey, R., Kloft, W. J. and Deloach, J. R. (1991a) Mechanism of hemolysis of erythrocytes by hemolytic factors from *Stomoxys calcitrans* (L) (Diptera, Muscidae). *J. Insect Phys.*, **37**, 851–61.

Kirch, H. J., Spates, G., Kloft, W. J. and Deloach, J. R. (1991b) The relationship of membrane-lipids to species-specific hemolysis by hemolytic factors from *Stomoxys calcitrans* (L) (Diptera, Muscidae). *Insect Biochem.*, **21**, 113.

Klein, T. A., Harrison, B. A., Andre, R. G., Whitmire, R. E. and Inlao, I. (1982) Detrimental effects of *Plasmodium cynomolgi* infections on the longevity of *Anopheles dirus*. *Mosq. News*, **42**, 265–71.

Kline, D. L. and Lemire, G. F. (1995) Field evaluation of heat as an added attractant to traps baited with carbon dioxide and octenol for *Aedes taeniorhynchus*. *J. Am. Mosq. Control Assoc.*, **11**, 454–6.

Klowden, M. J. (1993) Mating and nutritional state affect the reproduction of *Aedes albopictus* mosquitoes. *J. Am. Mosq. Control Assoc.*, **9**, 169–73.

Klowden, M. J., Davis, E. E. and Bowen, M. F. (1987) Role of the fat body in the regulation of host-seeking behaviour in the mosquito, *Aedes aegypti*. *J. Insect Physiol.*, **33**, 643–6.

Klowden, M. J., Kline, D. L., Takken, W., Wood, J. R. and Carlson, D. A. (1990) Field studies on the potential of butanone, carbon-dioxide, honey extract, 1-octen-3-ol, L-lactic acid and phenols as attractants for mosquitos. *Med. Vet. Entomol.*, **4**, 383–91.

Klowden, M. J. and Lea, A. O. (1979) Abdominal distention terminates subsequent host-seeking behavior of *Aedes aegypti* following a blood meal. *J. Insect Physiol.*, **25**, 583–5.

Knols, B. G. J., de Jong, R. and Takken, W. (1995) Differential attractiveness of isolated humans to mosquitoes in Tanzania. *Trans. R. Soc. Trop. Med. Hyg.*, **89**, 604–6.

Knols, B. G. J., Mboera, L. E. G. and Takken, W. (1998) Electric nets for studying odour-mediated host-seeking behaviour of mosquitoes. *Med. Vet. Entomol.*, **12**, 116–20.

Knols, B. G. J., van Loon, J. J. A., Cork, A. *et al.* (1997) Behavioural and electrophysiological responses of the female malaria mosquito *Anopheles gambiae* (Diptera: Culicidae) to Limburger cheese volatiles. *Bull. Ent. Res.*, **87**, 151–9.

Koella, J. C., Agnew, P. and Michalakis, Y. (1998a) Coevolutionary interactions between host life histories and parasite life cycles. *Parasitology*, **116**, S47–S55.

Koella, J. C. and Boete, C. (2002) A genetic correlation between age at pupation and melanization immune response of the yellow fever mosquito *Aedes aegypti*. *Evolution Int. J. Org. Evolution*, **56**, 1074–9.

Koella, J. C., Sorensen, F. L. and Anderson, R. A. (1998b) The malaria parasite, *Plasmodium falciparum*, increases the frequency of multiple feeding of its mosquito vector, *Anopheles gambiae*. *Proc. R. Soc. Lond. B Biol. Sci.*, **265**, 763–8.

Komano, H., Mizuno, D. and Natori, S. (1980) Purification of a lectin induced in the haemolymph of *Sarcophaga peregrina* larvae on injury. *J. Biol. Chem.*, **255**, 2919–24.

Kramer, L. D., Hardy, J. L., Presser, S. B. and Houk, E. G. (1981) Dissemination barriers for western equine encephalomyelitis virus in *Culex tarsalis* infected after ingestion of low viral doses. *Am. J. Trop. Med. Hyg.*, **30**, 190–7.

Krasnov, B. R., Khokhlova, I. S. and Shenbrot, G. I. (2003a) Density-dependent host selection in ectoparasites: an application of isodar theory to fleas parasitizing rodents. *Oecologia*, **134**, 365–72.

Krasnov, B. R., Sarfati, M., Arakelyan, M. S. *et al.* (2003b) Host specificity and foraging efficiency in blood-sucking parasite: feeding patterns of the flea *Parapulex chephrenis* on two species of desert rodents. *Parasitology Research*, **90**, 393–9.

Krynski, S., Kuchta, A. and Becla, E. (1952) Research on the nature of the noxious action of guinea-pig blood on the body louse (in Polish). *Bull. Inst. Mar. Med. Gdansk*, **4**, 104–7.

Ksiazkiewicz-Ilijewa, M. and Rosciszewska, E. (1979) Ultrastructure of the haemocytes of *Tetrodontophora bielanensis* Waga (Collembola). *Cytobios*, **26**, 113–21.

Kunz, S. E., Murrell, K. D., Lambert, G., James, L. F. and Terrill, C. E. (1991) Estimated losses of livestock to pests. In D. Pimentel (ed.), *CRC Handbook of*

Pest Management in Agriculture. Boca Raton, Florida: CRC Press, Vol. 1, pp. 69–98.

Kurata, S. (2004) Recognition of infectious non-self and activation of immune responses by peptidoglycan recognition protein (PGRP)-family members in *Drosophila. Dev. Comp. Immunol.*, **28**, 89–95.

Kurtz, J. and Franz, K. (2003) Evidence for memory in invertebrate immunity. *Nature*, **425**, 37–8.

La Breque, G. C., Meifert, D. W. and Rye, J. (1972) Experimental control of stable flies, *Stomoxys calcitrans* (Diptera: Muscidae), by the release of chemosterilized adults. *Can. Entomol.*, **104**, 885–7.

Laarman, J. J. (1958) The host-seeking behaviour of anopheline mosquitoes. *Trop. Geogr. Med.*, **10**, 293–305.

Lackie, A. M. (1986) Evasion of insect immunity by helminth larvae. In A. M. Lackie (ed.), *Immune Mechanisms in Invertebrate Vectors*, Oxford: Oxford University Press, Vol. 56, 161–78.

Lafond, M. M., Christensen, B. M. and Lasee, B. A. (1985) Defense reactions of mosquitoes to filarial worms: potential mechanisms for avoidance of the response by *Brugia pahangi* microfilaria. *J. Invert. Pathol.*, **46**, 26–30.

Lainson, R. and Shaw, J. J. (1987) Evolution, classification and geographical distribution. In W. Peters and R. Killick-Kendrick (eds.), *The Leishmaniases in Biology and Medicine.* New York: Academic Press.

Lall, S. B. (1969) Phagostimulants of haematophagous tabanids. *Entomol. Exp. Appl.*, **12**, 325–36.

Land, M. F., Gibson, G. and Horwood, J. (1997) Mosquito eye design: conical rhabdoms are matched to wide aperture lenses. *Proc. R. Soc. Lond. B. Biol. Sci.*, **264**, 1183–7.

Land, M. F., Gibson, G., Horwood, J. and Zeil, J. (1999) Fundamental differences in the optical structure of the eyes of nocturnal and diurnal mosquitoes. *J. Comp. Physiol.*, **185**, 91–103.

Lane, N. J. and Harrison, J. B. (1979) An unusual cell surface modification: a double plasma membrane. *J. Cell Sci.*, **39**, 355–72.

Langley, P. A. (1970) Post-teneral development of thoracic flight musculature in the tsetse flies *Glossina austeni* and *G. morsitans. Entomologia Exp. Appl.*, **13**, 133–40.

Langley, P. A. and Maly, H. (1969) Membrane feeding technique for tsetse flies (*Glossina* spp.). *Nature*, **221**, 855–6.

Lanzaro, G. C., Toure, Y. T., Carnahan, J. *et al.* (1998) Complexities in the genetic structure of *Anopheles gambiae* populations in west Africa as revealed by microsatellite DNA analysis. *Proceedings of the National Academy of Sciences of the United States of America*, **95**, 14260–5.

Larrivee, D. H., Benjamini, E., Feingold, B. F. and Shimuzu, M. (1964) Histologic studies of guinea pig skin: different stages of allergic reactivity to flea bites. *Exp. Parasitol.*, **15**, 491–502.

Laurence, B. R. (1966) Intake and migration of the microfilariae of *Onchocerca volvulus* (Leukart) in *Simulium damnosum* Theobald. *J. Helm.*, **40**, 337–42.

Laurence, B. R. and Pester, F. R. N. (1961) The ability of *Anopheles gambiae* Giles to transmit *Brugia patei* (Buckley, Nelson and Heisch). *J. Trop. Med. Hyg.*, **64**, 169–71.

(1967) Adaptation of a filarial worm, *Brugia patei*, to a new mosquito host, *Aedes togoi*. *J. Helminth.*, **41**, 365–92.

Lavine, M. D. and Strand, M. R. (2002) Insect hemocytes and their role in immunity. *Insect Biochem. Mol. Biol.*, **32**, 1295–309.

Lavoipierre, M. M. J., Dickerson, G. and Gordon, R. M. (1959) Studies on the methods of feeding of blood-sucking arthropods. I. The manner in which triatomine bugs obtain their blood meal as observed in the tissues of the living rodent, with some remarks on the effects of the bite on human volunteers. *Ann. Trop. Med. Parasitol.*, **53**, 235–50.

Lavoipierre, M. M. J. and Hamachi, M. (1961) An apparatus for observations on the feeding mechanism of the flea. *Nature*, **192**, 998–9.

Lazzari, C. R., Reiseman, C. E. and Insausti, T. C. (1998) The role of the ocelli in the phototactic behaviour of the haematophagous bug *Triatoma infestans*. *J. Insect Physiol.*, **44**, 1159–62.

Lebestky, T., Chang, T., Hartenstein, V. and Banerjee, U. (2000) Specification of *Drosophila* hematopoietic lineage by conserved transcription factors. *Science*, **288**, 146–9.

Lee, J. H., Rowley, W. A. and Platt, K. B. (2000) Longevity and spontaneous flight activity of *Culex tarsalis* (Diptera: Culicidae) infected with western equine encephalomyelitis virus. *J. Med. Ent.*, **37**, 187–93.

Lee, R. (1974) Structure and function of the fascicular stylets, and the labral and cibarial sense organs of male and female *Aedes aegypti* (L>). *Quest. Entomol.*, **10**, 187–215.

Lehane, M. J. (1976a) Digestive enzyme secretion in *Stomoxys calcitrans* (Diptera: Muscidae). *Tissue and Cell*, **170**, 275–87.

(1976b) The formation and histochemical structure of the peritrophic membrane in the stable fly, *Stomoxys calcitrans*. *J. Insect Physiol.*, **22**, 1551–7.

(1977a) An hypothesis of the mechanism controlling proteolytic digestive enzyme production levels in *Stomoxys calcitrans*. *J. Insect Physiol.*, **23**, 713–15.

(1977b) Transcellular absorption of lipids in the midgut of the stablefly, *Stomoxys calcitrans*. *J. Insect Physiol.*, **23**, 945–54.

(1985) Determining the age of an insect. *Parasitology Today*, **1**, 81–5.

(1987) Quantitative evidence for merocrine secretion in an insect midgut cell. *Tissue and Cell*, **19**, 451–561.

(1988) Evidence for secretion by the release of cytoplasmic extrusions from midgut cells of *Stomoxys calcitrans*. *J. Insect Physiol.*, **34**, 949–53.

(1989) The intracellular pathway and kinetics of digestive enzyme secretion in an insect midgut cell. *Tissue and Cell*, **21**, 101–11.

(1997) Peritrophic matrix structure and function. *Ann. Rev. Ent.*, **42**, 525–50.

Lehane, M. J., Aksoy, S., Gibson, W. (2003) Adult midgut EST from the tsetse fly *Glossina morsitans morsitans* and expression analysis of putative immune response genes. *Genome Biology*, **4** (10), R63.

Lehane, M. J., Aksoy, S. and Levashina, E. A. (2004) Blood-sucking insect immune responses and parasite transmission. *Trends in Parasitology*, in press.

Lehane, M. J., Allingham, P. G. and Weglicki, P. (1996) Peritrophic matrix composition of the tsetse fly, *Glossina morsitans morsitans*. *Cell and Tissue Research*, **272**, 158–62.

Lehane, M. J. and Billingsley, P. A. (eds.) (1996) *The Biology of the Insect Midgut.* London: Chapman and Hall.

Lehane, M. J., Crisanti, A. and Mueller, H. M. (1996) Mechanisms controlling the synthesis and secretion of digestive enzymes in insects. In M. J. Lehane (ed.), *The Insect Midgut.* London: Chapman and Hall.

Lehane, M. J. and Hargrove, J. (1988) Field experiments on a new method for determining age in tsetse flies (Diptera, Glossinidae). *Ecol. Entomol.,* **13,** 319–22.

Lehane, M. J. and Laurence, B. R. (1977) Flight muscle ultrastructure of susceptible and refractory mosquitoes parasitized by larval *Brugia pahangi. Parasitology,* **74,** 87–92.

Lehane, M. J. and Mail, T. S. (1985) Determining the age of adult male and female *Glossina morsitans morsitans* using a new technique. *Ecol. Entomol.,* **10,** 219–24.

Lehane, M. J. and Schofield, C. J. (1981) Field experiments of dispersive flight by *Triatoma infestans. Trans. R. Soc. Trop. Med. Hyg.,* **75,** 399–400.

(1982) Flight initiation in *Triatoma infestans* (Klug) (Hemiptera: Reduviidae). *Bull. Ent. Res.,* **72,** 497–510.

Lehane, M. J., Wu, D. and Lehane, S. M. (1997) Midgut-specific immune molecules are produced by the blood-sucking insect *Stomoxys calcitrans. Proceedings of the National Academy of Sciences of the United States of America,* **94,** 11502–7.

Lehane, S. M., Assinder, S. J. and Lehane, M. J. (1998) Cloning, sequencing, temporal expression and tissue-specificity of two serine proteases from the midgut of the blood-feeding fly *Stomoxys calcitrans. European Journal of Biochemistry,* **254,** 290–6.

Lemaitre, B., Reichhart, J. M. and Hoffmann, J. A. (1997) *Drosophila* host defense: differential induction of antimicrobial peptide genes after infection by various classes of microorganisms. *Proceedings of the National Academy of Sciences of the United States of America,* **94,** 14614–19.

Lemos, F. J. A., Cornel, A. J. and Jacobs Lorena, M. (1996) Trypsin and aminopeptidase gene expression is affected by age and food composition in *Anopheles gambiae. Insect Biochemistry and Molecular Biology,* **26,** 651–8.

Lester, H. M. O. and Lloyd, L. (1929) Notes on the process of digestion in tsetse flies. *Bull. Ent. Res.,* **19,** 39–60.

Leulier, F., Parquet, C., Pili-Floury, S. *et al.* (2003) The *Drosophila* immune system detects bacteria through specific peptidoglycan recognition. *Nature Immunology,* **4,** 478–84.

Levashina, E. A., Langley, E., Green, C. *et al.* (1999) Constitutive activation of toll-mediated antifungal defense in serpin-deficient *Drosophila. Science,* **285,** 1917–19.

Levashina, E. A., Moita, L. F., Blandin, S. *et al.* (2001) Conserved role of a complement-like protein in phagocytosis revealed by dsRNA knockout in cultured cells of the mosquito *Anopheles gambiae. Cell,* **104,** 709–18.

Lewis, D. J. (1953) *Simulium damnosum* and its relation to Onchocerciasis in the Anglo-Egyptian Sudan. *Bull. Ent. Res.,* **43,** 597–644.

Lewis, L. F., Christenson, D. M. and Eddy, G. W. (1967) Rearing the long-nosed cattle louse and cattle-biting louse on host animals in Oregon. *J. Econ. Entomol.,* **60,** 755–7.

Lewis, T. and Taylor, L. R. (1965) Diurnal periodicity of flight by insects. *Trans. R. Ent. Soc. Lond.,* **116,** 393–479.

Li, X., Sina, B. and Rossignol, P. A. (1992) Probing behaviour and sporozoite delivery by *Anopheles stephensi* infected with *Plasmodium berghei*. *Med. Vet. Entomol.*, **6**, 57–61.

Ligoxygakis, P., Pelte, N., Hoffmann, J. A. and Reichart, J. M. (2002) Activation of *Drosophila* toll during fungal infection by a blood serine protease. *Nature Reviews Immunology*, **2**, 545.

Lindsay, L. B. and Galloway, T. D. (1998) Reproductive status of four species of fleas (Insecta: Siphonaptera) on Richardson's ground squirrels (Rodentia: Sciuridae) in Manitoba, Canada. *J. Med. Ent.*, **35**, 423–30.

Lindsay, S. W., Adiamah, J. H., Miller, J. E., Pleass, R. J. and Armstrong, J. R. M. (1993) Variation in attractiveness of human subjects to malaria mosquitoes (Diptera, Culicidae) in the Gambia. *J. Med. Ent.*, **30**, 368–73.

Linley, J. R. and Davies, J. B. (1971) Sandflies and tourism in Florida and the Bahamas and Caribbean area. *J. Econ. Entomol.*, **64**, 264–78.

Linsenmair, K. E. (1973) Die Windorientierurung laufender Insekten. *Fortschr. Zool.*, **21**, 59–79.

Liu, C. T., Hou, R. F. and Chen, C. C. (1998) Formation of basement membrane-like structure terminates the cellular encapsulation of microfilariae in the haemocoel of *Anopheles quadrimaculatus*. *Parasitology*, **116** (Pt 6), 511–18.

Lochmiller, R. L. and Deerenberg, C. (2000) Trade-offs in evolutionary immunology: just what is the cost of immunity? *Oikos*, **88**, 87–98.

Loder, P. M. J., Hargrove, J. W. and Randolph, S. E. (1998) A model for blood meal digestion and fat metabolism in male tsetse flies (Glossinidae). *Phys. Ent.*, **23**, 43–52.

Lodmell, D. L., Bell, J. F., Clifford, C. M., Moore, G. J. and Raymond, G. (1970) Effects of limb disability on lousiness in mice. V. Hierarchy disturbance on mutual grooming and reproductive capacities. *Expl. Parasit.*, **27**, 184–92.

Loke, H. and Randolph, S. E. (1995) Reciprocal regulation of fat-content and flight activity in male tsetse-flies (*Glossina palpalis*). *Phys. Ent.*, **20**, 243–7.

Lord, W. D., DiZinno, J. A., Wilson, M. R. *et al.* (1998) Isolation, amplification, and sequencing of human mitochondrial DNA obtained from human crab louse, *Pthirus pubis* (L.), blood meals. *Journal of Forensic Sciences*, **43**, 1097–100.

Loudon, C. and McCulloh, K. (1999) Application of the Hagen-Poiseuille equation to fluid feeding through short tubes. *Ann. Ent. Soc. Am.*, **92**, 153–8.

Lowenberger, C. A., Ferdig, M. T., Bulet, P. *et al.* (1996) *Aedes aegypti* – induced antibacterial proteins reduce the establishment and development of *Brugia malayi*. *Experimental Parasitology*, **83**, 191–201.

Lowther, J. K. and Wood, D. M. (1964) Specificity of a black fly, *Simulium euryadmiculum* Davies, towards its host, the common loon. *Can. Entomol.*, **96**, 911–13.

Luckhart, S. and Rosenberg, R. (1999) Gene structure and polymorphism of an invertebrate nitric oxide synthase gene. *Gene*, **232**, 25–34.

Luckhart, S., Vodovotz, Y., Cui, L. W. and Rosenberg, R. (1998) The mosquito *Anopheles stephensi* limits malaria parasite development with inducible synthesis of nitric oxide. *Proceedings of the National Academy of Sciences of the United States of America*, **95**, 5700–5.

Lyman, D. E., Monteiro, F. A., Escalante, A. A. *et al.* (1999) Mitochondrial DNA sequence variation among triatomine vectors of Chagas' disease. *Am. J. Trop. Med. Hyg.*, **60**, 377–86.

Lythgoe, K. A. (2000) The coevolution of parasites with host-acquired immunity and the evolution of sex. *Evolution Int. J. Org. Evolution*, **54**, 1142–56.

Maa, T. C. and Marshall, A. G. (1981) Diptera Pupipara of the New Hebrides (South Pacific): taxonomy, zoogeography, host association and ecology. *Q. Jl. Taiwan Mus.*, **34**, 213–32.

MacCormack, C. P. (1984) Human ecology and behaviour in malaria control in tropical Africa. *Bull. WHO*, **62**, 81–7.

Macdonald, W. W. (1962a) The genetic basis of susceptibility to infection with semi-periodic *Brugia malayi* in *Aedes aegypti*. *Ann. Trop. Med. Parasit.*, **56**, 373–82.

  (1962b) The selection of a strain of *Aedes aegypti* susceptible to infection with semi-periodic *Brugia malayi*. *Ann. Trop. Med. Parasit.*, **56**, 368–72.

  (1963) Further studies on a strain of *Aedes aegypti* susceptible to infection with sub-periodic *Brugia malayi*. *Ann. Trop. Med. Parasit.*, **57**, 452–60.

Macdonald, W. W. and Ramachandran, A. (1965) The influence of the gene fm (filarial susceptibility, *Brugia malayi*) on the susceptibility of *Aedes aegypti* to several strains of *Brugia*, *Wuchereria* and *Dirofilaria*. *Annals of Tropical Medicine and Parasitology*, **59**, 64–73.

Mackie, F. P. (1907) The part played by *Pediculus corporis* in the transmission of relapsing fever. *British Medical Journal*, **2**, 1706–9.

Macvicker, J. A. K., Billingsley, P. F., Djamgoz, M. B. A. and Harrow, I. D. (1994) Ouabain-sensitive Na^+/K^+-ATPase activity in the reservoir zone of the midgut of *Stomoxys calcitrans* (Diptera, Muscidae). *Insect Biochemistry and Molecular Biology*, **24**, 151–9.

Maddrell, S. H. P. (1963) Control of ingestion in *Rhodnius prolixus* Stal. *Nature*, **198**, 210.

  (1980) Characteristics of epithelial transport in insect Malpighian tubules. *Curr. Topics Memb. Transport*, **14**, 427–63.

Magesa, S. M., Mdira, Y. K., Akida, J. A., Bygbjerg, I. C. and Jakobsen, P. H. (2000) Observations on the periodicity of *Plasmodium falciparum* gametocytes in natural human infections. *Acta Trop.*, **76**, 239–46.

Mahmood, F. (2000) Susceptibility of geographically distinct *Aedes aegypti* L. from Florida to *Dirofilaria immitis* (Leidy) infection. *J. Vector Ecol.*, **25**, 36–47.

Mahon, R. and Gibbs, A. (1982) Arbovirus-infected hens attract more mosquitoes. In J. S. MacKenzie (ed.), *Viral Diseases in South Esat Asia and the Western Pacific*. New York: Academic Press.

Maier, W. A. and Omer, O. (1973) Der einfluss von *Plasmodium cathemerium* auf den Aminosauregehalt und die eizahl von *Culex pipiens fatigans*. *Z. Parasit.*, **42**, 265–78.

Malhotra, I., Ouma, J. H., Wamachi, A. *et al.* (2003) Influence of maternal filariasis on childhood infection and immunity to *Wuchereria bancrofti* in Kenya. *Infection and Immunity*, **71**, 5231–7.

Mallon, E. B., Loosli, R. and Schmid-Hempel, P. (2003) Specific versus nonspecific immune defense in the bumblebee, *Bombus terrestris* L. *Evolution*, **57**, 1444–7.

Mans, B. J., Louw, A. I. and Neitz, A. W. H. (2002) Evolution of hematophagy in ticks: common origins for blood coagulation and platelet aggregation inhibitors from soft ticks of the genus *Ornithodoros*. *Molecular Biology and Evolution*, **19**, 1695–705.

Manson, P. (1878) On the development of *Filaria sanguinis hominis*, and on the mosquito considered as a nurse. *J. Linn. Soc. Zool. London*, **14**, 304–11.

Marchoux, E. and Salinberi, A. (1903) La spirillose des poules. *Annales de la Institut Pasteur*, **17**, 569–80.

Margalit, J., Galun, R. and Rice, M. J. (1972) Mouthpart sensilla of the tsetse fly and their function. I. Feeding patterns. *Ann. Trop. Med. Parasit.*, **66**, 525–36.

Marshall, A. G. (1981) *The Ecology of Ectoparasitic Insects*. New York: Academic Press.

Marx, R. (1955) Über die wirtsfindung und die Bedeutung de artspezifischen duftstoffes bei *Cimex lectularius* Linne. *Z. Parasit.*, **17**, 41–72.

Masaninga, F. and Mihok, S. (1999) Host influence on adaptation of *Trypanosoma congolense* metacyclics to vertebrate hosts. *Med. Vet. Entomol.*, **13**, 330–2.

Matsumoto, Y., Oda, Y., Uryu, M. and Hayakawa, Y. (2003) Insect cytokine growth-blocking peptide triggers a termination system of cellular immunity by inducing its binding protein. *J. Biol. Chem.*, **278**, 38579–85.

Matthysse, J. G. (1946) Cattle lice: their biology and control. *Cornell Agr. Exp. Sta. Bull*, **832**, 1–67.

Mattingley, P. F. (1965) The evolution of parasite–arthropod vector systems. In A. E. R. Taylor (ed.), *Symposium of the British Society for Parasitology*. Oxford: Blackwell, Vol. 3.

Maudlin, I. and Dukes, P. (1985) Extrachromosomal inheritance of susceptibility to trypanosome infection in tsetse flies. I. Selection of susceptible and refractory strains of *Glossina morsitans morsitans*. *Ann. Trop. Med. Parasit.*, **79**, 317–24.

Maudlin, I. and Ellis, D. (1985) Association between intracellular rickettsia-like infections of midgut cells and susceptibility to trypanosome infections in *Glossina* species. *Z. Parasit.*, **71**, 683–7.

Maudlin, I., Kabayo, J. P., Flood, M. E. T. and Evans, D. A. (1984) Serum factors and the maturation of *Trypanosoma congolense* infections in *Glossina morsitans*. *Z. Parasit.*, **70**, 11–19.

Maudlin, I. and Welburn, S. C. (1987) Lectin-mediated establishment of midgut infections of *Trypanosoma congolense* and *Trypanosoma bruce* in *Glossina morsitans*. *Tropical Medicine and Parasitology*, **38**, 167–70.

Maudlin, I., Welburn, S. C. and Milligan, P. J. M. (1998) Trypanosome infections and survival in tsetse. *Parasitology*, **116**, S23–S28.

Mayer, M. S. and James, J. D. (1969) Attraction of *Aedes aegypti* (L.): responses to human arms, carbon dioxide, and air currents in a new type of olfactometer. *Bull. Ent. Res.*, **58**, 629–42.

(1970) Attraction of *Aedes aegypti*. II. Velocity of reaction to host with and without additional carbon dioxide. *Ent. Exp. Appl.*, **13**, 47–53.

Mbow, M. L., Bleyenberg, J. A., Hall, L. R. and Titus, R. G. (1998) *Phlebotomus papatasi* sand fly salivary gland lysate down-regulates a Th1, but up-regulates a Th2, response in mice infected with *Leishmania major*. *Journal of Immunology*, **161**, 5571–7.

McCabe, C. T. and Bursell, E. (1975a) Interrelationships between amino acid and lipid metabolism in the tsetse fly, *Glossina morsitans*. *Insect Biochem.*, **5**, 781–9.

(1975b) Metabolism of digestive products in the tsetse fly, *Glossina morsitans*. *Insect Biochem.*, **5**, 769–79.

McCall, P. J. and Kelly, D. W. (2002) Learning and memory in disease vectors. *Trends in Parasitology*, **18**, 429–33.

McCall, P. J. and Lemoh, P. A. (1997) Evidence for the 'invitation effect' during blood-feeding by blackflies of the *Simulium damnosum* complex (Diptera, Simuliidae). *Journal of Insect Behavior*, **10**, 299–303.

McCall, P. J., Mosha, F. W., Njunwa, K. J. and Sherlock, K. (2001) Evidence for memorized site-fidelity in *Anopheles arabiensis*. *Trans. R. Soc. of Trop. Med. Hyg.*, **95**, 587–90.

McDermott, M. J., Weber, E., Hunter, S. *et al.* (2000) Identification, cloning, and characterization of a major cat flea salivary allergen (Cte f 1). *Molecular Immunology*, **37**, 361–75.

McGavin, G. C. (2001) *Essential Entomology: An Order by Order Introduction.* Oxford: Oxford University Press.

McGreevy, P. B., Bryan, J. H., Oothuman, P. and Kolstrup, N. (1978) The lethal effects of the cibarial and pharyngeal armatures of mosquitoes on microfilariae. *Trans. R. Soc. Trop. Med. Hyg.*, **74**, 361–8.

McGreevy, P. B., McClelland, G. A. H. and Lavoipierre, M. M. J. (1974) Inheritance of susceptibility to *Dirofilaria immitis* infection in *Aedes aegypti*. *Ann. Trop. Med. Parasit.*, **68**, 97–109.

McKeever, S. (1977) Observations of *Corethrella* feeding on tree frogs (*Hyla*). *Mosq. News*, **37**, 522.

McKeever, S. and French, F. E. (1991) *Corethrella* (Diptera, Corethrellidae) of Eastern North-America – Laboratory Life-History and Field Responses to Anuran Calls. *Ann. Ent. Soc. Am.*, **84**, 493–7.

McKelvey, J. J. (1973) *Man against Tsetse: Struggle for Africa.* Ithaca: Cornell University Press.

Mead-Briggs, A. R. (1964) The reproductive biology of the rabbit flea *Spilopsyllus cuniculi* (Dale) and the dependance of this species on the upon the breeding of its host. *J. Exp. Biol.*, **41**, 371–402.

Medvedev, S. I. and Skylar, V. Y. (1974) Beetles (Coleoptera) from nests of small mammals in Donotsk Province (in Russian). *Entomologicheskoe Obozrenie*, **53**, 561–71.

Meijerink, J., Braks, M. A. H., Brack, A. A. *et al.* (2000) Identification of olfactory stimulants for *Anopheles gambiae* from human sweat samples. *Journal of Chemical Ecology*, **26**, 1367–82.

Meister, M. and Lagueux, M. (2003) *Drosophila* blood cells. *Cell Microbiol.*, **5**, 573–80.

Mellink, J. J. (1981) Selections for blood-feeding efficiency in colonized *Aedes aegypti*. *Mosq. News*, **41**, 119–25.

Mellink, J. J. and Van Den Bovenkamp, W. (1981) Functional aspects of mosquito salivation in blood feeding in *Aedes aegypti*. *Mosq. News*, **41**, 110–15.

Mellor, P. S. and Boorman, J. (1980) Multiplication of the bluetongue virus in *Culicoides nubeculosus* (Meigen) simultaneously infected with the virus and microfilaria of *Onchocerca cervicalis* (Railliet and Henry). *Ann. Trop. Med. Parasit.*, **74**, 463–9.

Menezes, H. and Jared, C. (2002) Immunity in plants and animals: common ends through different means using similar tools. *Comp. Biochem. Physiol. C Toxicol. Pharmacol.*, **132**, 1–7.

Mews, A. R., Baumgartner, H., Luger, D. and Offori, E. D. (1976) Colonisation of *Glossina morsitans morsitans* Westw. in the laboratory using in vitro feeding techniques. *Bull. Ent. Res.*, **65**, 631–42.

Miall, R. C. (1978) The flicker fusion frequencies of six laboratory insects, and the response of the compound eye to mains fluorescent 'ripple'. *Phys. Ent.*, **3**, 99–106.

Michel, T., Reichhart, J. M., Hoffmann, J. A. and Royet, J. (2001) *Drosophila* toll is activated by Gram-positive bacteria through a circulating peptidoglycan recognition protein. *Nature*, **414**, 756–9.

Miller, N. and Lehane, M. J. (1990) In vitro perfusion studies on the peritrophic membrane of the tsetse fly *Glossina morsitans morsitans* (Diptera: Glossinidae). *J. Insect Phys.*, **36**, 813–18.

Minchella, D. J. (1985) Host life-history variation in response to parasitism. *Parasitol.*, **90**, 205–16.

Minchella, D. J. and Loverde, P. T. (1983) Laboratory comparison of the relative success of *Biomphalaria glabrata* stocks which are susceptible and insusceptible to infection with *Schistosoma mansoni*. *Parasitology*, **86**, 335–44.

Mitchell, B. K. and Reinouts van Haga-Kelker, H. A. (1976) A comparison of the feeding behaviour in teneral and post-teneral *Glossina morsitans* (Diptera, Glossinidae) using an artificial membrane. *Ent. Exp. Appl.*, **20**, 105–12.

Mockford, E. L. (1967) Some Psocoptera from the plumage of birds. *Proc. Ent. Soc. Washington*, **69**, 307–9.

  (1971) Psocoptera from the dusky-footed wood rat in southern California (Psocoptera: Atropidae, Psoguillidae, Liposcelidae). *Pan-Pacific Entomologist*, **47**, 127–40.

Moffatt, M. R., Blakemore, D. and Lehane, M. J. (1995) Studies on the synthesis and secretion of digestive trypsin in *Stomoxys calcitrans* (Insecta-Diptera). *Comp. Biochem. Phys. B*, **110B**, 291–300.

Mohr, C. O. (1943) Cattle droppings as ecological units. *Ecol. Monographs*, **13**, 275.

Moloo, S. K. (1983) Feeding behaviour of *Glossina morsitans morsitans* infected with *Trypanosoma vivax, T. congolense* or *T. brucei. Parasit.*, **86**, 51–6.

Moloo, S. K. and Dar, F. (1985) Probing by *Glossina morsitans centralis* infected with pathogenic *Trypanosoma* species. *Trans. R. Soc. of Trop. Med. Hyg.*, **79**, 119.

Moloo, S. K. and Kutuza, S. B. (1970) Feeding and crop-emptying in *Glossina brevipalpis* Newstead. *Acta Trop.*, **27**, 356–77.

Moloo, S. K., Sabwa, C. L. and Baylis, M. (2000) Feeding behaviour of *Glossina pallidipes* and *G. morsitans centralis* on Boran cattle infected with *Trypanosoma congolense* or *T. vivax* under laboratory conditions. *Med. Vet. Entomol.*, **14.**, 290–9.

Molyneux, D. H. (1984) Evolution of the Trypanosomatidae: considerations of polyphyletic origins of mammalian parasites. *CNRS/INSERM*, **1986**, 231–40.

Molyneux, D. H., Bradley, M., Hoerauf, A., Kyelem, D. and Taylor, M. J. (2003) Mass drug treatment for lymphatic filariasis and onchocerciasis. *Trends in Parasitology*, **19**, 516–22.

Molyneux, D. H. and Killick-Kendrick, R. (1987) Morphology, ultrastructure and life cycles. In W. Peters and R. Killick-Kendrick (eds.), *The Leishmaniases in Biology and Medicine*. New York: Academic Press.

Molyneux, D. H., Killick-Kendrick, R. and Ashford, R. W. (1975) *Leishmania* in phlebotomid sandflies. III. The ultrastructure of *Leishmania mexicana amazonensis* in the midgut and pharynx of *Lutzomyia longipalpis*. *Proc. R. Soc. B*, **190**, 341–57.

Montfort, W. R., Weichsel, A. and Andersen, J. F. (2000) Nitrophorins and related antihemostatic lipocalins from *Rhodnius prolixus* and other blood-sucking arthropods. *Biochimica et Biophysica Acta – Protein Structure and Molecular Enzymology*, **1482**, 110–18.

Mooring, M. S., Benjamin, J. E., Harte, C. R. and Herzog, N. B. (2000) Testing the interspecific body size principle in ungulates: the smaller they come, the harder they groom. *Anim. Behav.*, **60**, 35–45.

Mooring, M. S. and Hart, B. L. (1992) Animal grouping for protection from parasites – selfish herd and encounter-dilution effects. *Behaviour*, **123**, 173–93.

Morand, S. and Poulin, R. (1998) Density, body mass and parasite species richness of terrestrial mammals. *Evolutionary Ecology*, **12**, 717–27.

Moret, Y. and Schmid-Hempel, P. (2000) Survival for immunity: the price of immune system activation for bumblebee workers. *Science*, **290**, 1166–8.

Moro, O. and Lerner, E. A. (1997) Maxadilan, the vasodilator from sand flies, is a specific pituitary adenylate cyclase activating peptide type I receptor agonist. *J. Biol. Chem.*, **272**, 966–70.

Morris, R. V., Shoemaker, C. B., David, J. R., Lanzaro, G. C. and Titus, R. G. (2001) Sandfly maxadilan exacerbates infection with *Leishmania major* and vaccinating against it protects against *L. major* infection. *J. Immunol.*, **167**, 5226–30.

Moskalyk, L. A. and Friend, W. G. (1994) Feeding behavior of female *Aedes aegypti* – effects of diet, temperature, bicarbonate and feeding technique on the response to ATP. *Phys. Ent.*, **19**, 223–9.

Moyer, B. R., Gardiner, D. W. and Clayton, D. H. (2002) Impact of feather molt on ectoparasites: looks can be deceiving. *Oecologia*, **131**, 203–10.

Msangi, A. R., Whitaker, C. J. and Lehane, M. J. (1998) Factors influencing the prevalence of trypanosome infection of *Glossina pallidipes* on the Ruvu flood plain of Eastern Tanzania. *Acta Trop.*, **70**, 143–55.

Muir, L. E., Thorne, M. J. and Kay, B. H. (1992) *Aedes aegypti* (Diptera, Culicidae) vision – spectral sensitivity and other perceptual parameters of the female eye. *J. Med. Ent.*, **29**, 278–81.

Mukabana, W. R., Takken, W. and Knols, B. G. J. (2002a) Analysis of arthropod bloodmeals using molecular genetic markers. *Trends in Parasitology*, **18**, 505–9.

Mukabana, W. R., Takken, W., Seda, P. *et al.* (2002b) Extent of digestion affects the success of amplifying human DNA from blood meals of *Anopheles gambiae* (Diptera: Culicidae). *Bull. Ent. Res.*, **92**, 233–9.

Mukerji, D. and Sen-Sarma, P. (1955) Anatomy and affinity of the elephant louse, *Haematomyzus elephantis* Piaget (Insecta: Rhyncophthiraptera). *Parasitol.*, **45**, 5–30.

Mukwaya, L. G. (1977) Genetic control of feeding preference in the mosquitoes *Aedes* (*Stegomyia*) *simpsoni* and *aegypti*. *Phys. Ent.*, **2**, 133–45.

Mullens, B. A. and Gerhardt, R. R. (1979) Feeding behaviour of some Tennessee Tabanidae. *Environ. Ent.*, **8**, 1047–51.

Muller, H. M., Catteruccia, F., Vizioli, J., DellaTorre, A. and Crisanti, A. (1995) Constitutive and blood meal-induced trypsin genes in *Anopheles gambiae*. *Experimental Parasitology*, **81**, 371–85.

Muller, H. M., Crampton, J. M., Dellatorre, A., Sinden, R. and Crisanti, A. (1993) Members of a trypsin gene family in *Anopheles gambiae* are induced in the gut by blood meal. *EMBO J.*, **12**, 2891–900.

Mumcuoglu, Y. and Galun, R. (1987) Engorgement response of human body lice *Pediculus humanus* (Insecta: Anoplura) to blood fractions and their components. *Phys. Ent.*, **12**, 171–4.

Munstermann, L. E. and Conn, J. E. (1997) Systematics of mosquito disease vectors (Diptera, Culicidae): impact of molecular biology and cladistic analysis. *Ann. Rev. Ent.*, **42**, 351–69.

Murlis, J., Willis, M. A. and Carde, R. T. (2000) Spatial and temporal structures of pheromone plumes in fields and forests. *Phys. Ent.*, **25**, 211–22.

Murray, M. D. (1957) The distribution of the eggs of mammalian lice on their hosts. II. Analysis of the oviposition behaviour of *Damalinia ovis. Aust. J. Zool*, **5**, 19–29.

(1963) Influence of temperature on the reproduction of *Damalinia equi* (Denny). *Aust. J. Zool.*, **11**, 183–9.

(1987) Effects of host grooming on louse populations. *Parasitology Today*, **3**, 276–8.

Murray, M. D. and Nicholls, D. G. (1965) Studies on the ectoparasites of seals and penguins. I. The ecology of the louse *Lepidophthirus macrorhini* Enderlein on the southern elephant seal, *Mirounga leonina* (L.). *Aust. J. Zool.*, **13**, 437–54.

Mwandawiro, C., Boots, M., Tuno, N. *et al.* (2000) Heterogeneity in the host preference of Japanese encephalitis vectors in Chiang Mai, northern Thailand. *Trans. R. Soc. Trop. Med. Hyg.*, **94**, 238–42.

Naksathit, A. T., Edman, J. D. and Scott, T. W. (1999) Utilization of human blood and sugar as nutrients by female *Aedes aegypti* (Diptera: Culicidae). *J. Med. Ent*, **36**, 13–17.

Naksathit, A. T. and Scott, T. W. (1998) Effect of female size on fecundity and survivorship of *Aedes aegypti* fed only human blood versus human blood plus sugar. *J. Am. Mosq. Control Assoc.*, **14**, 148–52.

Napier Bax, S. (1937) The senses of smell and sight in *Glossina swynnertoni. Bull. Ent. Res.*, **28**, 539–82.

Nappi, A. J., Vass, E., Frey, F. and Carton, Y. (2000) Nitric oxide involvement in *Drosophila* immunity. *Nitric Oxide Biology and Chemistry*, **4**, 423–30.

Nasci, R. S. (1982) Differences in host choice between the sibling species of treehole mosquitoes *Aedes triseriatus* and *Aedes hendersoni. Am. J. Trop. Med. Hyg.*, **31**, 411–15.

Nelson, R. L. (1965) Carbon dioxide as an attractant for *Culicoides. J. Med. Ent.*, **2**, 56–7.

Nelson, W. A. (1987) Other blood-sucking and myiasis-producing arthropods. In E. J. L. Soulsby (ed.), *Immune Responses in Parasitic Infections: Immunology, Immunopathology and Immunoprophylaxis*. Boca Raton, Florida: CRC Press, Vol. IV.

Nelson, W. A., Bell, J. F., Clifford, C. M. and Keirans, A. J. (1977) Interaction of ectoparasites and their hosts. *J. Med. Ent.*, **13**, 389–428.

Nelson, W. A., Keirans, J. E., Bell, J. F. and Clifford, C. M. (1975) Host–ectoparasite relationships. *J. Med. Ent.*, **12**, 143–66.

Nelson, W. A. and Kozub, G. C. (1980) *Melophagus ovinus* (Diptera: Hippoboscidae): evidence of local mediation in acquired resistance of sheep to keds. *J. Med. Ent.*, **17**, 291–7.

Newson, R. M. and Holmes, R. G. (1968) Some ectoparasites of the coypu (*Myocastor coypus*) in eastern England. *J. Anim. Ecol.*, **37**, 471–81.

Nguu, E. K., Osir, E. O., Imbuga, M. O. and Olembo, N. K. (1996) The effect of host blood in the *in vitro* transformation of bloodstream trypanosomes by tsetse midgut homogenates. *Med. Vet. Entomol.*, **10**, 317–22.

Niare, O., Markianos, K., Volz, J. *et al.* (2002) Genetic loci affecting resistance to human malaria parasites in a west African mosquito vector population. *Science*, **298**, 213–16.

Nieves, E. and Pimenta, P. F. P. (2002) Influence of vertebrate blood meals on the development of *Leishmania* (Viannia) *braziliensis* and *Leishmania* (*Leishmania*) *amazonensis* in the sand fly *Lutzomyia migonei* (Diptera: Psychodidae). *Am. J. Trop. Med. Hyg.*, **67**, 640–7.

Nigam, Y. and Ward, R. D. (1991) The effect of male sandfly pheromone and host factors as attractants for female *Lutzomyia longipalpis* (Diptera, Psychodidae). *Phys. Ent.*, **16**, 305–12.

Nogge, G. (1978) Aposymbiotic tsetse flies, *Glossina morsitans morsitans* obtained by feeding on rabbits immunized specifically with symbionts. *J. Insect Physiol.*, **24**, 299–304.

(1981) Significance of symbionts for the maintenance of an optimal nutritional state for successful reproduction in haematophagous arthropods. *Parasitology*, **82**, 101–4.

Nogge, G. and Ritz, R. (1982) Number of symbionts and its regulation in tsetse flies, *Glossina* spp. *Ent. Exp. Appl.*, **31**, 249–54.

Noriega, F. G., Edgar, K. A., Bechet, R. and Wells, M. A. (2002) Midgut exopeptidase activities in *Aedes aegypti* are induced by blood feeding. *J. Insect Physiol.*, **48**, 205–12.

Noriega, F. G. and Wells, M. A. (1999) A molecular view of trypsin synthesis in the midgut of *Aedes aegypti*. *J. Insect Physiol.*, **45**, 613–20.

Nuttal, G. H. F. (1899) On the role of insects, arachnids, and myriapods as carriers in the spread of bacterial and parasitic disease of man and animals. A critical and historical study. *Johns Hopkins Hospital Reports*, **8**, 1–154.

Obiamiwe, B. A. and Macdonald, W. W. (1973) 1. The effect of heparin on the migration of *Brugia pahangi* microfilariae *Culex pipiens*. 2. The uptake of *B. pahangi* microfilariae in *C. pipiens* and the infectivity of *C. pipiens* in relation to microfilarial densities. 3. Evidence of a sex-linked recessive gene, sb, controlling susceptibility of *C. pipiens* to *B. pahangi*. *Trans. R. Soc. Trop. Med. Hyg.*, **67**, 32–3.

Ochiai, M., Niki, T. and Ashida, M. (1992) Immunocytochemical localization of beta-1,3-glucan recognition protein in the silkworm, *Bombyx mori*. *Cell and Tissue Research*, **268**, 431–7.

Ogston, C. W. and London, W. T. (1980) Excretion of hepatitis B surface antigen by the bedbug *Cimex hemipterus* Fabr. *Trans. R. Soc. Trop. Med. Hyg.*, **74**, 823–5.

Olubayo, R. O., Mihok, S., Munyoki, E. and Otieno, L. H. (1994) Dynamics of host blood effects in *Glossina morsitans* spp. infected with *Trypanosoma congolense* and *Trypanosoma brucei*. *Parasitology Research*, **80**, 177–81.

O'Meara, G. F. (1979) Variable expressions of autogeny in three mosquito species. *Int. J. Invert. Reprod.*, **1**, 253–61.

(1985) Ecology and autogeny in mosquitoes. In L. P. Lounibos, J. R. Rey and J. H. Frank,(eds.), *Ecology of Mosquitoes*. Florida: Florida Medical Laboratory.

(1987) Nutritional ecology of blood feeding diptera. In F. Slansky and J.G. Rodriguez (eds.), *Nutritional Ecology of Insects, Mites, Spiders and Related Invertebrates*. New York: Wiley.

O'Meara, G. F. and Edman, J. D. (1975) Autogenous egg production in the salt marsh mosquito, *Aedes taeniorrhynchus*. *Biol. Bull.*, **149**, 384–96.

O'Meara, G. F. and Evans, D. G. (1973) Blood-feeding requirements of the mosquito: geographical variation in *Aedes taeniorhynchus*. *Science*, **180**, 1291–3.

(1976) The influence of mating on autogenous egg development in the mosquito, *Aedes taeniorrhynchus*. *J. Insect Physiol.*, **22**, 613–17.

Omer, S. M. and Gillies, M. T. (1971) Loss of response to carbon dioxide in palpectomized female mosquitoes. *Ent. Exp. Appl.*, **14**, 251–2.

Osbrink, L. A. and Rust, M. A. (1985) Cat flea (Siphonaptera: Pulicidae): factors influencing host-finding behaviour in the laboratory. *Ann. Ent. Soc. Am.*, **78**, 29–34.

O'Shea, B., Rebollar-Tellez, E., Ward, R. D. *et al.* (2002) Enhanced sandfly attraction to *Leishmania*-infected hosts. *Trans. R. Soc. Trop. Med. Hyg.*, **96**, 117–18.

Overal, W. L. (1980) Biology and behaviour of North American *Trichobius* bat flies (Diptera: Streblidae). Ph.D. thesis, University of Kansas.

Overal, W. L. and Wingate, L. R. (1976) The biology of the batbug *Strictimex antennatus* (Hemiptera: Cimicidae) in South Africa. *Ann. Natal Mus.*, **22**, 821–8.

Owaga, M. L. and Challier, A. (1985) Catch composition of the tsetse *Glossina pallidipes* Austen in revolving and stationary traps with respect to age, sex ratio and hunger stage. *Insect Sci. Applic.*, **6**, 711–18.

Page, R. D. M., Clayton, D. H. and Paterson, A. M. (1996) Lice and cospeciation: a response to Barker. *Int. J. Parasit.*, **26**, 213–18.

Pagel, M. and Bodmer, W. (2003) A naked ape would have fewer parasites. *Proc. R. Soc. Lond. B Biol. Sci.*, **270**, Suppl 1, 117–19.

Pant, C. P., Houba, V. and Engers, H. D. (1987) Bloodmeal identification in vectors. *Parasitology Today*, **3**, 324–6.

Panton, L. J., Tesh, R. B., Nadeau, K. C. and Beverley, S. M. (1991) A test for genetic exchange in mixed infections of *Leishmania major* in the sand fly *Phlebotomus papatasi*. *J. Protozool*, **38**, 224–8.

Pappas, L. G., Pappas, C. D. and Grossman, G. L. (1986) Hemodynamics of human skin during mosquito (Diptera: Culicidae) blood feeding. *J. Med. Ent.*, **23**, 581–7.

Parker, K. R. and Gooding, R. H. (1979) Effects of host anaemia, local skin factors and circulating antibodies upon biology of laboratory reared *Glossina morsitans morsitans* (Diptera: Glossinidae). *Can. J. Zool.*, **57**, 2393–401.

Paskewitz, S. M., Brown, M. R., Lea, A. O. and Collins, F. H. (1988) Ultrastructure of the encapsulation of *Plasmodium cynomolgi* (B-strain) on the midgut of a refractory strain of *Anopheles gambiae*. *Journal of Parasitology*, **74**, 432–9.

Patton, W. S. and Craig, F. W. (1913) On certain haematophagous species of the genus *Musca*, with descriptions of two new species. *Indian Journal of Medical Research*, **1**, 13–25.

Peacock, A. J. (1981) Distribution of ($Na^+ K^+$)-ATPase activity in the mid-guts and hind-guts of adult *Glossina morsitans* and *Sarcophaga nodosa* and the hind-gut of *Bombyx mori* larvae. *Comp. Biochem. Physiol. A*, **69**, 133–6.

(1982) Effects of sodium transport inhibitors on diuresis and midgut (Na^+ and K^+) ATPase in the tsetse fly *Glossina morsitans*. *J. Insect Physiol.*, **28**, 553–8.

Pearman, J. V. (1960) Some African psocoptera found on rats. *Entomologist*, **93**, 246–50.

Pearson, T. W., Beecroft, R. P., Welburn, S. C. *et al.* (2000) The major cell surface glycoprotein procyclin is a receptor for induction of a novel form of cell death in African trypanosomes in vitro. *Molecular and Biochemical Parasitology*, **111**, 333–49.

Pell, P. E. and Southern, D. I. (1976) Effect of the coccidiostat, sulphaquinoxline, on symbiosis in the tsetse fly, *Glossina* species. *Microbios Letters*, **2**, 203–11.

Pereira, H., Penido, C. M., Martins, M. S. and Diotaiuti, L. (1998) Comparative kinetics of bloodmeal intake by *Triatoma infestans* and *Rhodnius prolixus*, the two principal vectors of Chagas disease. *Med. Vet. Entomol.*, **12**, 84–8.

Pereira, M. E. A., Andrade, A. F. B. and Ribeiro, J. M. C. (1981) Lectins of distinct specificity in *Rhodnius prolixus* interact selectively with *Trypanosoma cruzi*. *Science*, **211**, 597–9.

Perrin, N., Christe, P. and Richner, H. (1996) On host life-history response to parasitism. *Oikos*, **75**, 317–20.

Peschken, D. P. and Thorsteinson, A. J. (1965) Visual orientation of black flies (Simuliidae: Diptera) to colour, shape and movement of targets. *Ent. Exp. Appl.*, **8**, 282–8.

Peters, W. (1968) Vorkommen, Zusammensetzung und feinstruktur peritrophischer membranen im tierreich. *Zeit. Morph. Okol. Tiere.*, **64**, 21–58.

Peters, W., Kolb, H. and Kolb-Bachofen, V. (1983) Evidence for a sugar receptor (lectin) in the peritrophic membrane of the blowfly larva, *Calliphora erythrocephala* Mg. (Diptera). *J. Insect Physiol.*, **29**, 275–80.

Peters, W., Zimmermann, U. and Becker, B. (1973) Investigations on the transport function and structure of peritrophic membranes. IV. Anisotropic cross bands in peritrophic membranes of Diptera. *J. Insect Physiol.*, **19**, 1067–77.

Peterson, D. G. and Brown, A. W. A. (1951) Studies of the responses of female *Aedes* mosquito. III. The response of *Aedes aegypti* (L.) to a warm body and its radiation. *Bull. Ent. Res.*, **42**, 535–41.

Phelps, R. J. and Vale, G. A. (1976) Studies on the local distribution and on the methods of host location of some Rhodesian Tabanidae (Diptera). *J. Ent. Soc. S. Afr.*, **39**, 67–81.

Pichon, G., Awono-Ambene, H. P. and Robert, V. (2000) High heterogeneity in the number of *Plasmodium falciparum* gametocytes in the bloodmeal of mosquitoes fed on the same host. *Parasitology*, **121**, 115–20.

Piot, P. and Schofield, C. J. (1986) No evidence for arthropod transmission of AIDS. *Parasitology Today*, **2**, 294–5.

Platt, K. B., Linthicum, K. J., Myint, K. S. A. *et al.* (1997) Impact of dengue virus infection on feeding behavior of *Aedes aegypti*. *Am. J. Trop. Med. Hyg.*, **57**, 119–25.

Politzar, H. and Merot, P. (1984) Attraction of the tsetse fly *Glossina morsitans submorsitans* to acetone, 1 octen-3-ol, and the combination of these compounds in west Africa. *Rev. Elev. Med. Vet. Pays Trop.*, **37**, 468–73.

Ponnudurai, T., Billingsley, P. F. and Rudin, W. (1988) Differential infectivity of *Plasmodium* for mosquitoes. *Parasitology Today*, **4**, 319–21.

Port, G. R., Bateham, P. F. L. and Bryan, J. H. (1980) The relationship of host size to feeding by mosquitoes of the *Anopheles gambiae* complex (Diptera: Culicidae). *Bull. Ent. Res.*, **70**, 133–44.

Pospisil, J. and Zdarek, J. (1965) On the visual orientation of the stable fly (*Stomoxys calcitrans* L.) to colours. *Acta Entomol. Bohemoslov.*, **62**, 85–91.

Powell, J. R., Petrarca, V., Della Torre, A., Caccone, A. and Coluzzi, M. (1999) Population structure, speciation, and introgression in the *Anopheles gambiae* complex. *Parasitologia*, **41**, 101–13.

Price, G. D., Smith, N. and Carlson, D. A. (1979) The attraction of female mosquitoes (*Anopheles quadrimaculatus* Say) to stored human emanations in conjunction with adjusted levels of relative humidity, temperature and carbon dioxide. *J. Chem. Ecol.*, **5**, 383–95.

Price, R. D. (1975) The *Menacanthus eurysternus* complex (Mallophaga: Menoponidae) of the Passeriformes and Piciformes (Aves). *Ann. Ent. Soc. Am.*, **68**, 617–22.

Prior, A. and Torr, S. J. (2002) Host selection by *Anopheles arabiensis* and *An. quadriannulatus* feeding on cattle in Zimbabwe. *Med. Vet. Entomol.*, **16**, 207–13.

Raikhel, A. S., Kokoza, V. A., Zhu, J. *et al.* (2002) Molecular biology of mosquito vitellogenesis: from basic studies to genetic engineering of antipathogen immunity. *Insect Biochem. Mol. Biol.*, **32**, 1275–86.

Ramet, M., Lanot, R., Zachary, D. and Manfruelli, P. (2002a) JNK signaling pathway is required for efficient wound healing in *Drosophila*. *Dev. Biol.*, **241**, 145–56.

Ramet, M., Manfruelli, P., Pearson, A., Mathey-Prevot, B. and Ezekowitz, R. A. B. (2002b) Functional genomic analysis of phagocytosis and identification of a *Drosophila* receptor for E-coli. *Nature*, **416**, 644–8.

Ratcliffe, N. A. and Rowley, A. F. (1979) Role of hemocytes in defence against biological agents. In A. P. Gupta (ed.), *Insect Hemocytes*. Cambridge: Cambridge University Press, 331–422.

Ratzlaff, R. E. and Wikel, S. K. (1990) Murine immune responses and immunization against *Polyplax serrata* (Anoplura: Polyplacidae). *J. Med. Ent.*, **27**, 1002–7.

Ready, P. D. (1978) The feeding habits of laboratory bred *Lutzomyia longipalpis* (Diptera: Psychodidae). *J. Med. Ent.*, **14**, 545–552.

Reddy, V. B., Kounga, K., Mariano, F. and Lerner, E. A. (2000) Chrysoptin is a potent glycoprotein IIb/IIIa fibrinogen receptor antagonist present in salivary gland extracts of the deerfly. *Journal of Biological Chemistry*, **275**, 15861–7.

Read, W., Carrall, J., Agramonte, A. and Lazear, J. (1900) The etiology of yellow fever: a preliminary note. *The Philadelphia Medical Journal*, **6**, 790–3.

Reichardt, T. R. and Galloway, T. D. (1994) Seasonal occurrence and reproductive status of *Opisocrostis bruneri* (Siphonaptera, Ceratophyllidae), a flea on franklin ground-squirrel, *Spermophilus franklinii* (Rodentia, Sciuridae) near Birds Hill Park, Manitoba. *J. Med. Ent.*, **31**, 105–13.

Reid, G. D. F. and Lehane, M. J. (1984) Peritrophic membrane formation in three temperate simuliids, *Simulium ornatum*, *S. equinum* and *S. lineatum* with respect to the migration of Onchocercal microfilariae. *Ann. Trop. Med. Parasit.*, **78**, 527–39.

Reinouts van Haga, H. A. and Mitchell, B. K. (1975) Temperature receptors on tarsi of the tsetse fly *Glossina morsitans* West. *Nature*, **255**, 225–6.

Reunala, T., Brummer-Korvenkontio, H., Lappalainen, P., Rasanen, L. and Palosuo, T. (1990) Immunology and treatment of mosquito bites. *Clin. Exp. Allergy*, **20**, Suppl 4, 19–24.

Ribeiro, J. M. C. (1982) The anti-serotonin and antihistamine activities of salivary secretion of *Rhodnius prolixus*. *J. Insect Physiol.*, **28**, 69–75.

(1987) Role of saliva in blood-feeding by arthropods. *Ann. Rev. Ent.*, **32**, 463–78.

(1995) Blood-feeding arthropods – live syringes or invertebrate pharmacologists. *Infectious Agents and Disease – Reviews Issues and Commentary*, **4**, 143–52.

(1998) *Rhodnius prolixus* salivary nitrophorins display heme-peroxidase activity. *Insect Biochemistry and Molecular Biology*, **28**, 1051–7.

Ribeiro, J. M. C., Charlab, R., Rowton, E. D. and Cupp, E. W. (2000) *Simulium vittatum* (Diptera: Simuliidae) and *Lutzomyia longipalpis* (Diptera: Psychodidae) salivary gland hyaluronidase activity. *J. Med. Ent.*, **37**, 743–7.

Ribeiro, J. M. C., Charlab, R. and Valenzuela, J. G. (2001) The salivary adenosine deaminase activity of the mosquitoes *Culex quinquefasciatus* and *Aedes aegypti*. *J. Exp. Biol.*, **204**, 2001–10.

Ribeiro, J. M. C. and Francischetti, I. M. B. (2003) Role of arthropod saliva in blood feeding: sialome and post-sialome perspectives. *Ann. Rev. Ent.*, Vol. 48, 73–88.

Ribeiro, J. M. C. and Garcia, E. S. (1981a) Platelet antiaggregating activity in the salivary secretion of the blood-sucking bug *Rhodnius prolixus*. *Experientia*, **37**, 384–5.

(1981b) The role of saliva in feeding in *Rhodnius prolixus*. *J. Exp. Biol.*, **94**, 219–30.

Ribeiro, J. M. C., Katz, O., Pannell, L. K., Waitumbi, J. and Warburg, A. (1999) Salivary glands of the sand fly *Phlebotomus papatasi* contain pharmacologically active amounts of adenosine and 5-AMP. *J. Exp. Biol.*, **202**, 1551–9.

Ribeiro, J. M. C., Rossignol, P. A. and Spielman, A. (1985) Salivary gland apyrase determines probing time in anopheline mosquitoes. *J. Insect Physiol.*, **9**, 689–92.

Ribeiro, J. M. C., Schneider, M. and Guimaraes, J. A. (1995) Purification and characterization of prolixin-S (nitrophorin-2), the salivary anticoagulant of the blood-sucking bug *Rhodnius prolixus*. *Biochemical Journal*, **308**, 243–9.

Ribeiro, J. M. C. and Valenzuela, J. G. (1999) Purification and cloning of the salivary peroxidase/catechol oxidase of the mosquito *Anopheles albimanus*. *J. Exp. Biol.*, **202**, 809–16.

Rice, M. J., Galun, R. and Margalit, J. (1973) Mouthpart sensilla of the tsetse fly and their function. II. Labial sensilla. *Ann. Trop. Med. Parasit.*, **67**, 101–7.

Richman, A. M., Dimopoulos, G., Seeley, D. and Kafatos, F. C. (1997) *Plasmodium* activates the innate immune response of *Anopheles gambiae* mosquitoes. *EMBO J.*, **16**, 6114–9.

Roberts, L. W. (1981) Probing by *Glossina morsitans morsitans* and transmission of *Trypanosoma* (Nannomonas) *congolense*. *Am. J. Trop. Med. Hyg.*, **30**, 948–51.

Roberts, R. H. (1972) Relative attractiveness of CO_2 and a steer to Tabanidae, Culicidae, and *Stomoxys calcitrans*. *Mosq. News*, **32**, 208–11.

(1977) Attractancy of two black decoys and CO_2 to tabanids (Diptera: Tabanidae). *Mosq. News*, **37**, 169–72.

Robinson, A. (1939) The mouthparts and their function in the female mosquito, *Anopheles maculipennis*. *Parasitol.*, **31**, 212–42.

Rogers, K. A. and Titus, R. G. (2003) Immunomodulatory effects of Maxadilan and *Phlebotomus papatasi* sand fly salivary gland lysates on human primary in vitro immune responses. *Parasite Immunol.*, **25**, 127–34.

Rogers, M. E., Chance, M. L. and Bates, P. A. (2002) The role of promastigote secretory gel in the origin and transmission of the infective stage of *Leishmania mexicana* by the sandfly *Lutzomyia longipalpis*. *Parasitology*, **124**, 495–507.

Rosenfeld, A. and Vanderberg, J. P. (1998) Identification of electrophoretically separated proteases from midgut and hemolymph of adult *Anopheles stephensi* mosquitoes. *Journal of Parasitology*, **84**, 361–5.

Ross, R. (1897) On same peculier pigmented cells found in two mosquitoes fed on malaria blood. *British Medical Journal*, **2**, 1786–8.

Ross, R. (1898) Report on the cultivation of protessoma, Labb, in grey mosquitoes. *Indian Med. Gaz.*, **33**, 401–8.

Rossignol, P. A., Ribeiro, J. M. C., Jungery, M., Turell, M. J., Spielman, A. and Bailey, C. L. (1985) Enhanced mosquito blood-finding success on parasitaemic hosts: Evidence for vector-parasite mutualism. *Proc. Nat. Acad. Sci.*, **82**, 7725–7.

Rossignol, P. A., Ribeiro, J. M. C. and Spielman, A. (1984) Increased intradermal probing time in sporozoite-infected mosquitoes. *Am. J. Trop. Med. Hyg.*, **33**, 17–20.

(1986) Increased biting rate and reduced fertility in sporozoite-infected mosquitoes. *Am. J. Trop. Med. Hyg.*, **35**, 277–9.

Rossignol, P. A. and Rossignol, A. M. (1988) Simulations of enhanced malaria transmission and host bias induced by modified vector blood location behaviour. *Parasitol.*, **97**, 363–72.

Rothschild, M. (1975) Recent advances in our knowledge of the Siphonoptera. *Ann. Rev. Ent.*, **20**, 241–59.

Rothschild, M. and Clay, T. (1952) *Fleas, Flukes and Cuckoos*. New York: Philosophical Library.

Rothschild, M. and Ford, B. (1973) Factors influencing the breeding of the rabbit flea (*Spilopsyllus cuniculi*): a spring-time accelerator and a kairomone in nestling rabbit urine (with notes on *Cediopsylla simplex*, another 'hormone bound' species). *J. Zool.*, **170**, 87–137.

Rothschild, M., Schlein, Y., Parker, K. and Sternberg, S. (1972) Jump of the oriental rat flea *Xenopsylla cheopis* (Roths.). *Nature*, **239**, 45–8.

Rowland, M. and Boersma, E. (1988) Changes in the spontaneous flight activity of the mosquito *Anopheles stephensi* by parasitization with the rodent malaria *Plasmodium yoelii*. *Parasitology*, **97**, 221–7.

Rowland, M. W. and Lindsay, S. L. (1986) The circadian flight activity of *Aedes aegypti* parasitized with the filarial nematode *Brugia pahangi*. *Phys. Ent.*, **11**, 325–34.

Roy, D. N. (1936) On the role of blood in ovulation in *Aedes aegypti*, Linn. *Bull. Ent. Res.*, **27**, 423–9.

Royet, J., Meister, M. and Ferrandon, D. (2003) Humoral and cellular responses in *Drosophila* innate immunity. In R. A. Ezekowitz and J. A. Hoffman (eds.), *Infectious Disease: Innate Immunity*. Totowa, NJ: Humana Press 137–53.

Rubenstein, D. I. and Hohmann, M. E. (1989) Parasites and social-behavior of island feral horses. *Oikos*, **55**, 312–20.

Rudin, W. and Hecker, H. (1979) Functional morphology of the midgut of *Aedes aegypti* L. (Insecta; Diptera) during blood digestion. *Cell*, **200**, 193–203.

(1982) Functional morphology of the midgut of a sandfly as compared to other haematophagous nematocera. *Tissue and Cell*, **14**, 751–8.

Rutberg, A. T. (1987) Horse fly harassment and the social-behavior of feral ponies. *Ethology*, **75**, 145–54.

Sabelis, M. W. and Schippers, P. (1984) Variable wind direction and anemotactic strategies of searching for an odour plume. *Oecologia*, **63**, 225–8.

Sacks, D. L. (1989) Metacyclogenesis in *Leishmania* promastigotes. *Exp. Parasitol.*, **69**, 100–3.

Sacks, D. L. and Kamhawi, S. (2001) Molecular aspects of parasite–vector and vector–host interactions in *Leishmania*sis. *Ann. Rev. Microbiol.*, **55**, 453–83.

Sallum, M. A. M., Schultz, T. R., Foster, P. G. *et al.* (2002) Phylogeny of Anophelinae (Diptera: Culicidae) based on nuclear ribosomal and mitochondrial DNA sequences. *Systematic Entomology*, **27**, 361–82.

Samuel, W. M. and Trainer, D. O. (1972) *Lipoptena mazamae* Rondani, 1878 (Diptera: Hippoboscidae) on white-tailed deer in southern Texas. *J. Med. Ent.*, **9**, 104–6.

Sandeman, R. M. (1996) Immune rsponses to mosquitoes and flies. In S. K. Wikel (ed.), *The Immunology of Host–Ectoparasitic Arthropod Interactions*. Wallingford: CAB International 175–203.

Sangiorgi, G. and Frosini, D. (1940) Di un principio emolitico ('Cimicina') nella saliva del *Cimex lectularius*. *Pathologica*, **32**, 189–91.

Saraiva, E. M., Pimenta, P. F., Brodin, T. N. (1995) Changes in lipophosphoglycan and gene expression associated with the development of *Leishmania major* in *Phlebotomus papatasi*. *Parasitology*, **111**, (Pt 3), 275–87.

Sarkis, J. J. F., Guimaraes, J. A. and Ribeiro, J. M. C. (1986) Salivary apyrase of *Rhodnius prolixus*: kinetics and purification. *Biochem. J.*, **233**, 885–91.

Scaraffia, P. Y. and Wells, M. A. (2003) Proline can be utilized as an energy substrate during flight of *Aedes aegypti* females. *J. Insect Physiol.*, **49**, 591–601.

Schall, J. J. (2002) Parasite virulence. In E. E. Lewis, J. F. Campbell and M. D. K. Sukdheo (eds.), *The Behavioural Ecology of Parasites*. Wallingford: CAB International.

Schiefer, B. A. *et al.* (1977) *Plasmodium cynomolgi*: effects of malaria infection on laboratory flight performance of *Anopheles stephensi* mosquiotoes. *Exp. Parasitol.*, **41**(2), 397–404.

Schlein, Y. (1977) Lethal effect of tetracycline on tsetse flies following damage to bacteroid symbionts. *Experimentia*, **33**, 450–1.

Schlein, Y. and Jacobson, R. L. (1998) Resistance of *Phlebotomus papatasi* to infection with *Leishmania donovani* is modulated by components of the infective bloodmeal. *Parasitology*, **117**, 467–73.

Schlein, Y., Warburg, A., Schnur, L. F. and Shlomai, J. (1983) Vector compatibility of *Phlebotomus papatasi* dependent on differentially induced digestion. *Acta Trop.*, **40**, 65–70.

Schlein, Y., Yuval, B. and Warburg, A. (1984) Aggregation pheromone released from the palps of feeding female *Phlebotomus papatasi* (Psychodidae). *J. Insect Physiol.*, **30**, 153–6.

Schmid-Hempel, P. (2003) Immunology and evolution of infectious disease. *Science*, **300**, 254.

Schmitz, H., Trenner, S., Hofmann, M. H. and Bleckmann, H. (2000) The ability of *Rhodnius prolixus* (Hemiptera; Reduviidae) to approach a thermal source solely by its infrared radiation. *J. Insect Physiol.*, **46**, 745–51.

Schoeler, G. B. and Wikel, S. K. (2001) Modulation of host immunity by haematophagous arthropods. *Annals of Tropical Medicine and Parasitology*, **95**, 755–71.

Schofield, C. J. (1981) Chagas disease, triatomine bugs, and blood loss. *Lancet*, **1**, 1316.

(1982) The role of blood intake in density regulation of populations of *Triatoma infestans* (Klug) (Hemiptera: Reduviidae). *Bull. Ent. Res.*, **72**, 617–29.

(1985) Population dynamics and control of *Triatoma infestans*. *Ann. Soc. Belge, Med. Trop.*, **65**, 149–64.

(1988) Biosystematics of the triatominae. In M. W. Service (ed.), *Biosystematics of Haematophagous Insects*. Oxford: Clarendon Press.

Schofield, S. and Sutcliffe, J. F. (1996) Human individuals vary in attractiveness for host-seeking black flies (Diptera: Simuliidae) based on exhaled carbon dioxide. *J. Med. Ent.*, **33**, 102–8.

(1997) Humans vary in their ability to elicit biting responses from *Simulium venustum* (Diptera: Simuliidae). *J. Med. Ent.*, **34**, 64–7.

Schofield, S. and Torr, S. J. (2002) A comparison of the feeding behaviour of tsetse and stable flies. *Med. Vet. Entomol.*, **16**, 177–85.

Senghor, J. E. and Samba, E. M. (1988) Onchocerciasis control program – the human perspective. *Parasitology Today*, **4**, 332–3.

Severson, D. W., Brown, S. E. and Knudson, D. L. (2001) Genetic and physical mapping in mosquitoes: molecular approaches. *Ann. Rev. Ent.*, **46**, 183–219.

Severson, D. W., Mori, A., Zhang, Y. and Christensen, B. M. (1994) Chromosomal mapping of two loci affecting filarial worm susceptibility in *Aedes aegypti*. *Insect Mol. Biol.*, **3**, 67–72.

Severson, D. W., Thathy, V., Mori, A., Zhang, Y. and Christensen, B. M. (1995) Restriction-fragment-length-polymorphism mapping of quantitative trait loci for malaria parasite susceptibility in the mosquito *Aedes aegypti*. *Genetics*, **139**, 1711–17.

Shahabuddin, M. (1998) *Plasmodium* ookinete development in the mosquito midgut: a case of reciprocal manipulation. *Parasitology*, **116**, S83–S93.

Shahabuddin, M., Fields, I., Bulet, P., Hoffmann, J. A. and Miller, L. H. (1998) *Plasmodium gallinaceum*: differential killing of some mosquito stages of the parasite by insect defensin. *Experimental Parasitology*, **89**, 103–12.

Shahan, M. S. and Giltner, L. T. (1945) A review of the epizootiology of equine encephalomyelitis in the United States. *J. Am. Vet. Med. Assoc.*, **107**, 279–88.

Shin, S. W., Kokoza, V., Lobkov, I. and Raikhel, A. S. (2003) Relish-mediated immune deficiency in the transgenic mosquito *Aedes aegypti*. *Proc. Natl. Acad. Sci. USA*, **100**, 2616–21.

Sieber, K. P., Huber, M., Kaslow, D. *et al.* (1991) The peritrophic membrane as a barrier. Its penetration by *Plasmodium gallinaceum* and the effect of a monoclonal antibody to ookinetes. *Experimental Parasitology*, **72**, 145–56.

Silva, C. P., Ribeiro, A. F., Gulbenkian, S. and Terra, W. R. (1995) Organization, origin and function of the outer microvillar (perimicrovillar) membranes of *Dysdercus peruvianus* (Hemiptera) midgut cells. *J. Insect Physiol.*, **41**, 1093–103.

Silverman, N., Zhou, R., Stoven, S., Pandey, N., Hultmark, D. and Maniatis, T. (2000) A *Drosophila* IkappaB kinase complex required for Relish cleavage and antibacterial immunity. *Genes Dev.*, **14**, 2461–71.

Simond, P. L. (1898) La propagation de la peste. *Annales de la Institut Pasteur*, **12**, 625.

Sippel, W. L. and Brown, A. W. A. (1953) Studies on the responses of the female *Aedes* mosquito. Part V. The role of visual factors. *Bull. Ent. Res.*, **43**, 567–74.

Smit, F. G. A. M. (1972) On some adaptive structures in Siphonaptera. *Folia Parasit.*, **19**, 5–17.

Smith, C. N., Smith, N., Gouck, H. K. *et al.* (1970) L-lactic acid as a factor in the attraction of *Aedes aegypti* to human hosts. *Ann. Ent. Soc. Am.*, **63**, 760–70.

Smith, H. V. and Titchener, R. N. (1980) Mouthparts of ectoparasites and host damage. *Proc. R. Soc. Edin. B-Biol. Sci.*, **79**, 139.

Smith, J. J. B. (1979) Effect of diet viscosity on the operation of the pharyngeal pump in the blood-feeding bug *Rhodnius prolixus*. *J. Exp. Biol.*, **82**, 93–104.

(1984) Feeding mechanisms. In G. A. Kerkut and L. I. Gilbert (eds.), *Comprehensive Insect Physiology, Biochemistry and Pharmacology*. Oxford: Pergamon.

Smith, J. J. B. and Friend, W. G. (1970) Feeding in *Rhodnius prolixus*: responses to artificial diets as revealed by changes in electrical resistance. *J. Insect Physiol.*, **16**, 1709–20.

(1982) Feeding behaviour in response to blood fractions and chemical phagostimulants in the blackfly, *Simulium venustum*. *Phys. Ent.*, **7**, 219–26.

Smith, K. G. V. (ed.) (1973) *Insects and Other Arthropods of Medical Importance*. London: British Museum (Natural History).

Smith, T. and Kilbourne, F. L. (1893) Investigations into the nature, causation and prevention of Texas or Southern cattle fever. *U.S. Dept. Agric. Bur. Animal. Indust. Bull.*, Vol. 1, 301.

Snodgrass, R. E. (1944) The anatomy of the Mallophaga. *Occ. Pap. Calif. Acad. Sci.*, **6**, 145–229.

Soares, M. B., Titus, R. G., Shoemaker, C. B., David, J. R. and Bozza, M. (1998) The vasoactive peptide maxadilan from sand fly saliva inhibits TNF-alpha and induces IL-6 by mouse macrophages through interaction with the pituitary adenylate cyclase-activating polypeptide (PACAP) receptor. *J. Immunol.*, **160**, 1811–16.

Soderhall, K. and Cerenius, L. (1998) Role of the prophenoloxidase-activating system in invertebrate immunity. *Current Opinion in Immunology*, **10**, 23–8.

Soldatos, A. N., Metheniti, A., Mamali, I., Lambropoulou, M. and Marmaras, V. J. (2003) Distinct LPS-induced signals regulate LPS uptake and morphological changes in medfly hemocytes. *Insect Biochem. Mol. Biol.*, **33**, 1075–84.

Sorci, G., de Fraipont, M. and Clobert, J. (1997) Host density and ectoparasite avoidance in the common lizard (*Lacerta vivipara*). *Oecologia*, **111**, 183–8.

Soulsby, E. J. L. (1982) *Helminths, Arthropods and Protozoa of Domesticated Animals*. London: Bailliere Tindall.

Southwood, T. R. E., Khalaf, S. and Sinden, R. E. (1975) The micro-organisms of tsetse flies. *Acta Trop.*, **32**, 259–66.

Southworth, G. C., Mason, G. and Seed, J. R. (1968) Studies in frog trypanosomiasis. I. A 24-hour cycle in the parasitaemia level of *Trypanosoma rotatorium* in *Rana clamitans* from Louisiana. *J. Parasit.*, **54**, 255–8.

Spates, G. E. (1981) Proteolytic and haemolytic activity in the midgut of the stablefly *Stomoxys calcitrans* (L.): partial purification of the haemolysin. *Insect Biochem.*, **11**, 143–7.

Spates, G. E., Stipanovic, R. D., Williams, H. and Holman, G. M. (1982) Mechanisms of haemolysis in a blood-sucking dipteran, *Stomoxys calcitrans*. *Insect Biochem.*, **12**, 707–12.

Spindler, K. (2001) The man in the ice under special consideration of paleopathological evidence [in German]. *Verhandlungen der Deutschen Gesellschaft für Pathologie*, **85**, 229–36.

Stange, G. (1981) The ocellar component of flight equilibrium control in dragonflies. *J. Comp. Physiol.*, **141**, 335–47.

Stanko, M., Miklisova, D., De Bellocq, J. G. and Morand, S. (2002) Mammal density and patterns of ectoparasite species richness and abundance. *Oecologia*, **131**, 289–95.

Stark, K. R. and James, A. A. (1995) A factor Xa-directed anticoagulant from the salivary glands of the yellow fever mosquito *Aedes aegypti*. *Experimental Parasitology*, **81**, 321–31.

Steelman, C. D. (1976) Effects of external and internal arthropod parasites on domestic livestock production. *Ann. Rev. Ent*, **21**, 155–78.

Stevens, J. R., Noyes, H. A., Schofield, C. J. and Gibson, W. (2001) The molecular evolution of Trypanosomatidae. *Advances in Parasitology*, Vol. 48, 1–56.

Stierhof, Y. D., Bates, P. A., Jacobson, R. L. (1999) Filamentous proteophosphoglycan secreted by *Leishmania* promastigotes forms gel-like three-dimensional networks that obstruct the digestive tract of infected sandfly vectors. *European Journal of Cell Biology*, **78**, 675–89.

Stojanovich, C. J. (1945) The head and mouthparts of the sucking lice (Insecta: Anoplura). *Microentomology*, **10**, 1–46.

Stoven, S., Silverman, N., Junell, A. *et al.* (2003) Caspase-mediated processing of the *Drosophila* NF-kappaB factor Relish. *Proc. Natl. Acad. Sci. USA*, **100**, 5991–6.

Strand, M. R. and Clark, K. D. (1999) Plasmatocyte spreading peptide induces spreading of plasmatocytes but represses spreading of granulocytes. *Arch. Insect Biochem. Physiol.*, **42**, 213–23.

Strand, M. R. and Pech, L. L. (1995) Immunological basis for compatibility in parasitoid host relationships. *Ann. Rev. Ent.*, **40**, 31–56.

Stys, P. and Daniel, M. (1957) *Lyctocoris compestris* F. (Heteroptera: Anthocoridae) as a human facultative ectoparasite. *Acta Societatis Entomologicae Cechoslovenicae*, **54**, 1–10.

Sutcliffe, J. F. (1986) Black fly host location: a review, *Can J. Zool*, **64**(4), 1041–53.

Sutcliffe, J. F. (1987) Distance orientation of biting flies to their hosts. *Insect Sci. Applic.*, **8**, 611–16.

Sutcliffe, J. F. and McIver, S. B. (1975) Artificial feeding of simuliids (*Simulium venustum*), factors associated with probing and gorging. *Experientia*, **31**, 694–5.

Sutcliffe, J. F., Steer, D. J. and Beardsall, D. (1995) Studies of host location behavior in the black fly *Simulium arcticum* (Iis-10.11) (Diptera, Simuliidae) – aspects of close range trap orientation. *Bull. Ent. Res.*, **85**, 415–24.

Sutherland, D. R., Christensen, B. M. and Lasee, B. A. (1986) Midgut barrier as a possible factor in filarial worm vector competency in *Aedes trivittatus*. *J. Invert. Path.*, **47**, 1–7.

Sutherst, R. W., Ingram, J. S. I. and Scherm, H. (1998) Global change and vector-borne diseases. *Parasitology Today*, **14**, 297–9.

Sutton, O. G. (1947) The problem of diffusion in the lower atmosphere. *Quart. J. Roy. Meteorol. Soc.*, **73**, 257–81.

Swellengrebel, N. H. (1929) La dissociation des fonctions sexuelles de nutritives (dissociation gonotrophique) d'*Anopheles maculipennis* comme cause du paludisme dans les Pays-Bas et ses rapports avec 'l'infection domiciliare'. *Ann. Inst. Pasteur*, **43**, 1370–89.

Takehana, A., Katsuyama, T., Yano, T. *et al.* (2002) Overexpression of a pattern-recognition receptor, peptidoglycan-recognition protein-LE, activates imd/relish-mediated antibacterial defense and the prophenoloxidase cascade in *Drosophila* larvae. *Proc. Natl. Acad. Sci. USA*, **99**, 13705–10.

Takken, W. (1996) Synthesis and future challenges: the response of mosquitoes to host odours. In G. Cardew (ed.), *Olfaction in Mosquito–Host Interactions*. Chichester: Wiley 302–20.

Takken, W., Kager, P. A., and Kaay, H. J., (1999) Endemische malaria terug in Nederland? *Nederlands Tijdschrift voor Geneeskunde*, **143**, 836–8.

Takken, W., Klowden, M. J. and Chambers, G. M. (1998) Effect of body size on host seeking and blood meal utilization in *Anopheles gambiae sensu stricto* (Diptera: Culicidae): the disadvantage of being small. *J. Med. Ent.*, **35**, 639–45.

Takken, W. and Knols, B. G. J. (1999) Odor-mediated behaviour of afrotropical malaria mosquitoes. *Ann. Rev. Ent.*, **44**, 131–57.

Takken, W., van Loon, J. J. A. and Adam, W. (2001) Inhibition of host-seeking response and olfactory responsiveness in *Anopheles gambiae* following blood feeding. *J. Insect Physiol.*, **47**, 303–10.

Tashiro, H. and Schwardt, H. H. (1953) Biological studies of horse flies in New York. *J. Econ. Ent.*, **46**, 813–22.

Tawfik, M. S. (1968) Feeding mechanisms and the forces involved in some bloodsucking insects. *Quaes. Ent.*, **4**, 92–111.

Taylor, P. J. and Hurd, H. (2001) The influence of host haematocrit on the blood feeding success of *Anopheles stephensi*: implications for enhanced malaria transmission. *Parasitology*, **122**, 491–6.

Teesdale, C. (1955) Studies on the bionomics of *Aedes aegypti* L. in its natural habitats in a coastal region of Kenya. *Bull. Ent. Res.*, **46**, 711–42.

Tempelis, C. H. and Washino, R. K. (1967) Host feeding patterns of *Culex tarsalis* in the Sacramento Valley, California, with notes on other species. *J. Med. Ent.*, **4**, 315–18.

Terra, W. R. (1988a) Physiology and biochemistry of insect digestion: an evolutionary perspective. *Braz. J. Med. Biol. Res.*, **21**, 675–734.

(2001) The origin and functions of the insect peritrophic membrane and peritrophic gel. *Arch. Insect Biochem. Physiol.*, **47**, 47–61.

Terra, W. R. and Ferreira, C. (1994) Insect digestive enzymes – properties, compartmentalization and function. *Comp. Biochem. Physiol. B*, **109**, 1–62.

Terra, W. R., Ferreira, C. and Garcia, E. S. (1988b) Origin, distribution, properties and functions of the major *Rhodnius prolixus* midgut hydrolases. *Insect Biochem.*, **18**, 423–34.

Thathy, V., Severson, D. W. and Christensen, B. M. (1994) Reinterpretation of the genetics of susceptibility of *Aedes aegypti* to *Plasmodium gallinaceum*. *J. Parasitol.*, **80**, 705–12.

Theodor, O. (1967) *An Illustrated Catalogue of the Rothschild Collection of Nycteribiidae (Diptera) in the British Museum (Natural History)*. London: British Museum.

Theodos, C. M., Ribeiro, J. M. and Titus, R. G. (1991) Analysis of enhancing effect of sand fly saliva on *Leishmania* infection in mice. *Infection and Immunity*, **59**, 1592–8.

Theodos, C. M. and Titus, R. G. (1993) Salivary gland material from the sand fly *Lutzomyia longipalpis* has an inhibitory effect on macrophage function in vitro. *Parasite Immunology*, **15**, 481–7.

Thompson, B. H. (1976) Studies on the attraction of *Simulium damnosum* s.l. (Diptera: Simuliidae) to its hosts. I. The relative importance of sight, exhaled breath and smell. *Tropenmed. Parasitol.*, **27**, 455–73.

Thompson, W. H. and Beattey, B. J. (1977) Veneral transmission of La Crosse (California encephalitis) arbovirus in *Aedes triseriatus* mosquitoes. *Science*, **196**, 530–1.

Thorsteinson, A. J. and Bracken, G. K. (1965) The orientation behavior of horse flies and deer flies (Tabanidae: Diptera). III. The use of traps in the study of orientation of tabanids in the field. *Ent. Exp. Appl.*, **8**, 189–92.

Thorsteinson, A. J., Bracken, G. K. and Tostawaryk, W. (1966) The orientation behaviour of horse flies and deer flies (Tabanidae: Diptera). VI. The influence of the number of reflecting surfaces on attractiveness to tabanids of glossy black polyhedra. *Can. J. Zool.*, **44**, 275–9.

Tillyard, R. J. (1935) The evolution of scorpion flies and their derivatives (order Mecoptera). *Ann. Ent. Soc. Am.*, **28**, 37–45.

Titus, R. G. (1998) Salivary gland lysate from the sand fly *Lutzomyia longipalpis* suppresses the immune response of mice to sheep red blood cells in vivo and concanavalin A in vitro. *Exp. Parasitol.*, **89**, 133–6.

Titus, R. G. and Ribeiro, J. M. C. (1990) The role of vector saliva in transmission of arthropod-borne disease. *Parasitology Today*, **6**, 157–60.

Tobe, S. S. and Davey, K. G. (1972) Volume relationships during the pregnancy cycle of the tsetse fly *Glossina austeni*. *Can. J. Zool.*, **50**, 999–1010.

Torr, S. J. (1989) The host-orientated behaviour of tsetse flies (*Glossina*): the interaction of visual and olfactory stimuli. *Phys. Ent.*, **14**, 325–40.

Torr, S. J., Hall, D. R. and Smith, J. L. (1995) Responses of tsetse-flies (Diptera, Glossinidae) to natural and synthetic ox odors. *Bull. Ent. Res.*, **85**, 157–66.

Torr, S. J. and Mangwiro, T. N. C. (2000) Interactions between cattle and biting flies: effects on the feeding rate of tsetse. *Med. Vet. Ent.*, **14**, 400–9.

Torr, S. J., Wilson, P. J., Schofield, S. *et al.* (2001) Application of DNA markers to identify the individual-specific hosts of tsetse feeding on cattle. *Med. Vet. Ent.*, **15**, 78–86.

Traub, R. (1985) Coevolution of fleas and mammals. In K. C. Kim (ed.), *Coevolution of Parasitic Arthropods and Mammals*. New York: Wiley.

Trpis, M., Duhrkopf, R. E. and Parker, K. L. (1981) Non-Mendelian inheritance of mosquito susceptibility to infection with *Brugia malayi* and *Brugia pahangi*. *Science*, **211**, 1435–7.

Trudeau, W. L., Fernandez-Caldas, E., Fox, R. W. (1993) Allergenicity of the cat flea (*Ctenocephalides felis felis*). *Clinical and Experimental Allergy: Journal of the British Society for Allergy and Clinical Immunology*, **23**, 377–83.

Turell, M. J., Bailey, C. L. and Rossi, C. A. (1984a) Increased mosquito feeding on Rift Valley fever virus-infected lambs. *Am. J. Trop. Med. Hyg.*, **33**, 1232–8.

Turell, M. J., Rossignol, P. A., Spielman, A., Rossi, C. A. and Bailey, C. L. (1984b) Enhanced arboviral transmission by mosquitoes that concurrently ingested microfilaria. *Science*, **225**, 1039–41.

Turner, D. A. and Invest, J. F. (1973) Laboratory analyses of vision in tsetse flies (Dipt., Glossinidae). *Bull. Ent. Res.*, **62**, 343–57.

Tzou, P., De Gregorio, E. and Lemaitre, B. (2002) How *Drosophila* combats microbial infection: a model to study innate immunity and host-pathogen interactions. *Curr. Opin. Microbiol.*, **5**, 102–10.

Tzou, P., Ohresser, S., Ferrandon, D. *et al.* (2000) Tissue-specific inducible expression of antimicrobial peptide genes in *Drosophila* surface epithelia. *Immunity*, **13**, 737–48.

Underhill, G. W. (1940) Some factors influencing feeding activity of simuliids in the field. *J. Econ. Entomol.*, **33**, 915–17.

Vale, G. A. (1974a) New field methods for studying the response of tsetse flies (Diptera, Glossinidae) to hosts. *Bull. Ent. Res.*, **64**, 199–208.

(1974b) The response of tsetse flies (Diptera, Glossinidae) to mobile and stationary baits. *Bull. Ent. Res.*, **64**, 545–88.

(1977) Feeding responses of tsetse flies (Diptera: Glossinidae) to stationary hosts. *Bull. Ent. Res.*, **67**, 635–49.

(1980) Flight as a factor in host-finding behaviour of tsetse flies (Diptera: Glossinidae). *Bull. Ent. Res.*, **70**, 299–307.

(1982) The trap-orientated behaviour of tsetse flies (Glossinidae) and other Diptera. *Bull. Ent. Res.*, **72**, 71–93.

(1983) The effects of odours, wind direction and wind speeds on the distribution of *Glossina* (Diptera: Glossinidae) and other insects near stationary targets. *Bull. Ent. Res.*, **73**, 53–64.

Vale, G. A. and Hall, D. R. (1985a) The role of 1-octen-3-ol, acetone and carbon dioxide in the attraction of tsetse flies, *Glossina* spp. (Diptera: Glossinidae), to ox odour. *Bull. Ent. Res.*, **75**, 209–17.

(1985b) The use of 1-octen-3-ol, acetone and carbon dioxide to improve baits for tsetse flies, *Glossina* spp. (Diptera: Glossinidae). *Bull. Ent. Res.*, **75**, 219–31.

Vale, G. A., Hall, D. R. and Gough, A. J. E. (1988) The olfactory responses of tsetse flies, *Glossina* spp. (Diptera: Glossinidae), to phenols and urine in the field. *Bull. Ent. Res.*, **78**, 293–300.

Valenzuela, J. G., Belkaid, Y., Rowton, E. and Ribeiro, J. M. C. (2001) The salivary apyrase of the blood-sucking sand fly *Phlebotomus papatasi* belongs to the novel *Cimex* family of apyrases. *J. Exp. Biol.*, **204**, 229–37.

Valenzuela, J. G., Francischetti, I. M. B. and Ribeiro, J. M. C. (1999) Purification, cloning, and synthesis of a novel salivary anti-thrombin from the mosquito *Anopheles albimanus*. *Biochemistry*, **38**, 11209–15.

Valenzuela, J. G., Pham, V. M., Garfield, M. K., Francischetti, I. M. B. and Ribeiro, J. M. C. (2002) Toward a description of the sialome of the adult female mosquito *Aedes aegypti*. *Insect Biochemistry and Molecular Biology*, **32**, 1101–22.

Valenzuela, J. G. and Ribeiro, J. M. C. (1998) Purification and cloning of the salivary nitrophorin from the hemipteran *Cimex lectularius*. *J. Exp. Biol.*, **201**, 2659–64.

Van Handel, E. (1984) Metabolism of nutrients in the adult mosquito. *Mosq. News*, **44**, 573–9.

Van Naters, W. M. V., Den Otter, C. J. and Cuisance, D. (1998) The interaction of taste and heat on the biting response of the tsetse fly *Glossina fuscipes fuscipes*. *Phys. Ent.*, **23**, 285–8.

Vargaftig, B. B., Chignard, M. and Benveniste, J. (1981) Present concepts on the mechanism of platelet aggregation. *Biochem. Pharmacol.*, **30**, 263–71.

Vaughan, J. A. and Turell, M. J. (1996) Facilitation of Rift Valley fever virus transmission by *Plasmodium berghei* sporozoites in *Anopheles stephensi* mosquitoes. *Am. J. Trop. Med. Hyg.*, **55**, 407–9.

Venkatesh, K. and Morrison, P. E. (1982) Blood meal as a regulator of triacylglycerol synthesis in the haematophagous stable fly, *Stomoxys calcitrans*. *J. Comp. Physiol.*, **147**, 49–52.

Venkatesh, K., Morrison, P. E. and Kallapur, V. L. (1981) Influence of blood meals on the conversion of D-(U-14C)-glucose to lipid in the fat body of the haematophagous stablefly, *Stomoxys calcitrans*. *Comp. Biochem. Physiol.*, **68**, 425–9.

Vernick, K. D., Fujioka, H., Seeley, D. C. *et al.* (1995) *Plasmodium gallinaceum* – a refractory mechanism of ookinete killing in the mosquito, *Anopheles gambiae*. *Experimental Parasitology*, **80**, 583–95.

Victoir, K. and Dujardin, J. C. (2002) How to succeed in parasitic life without sex? Asking *Leishmania*. *Trends Parasitol.*, **18**, 81–5.

Voskamp, K. E., Den Otter, C. J. and Noorman, N. (1998) Electroantennogram responses of tsetse flies (*Glossina pallidipes*) to host odours in an open field and riverine woodland. *Phys. Ent.*, **23**, 176–83.

Waage, J. K. (1979) The evolution of insect/vertebrate associations. *Biological Journal of the Linnaeon Society*, **12**, 187–224.

(1981) How the zebra got its stripes – biting flies as selective agents in the evolution of zebra coloration. *J. Ent. Soc. Sth. Afr.*, **44**, 351–8.

Waage, J. K. and Davies, C. R. (1986) Host-mediated competition in a bloodsucking insect community. *Journal of Animal Ecology*, **55**, 171–80.

Waage, J. K. and Nondo, J. (1982) Host behaviour and mosquito feeding success: an experimental study. *Trans. R. Soc. Trop. Med. Hyg.*, **76**, 119–22.

Wahid, I., Sunahara, T. and Mogi, M. (2003) Maxillae and mandibles of male mosquitoes and female autogenous mosquitoes (Diptera: Culicidae). *J. Med. Ent.*, **40**, 150–8.

Ward, R. A. (1963) Genetic aspects of the susceptibility of mosquitoes to malaria infections. *Exp. Parasit.*, **13**, 328–41.

Warnes, M. L. (1995) Field studies on the effect of cattle skin secretion on the behavior of tsetse. *Med. Vet. Ent.*, **9**, 284–8.

Warnes, M. L. and Finlayson, L. H. (1985) Responses of the stable fly, *Stomoxys calcitrans* (L.) (Diptera: Muscidae), to carbon dioxide and host odours. I. Activation. *Bull. Ent. Res.*, **75**, 519–27.

(1986) Electroantennogram responses of the stable fly, *Stomoxys calcitrans*, to carbon dioxide and other odours. *Phys. Ent.*, **11**, 469–73.

(1987) Effect of host behaviour on host preference in *Stomoxys calcitrans*. *Med. Vet. Ent.*, **1**, 53–7.

Waterhouse, D. F. (1953) The occurrence and significance of the peritrophic membrane, with special reference to adult Lepidoptera and Diptera. *Aust. J. Zool.*, **1**, 299–318.

Watts, D. M., Pantuwatana, S., Defoliart, G. S., Yuill, T. M. and Thompson, W. H. (1973) Transovarial transmission of La Crosse virus (California encephalitis group) in the mosquito, *Aedes triseriatus*. *Science*, **182**, 1140–1.

Webb, P. A., Happ, C. M., Maupin, G. O. *et al.* (1989) Potential for insect transmission of HIV: experimental exposure of *Cimex hemipterus* and *Toxorhynchites amboinensis* to human immunodeficiency virus. *J. Infect Dis.*, **160**, 970–7.

Webber, L. A. and Edman, J. D. (1972) Anti-mosquito behaviour of ciconiiform birds. *Animal Behaviour*, **20**, 228–32.

Webster, J. P. and Woolhouse, M. E. J. (1999) Cost of resistance: relationship between reduced fertility and increased resistance in a snail-schistosome host-parasite system. *Proc. R. Soc. Lond. B Sci.*, **266**, 391–6.

Wee, W. L. and Anderson, J. R. (1995) Tethered flight capabilities and survival of *Lambornella clarki*-infected, blood-fed, and gravid *Aedes sierrensis* (Diptera, Culicidae). *J. Med. Ent.*, **32**, 153–60.

Weitz, B. (1963) The feeding habits of *Glossina*. *Bull. WHO*, **28**, 711–29.

Wekesa, J. W., Copeland, R. S. and Mwangi, R. W. (1992) Effect of *Plasmodium falciparum* on blood feeding behavior of naturally infected *Anopheles* mosquitoes in western Kenya. *Am. J. Trop. Med. Hyg.*, **47**, 484–8.

Welburn, S. C., *et al.* (1987) In vitro cultivation of rickettsia-like organisms from *Glossina* spp. *Ann. Trop. Med. Parasit.*, **81**(4), 331–5.

Welburn, S. C., Arnold, K., Maudlin, I. and Gooday, G. W. (1993) Rickettsialike organisms and chitinase production in relation to transmission of trypanosomes by tsetse-flies. *Parasitology*, **107**, 141–5.

Welburn, S. C. and Murphy, N. B. (1998) Prohibitin and RACK homologues are upregulated in trypanosomes induced to undergo apoptosis and in naturally occurring terminally differentiated forms. *Cell Death and Differentiation*, **5** (7), 615–22.

Wells, E. A. (1982) Trypanosomiasis in the absence of tsetse. In J. R. Baker (ed.), *Perspectives in Trypansosomiasis Research*. Chichester: Research Studies Press.

Wen, Y., Muir, L. E. and Kay, B. H. (1997) Response of *Culex quinquefasciatus* to visual stimuli. *J. Am. Mosq. Control Assoc.*, **13**, 150–2.

Wenk, P. (1962) Anatomie des Kopfes von *Wilhelmia equina* (Simuliidae syn. Melusinidae, Diptera). *Zool. Jahrb. Abt. Ontog. Tiere*, **80**, 81–134.

Wenk, P. and Schlorer, G. (1963) Wirtsorientierung und Kopulation bei blutsaugenden Simuliiden (Diptera). *Z. Tropenmed. Parasitol.*, **14**, 177–91.

Werner, T., Liu, G., Kang, D. *et al.* (2000) A family of peptidoglycan recognition proteins in the fruit fly *Drosophila* melanogaster. *Proc. Natl. Acad. Sci. USA*, **97**, 13772–7.

Werner-Reiss, U., Galun, R., Crnjar, R. and Liscia, A. (1999) Factors modulating the blood feeding behavior and the electrophysiological responses of labral apical chemoreceptors to adenine nucleotides in the mosquito *Aedes aegypti* (Culicidae). *J. Insect Physiol.*, **45**, 801–8.

Weyer, F. (1960) Biological relationships between lice (Anoplura) and microbial agents. *Ann. Rev. Ent.*, **5**, 405–20.

Wharton, R. H. (1957) Studies on filariasis in Malaya: observations on the development of *Wuchereria malayi* in *Mansonia* (*Mansonioides*) *longipalpis*. *Ann. Trop. Med. Parasit.*, **51**, 278–96.

White, G. B. (1974) *Anopheles gambiae* complex and disease transmission in Africa. *Trans. R. Soc. Trop. Med. Hyg.*, **4**, 278–98.

White, G. B., Magayuka, S. A. and Boreham, P. F. L. (1972) Comparative studies on sibling species of the *Anopheles gambiae* Giles complex (Dipt. Culicidae): bionomics and vectorial capacity of species A and species B at Segera, Tanzania. *Bull. Ent. Res.*, **62**, 295–317.

White, G. B. and Rosen, B. (1973) Comparative studies on sibling species of the *Anopheles gambiae* Giles complex (Dipt. Culicidae). II. Ecology of species A and B in savanna around Kaduna, Nigeria, during transition from wet to dry season. *Bull. Ent. Res.*, **62**, 613–25.

Whiting, M. F. (2002) Mecoptera is paraphyletic: multiple genes and phylogeny of Mecoptera and Siphonaptera. *Zoologica Scripta*, **31**, 93–104.

Whitten, M. M. and Ratcliffe, N. A. (1999) *In vitro* superoxide activity in the haemolymph of the West Indian leaf cockroach, *Blaberus discoidalis*. *J. Insect Physiol.*, **45**, 667–75.

Wigglesworth, V. B. (1941) The sensory physiology of the human louse *Pediculus humanus corporis* de Greer (Anoplura). *Parasitology*, **33**, 67–109.

(1979) Secretory activities of plasmatocytes and oenocytoids during the moulting cycle in an insect (*Rhodnius*). *Tissue and Cell*, **11**, 69–78.

Wigglesworth, V. B. and Gillett, J. D. (1934) The function of antennae in *Rhodnius prolixus* and the mechanism of orientation of the host. *J. Exp Biol.*, **11**, 120–39.

Williams, B. (1994) Models of trap seeking by tsetse-flies – anemotaxis, klinokinesis and edge-detection. *Journal of Theoretical Biology*, **168**, 105–15.

Williams, P. D. and Day, T. (2001) Interactions between sources of mortality and the evolution of parasite virulence. *Proc. R. Soc. Lond. B Sci.*, **268**, 2331–7.

Wilson, J. J., Neame, P. V. and Kelton, J. G. (1982) Infection induced thrombocytopaenia. *Seminars in Thrombosis and Haemostasis*, **8**, 217–33.

Wilson, M. (1978) The functional organisation of locust ocelli. *J. Comp. Physiol.*, **124**, 297–316.

Wilson, R., Chen, C. W. and Ratcliffe, N. A. (1999) Innate immunity in insects: the role of multiple, endogenous serum lectins in the recognition of foreign invaders in the cockroach, *Blaberus discoidalis*. *Journal of Immunology*, **162**, 1590–6.

Woke, P. A. (1937) Comparative effects of the blood of man and of canary on egg production of *Culex pipiens* Linn. *J. Parasit.*, **23**, 311–13.

Wood, D. M. (1964) Studies on the beetles *Leptinillus validus* (Horn) and *Platypsyllus castoris* Rissema (Coleoptera: Leptinidae) from beaver. *Proceedings of the Entomological Society of Ontario*, **95**, 33–63.

Wood, S. F. (1942) Observations on vectors of Chagas' disease in the United States. I. California. *Bull. Calif. Acad. Sci.*, **41**, 61–9.

Worms, M. J. (1972) Circadian and seasonal rhythms in blood parasites. In E. U. Canning and C. A. Wright (eds.), *Behavioural Aspects of Parasite Transmission*. London: Linnean Society.

Wright, R. H. (1958) The olfactory guidance of flying insects. *Can. Entomol.*, **90**, 81–9.

(1968) Tunes to which mosquitoes dance. *New Sci.*, **37**, 694–7.

Wright, R. H. and Kellogg, F. E. (1962) Response of *Aedes aegypti* to moist convection currents. *Nature*, **194**, 402–3.

Xie, H., Bain, O. and Williams, S. A. (1994) Molecular phylogenetic studies on filarial parasites based on 5s ribosomal spacer sequences. *Parasite-Journal de la Societe Française de Parasitologie*, **1**, 141–51.

Xu, P. X., Zwiebel, L. J. and Smith, D. P. (2003) Identification of a distinct family of genes encoding atypical odorant-binding proteins in the malaria vector mosquito, *Anopheles gambiae*. *Insect Mol. Biol.*, **12**, 549–60.

Yajima, M., Takada, M., Takahashi, N. *et al.* (2003) A newly established in vitro culture using transgenic *Drosophila* reveals functional coupling between the phospholipase A2-generated fatty acid cascade and lipopolysaccharide-dependent activation of the immune deficiency (imd) pathway in insect immunity. *Biochem. J.*, **371**, 205–10.

Yu, X. Q. and Kanost, M. R. (2000) Immulectin-2, a lipopolysaccharide specific lectin from an insect, *Manduca sexta*, is induced in response to gram-negative bacteria. *J. Biol. Chem.*, **275**, 37373–81.

Yuill, T. M. (1983) The role of mammals in the maintainence and dissemination of La Crosse virus. In C. H. Calisher and W. H. Thompson (eds.), *California Serogroup Viruses*. New York: Alan R. Liss.

Zahedi, M. (1994) The fate of *Brugia pahangi* microfilariae in *Armigeres subalbatus* during the first 48 hours post ingestion. *Tropical Medicine and Parasitology*, **45**, 33–5.

Zhang, D., Cupp, M. S. and Cupp, E. W. (2002) Thrombostasin: purification, molecular cloning and expression of a novel anti-thrombin protein from horn fly saliva. *Insect Biochemistry and Molecular Biology*, **32**, 321–30.

Zhang, Y., Ribeiro, J. M. C., Guimaraes, J. A. and Walsh, P. N. (1998) Nitrophorin-2: a novel mixed-type reversible specific inhibitor of the intrinsic factor-X activating complex. *Biochemistry*, **37**, 10681–90.

Zheng, L. (1999) Genetic basis of encapsulation response in *Anopheles gambiae*. *Parasitologia*, **41**, 181–4.

Zheng, L., Cornel, A. J., Wang, R. (1997) Quantitative trait loci for refractoriness of *Anopheles gambiae* to *Plasmodium cynomolgi* B. *Science*, **276**, 425–8.

Zheng, L., Wang, S., Romans, P., *et al.* (2003) Quantitative trait loci in *Anopheles gambiae* controlling the encapsulation response against *Plasmodium cynomolgi* Ceylon. *BMC Genet*, **4**, 16.

Zieler, H., Garon, C. F., Fischer, E. R. and Shahabuddin, M. (2000) A tubular network associated with the brush-border surface of the *Aedes aegypti* midgut: implications for pathogen transmission by mosquitoes. *J. Exp. Biol.*, **203**, 1599–611.

Index